OCCUPYING IRAQ
A HISTORY OF THE COALITION PROVISIONAL AUTHORITY

JAMES DOBBINS, SETH G. JONES,
BENJAMIN RUNKLE, SIDDHARTH MOHANDAS

Sponsored by the Carnegie Corporation of New York

Approved for public release; distribution unlimited

NATIONAL SECURITY
RESEARCH DIVISION

The research described in this report was sponsored by the Carnegie Corporation of New York and conducted within the International Security and Defense Policy Center of the RAND National Security Research Division (NSRD). NSRD conducts research and analysis for the Office of the Secretary of Defense, the Joint Staff, the Unified Combatant Commands, the defense agencies, the Department of the Navy, the Marine Corps, the U.S. Coast Guard, the U.S. Intelligence Community, allied foreign governments, and foundations.

Library of Congress Cataloging-in-Publication Data

Occupying Iraq : a history of the Coalition Provisional Authority / James Dobbins
    ... [et al.].
        p. cm.
    Includes bibliographical references.
    ISBN 978-0-8330-4665-9 (pbk. : alk. paper)
    1. Coalition Provisional Authority. 2. Postwar reconstruction—Iraq.
    3. Bremer, L. Paul. 4. Iraq—Politics and government—2003– I. Dobbins, James.
    II. Coalition Provisional Authority.

    DS79.769.O33 2009
    956.7044'31—dc22

                                                            2009007507

Cover photo credits (clockwise from top left):
AP Photo/Jerome Delay; AP Photo/Dusan Vranic;
AP Photo/Khalid Mohammed; AP Photo/Hussein Malla

The RAND Corporation is a nonprofit research organization providing objective analysis and effective solutions that address the challenges facing the public and private sectors around the world. RAND's publications do not necessarily reflect the opinions of its research clients and sponsors. **RAND**® is a registered trademark.

Cover design by Carol Earnest

Published 2009 by the RAND Corporation
1776 Main Street, P.O. Box 2138, Santa Monica, CA 90407-2138
1200 South Hayes Street, Arlington, VA 22202-5050
4570 Fifth Avenue, Suite 600, Pittsburgh, PA 15213-2665
RAND URL: http://www.rand.org/
To order RAND documents or to obtain additional information, contact
Distribution Services: Telephone: (310) 451-7002;
Fax: (310) 451-6915; Email: order@rand.org

Preface

The American engagement in Iraq has been looked at from many perspectives, including the flawed intelligence that provided the war's rationale, the failed effort to secure an international mandate, the rapid success of the invasion, and the long ensuing counterinsurgency campaign. This book focuses on the activities of the Coalition Provisional Authority (CPA) and its administrator, L. Paul Bremer, who governed Iraq from his arrival on May 12, 2003, to his departure on June 28 of the following year. It is an account of that occupation, seen largely from American eyes—mostly from Americans working in Baghdad for the CPA. It is based on interviews with many of those in Baghdad and Washington responsible for setting and implementing occupation policy, on the memoirs of American and Iraqi officials who have since left office, on journalists' accounts of the period, and on nearly 100,000 internal CPA documents to which the authors were allowed access.

This book recounts and evaluates the efforts of the United States and its coalition partners to restore public services; reform the judicial and penal systems; fight corruption; reduce inflation; expand the economy; and create the basis for a democratic constitution, free elections, and representative government. It also addresses the occupation's most striking failure: the inability of the United States and its coalition partners to protect the Iraqi people from the criminals and extremists in their midst.

This account is based largely on primary sources that include, in particular, the unclassified archives of the CPA. Because the CPA was a hastily improvised multinational organization, an unusually high portion of its work was, in fact, done on an unclassified basis. Nevertheless, a fuller history of the period will have to await the future release not just of classified CPA documents, but of the much more voluminous material held in Washington and by the U.S. military. A comparable history of Combined Joint Task Force-7 (CJTF-7), the CPA's military counterpart, would shed further valuable light on this critical period. Perhaps even more important to a fully rounded account of the period will be the development and exploration of Iraqi sources.

In its occupation of Iraq, the United States fell far short of the ambitious objectives set out by the Bush administration. This book illustrates how and why. It seeks to evaluate the CPA's performance not just against the benchmarks set in administration rhetoric but also against the record of numerous other, more or less contemporaneous, efforts at postwar reconstruction and reform. Iraq was, after all, not the first, but the seventh society that the United States had helped liberate and then tried to rebuild in little more than a decade, the others being Kuwait, Somalia, Haiti, Bosnia, Kosovo, and Afghanistan. The United Nations conducted an even larger number of nation-building missions over this same period. Iraq was among the largest and most challenging of these efforts, but it was not the first such attempt and will not be the last. It is useful, therefore, to judge how American efforts in Iraq stack up against other attempts to reform and reconstruct societies emerging from conflict.

The authors would like to thank all those who participated in interviews, reviewed early drafts of this work, and, in many cases, did both. These include Robert Blackwill, Lakhdar Brahimi, Douglas Brand, David Brannan, L. Paul Bremer, Andrew Card, Scott Carpenter, Keith Crane, Catherine Dale, Douglas Feith, David Gompert, Jeremy Greenstock, Terry Kelly, Patrick Kennedy, Roman Martinez, Clayton McManaway, Frank Miller, Meghan O'Sullivan, Joshua Paul, Andrew Rathmell, Charles Reis, Ricardo Sanchez, Omar al-Shahery, Dan Senor, Matt Sherman, and Olin Wethington. The authors would also like to thank Nora Bensahel, Steve Simon, and Dov Zakheim for their careful and thoughtful reviews.

This research was sponsored by the Carnegie Corporation of New York and conducted within the International Security and Defense

Policy Center (ISDP) of the RAND National Security Research Division (NSRD). NSRD conducts research and analysis for the Office of the Secretary of Defense, the Joint Staff, the Unified Combatant Commands, the defense agencies, the Department of the Navy, the Marine Corps, the U.S. Coast Guard, the U.S. Intelligence Community, allied foreign governments, and foundations.

For more information on RAND's International Security and Defense Policy Center, contact the Director, James Dobbins. He can be reached by email at James_Dobbins@rand.org; by phone at 703-413-1100, extension 5134; or by mail at the RAND Corporation, 1200 S. Hayes Street, Arlington, VA 22202. More information about RAND is available at www.rand.org.

Contents

Figures

Summary

L. Paul Bremer arrived in Baghdad on May 12, 2003, with a broad mandate and plenary powers. As administrator of the Coalition Provisional Authority, he was charged with governing Iraq and promoting the development of a functioning democracy that, it was hoped, would serve as a model for the entire Middle East. Bremer could dispose of all Iraqi state assets and direct all Iraqi government officials. He possessed full executive, legislative, and judicial authority. His instructions from Washington were quite general, and for the most part oral. Over the next several months he received plentiful advice but little further direction.

As a practical matter, Bremer's powers were much more limited than they appeared. He had no direct authority over 98 percent of official American personnel in Iraq. They were under military command. Most Iraqi officials had abandoned their offices, which had in turn been ransacked in rampant looting that had stripped most public facilities throughout the country to the bare walls, and beyond. The Iraqi army had deserted en masse, as had much of the police force. Several billion dollars in Iraqi funds were immediately available, but beyond this ready cash, the state was basically broke and producing no further revenue. Washington was still under the impression that the occupation would largely pay for itself and had made provision for only limited financial support to reconstruction. As a result, the CPA relied, throughout its lifespan, principally on Iraqi money to fund both reconstruction and Iraqi government operations.

Neither could Bremer count on much help from the rest of the world. The invasion had been launched against the advice of several of America's most important allies. Many of Iraq's neighbors, including Iran and Syria, were hostile to U.S. efforts and suspicious that the United States might eventually want to overthrow their regimes as well. The decision to treat Iraq, for legal purposes, as a conquered nation further increased the controversy associated with the enterprise. The occupations with which most Iraqis were familiar were the British control of their country after World War I and Israel's occupation of the West Bank and Gaza, then in its fourth decade. These were not reassuring precedents. An alternative to formal occupation would have been a UN-authorized "peace enforcement operation," as in Bosnia or Kosovo. That sort of arrangement might have attenuated, but not eliminated, Iraqi and regional resistance to the American presence. The price for such an international endorsement would have been some level of international oversight. In the bitter aftermath of the failed attempt to gain United Nations Security Council approval of the invasion, neither the United States nor the UN was interested in having the latter assume such a role in Iraq's governance.

On May 22, 2003, the UN Security Council formally recognized but did not endorse the United States and the United Kingdom as occupying powers. Attempts were made to enlist as many coalition countries as possible, but with limited success. The United Kingdom had contributed a large contingent of troops for the invasion but soon scaled back its contribution to the occupation to less than 10 percent of the total. Other allied contingents were even smaller and generally less capable. Unlike the Balkans, where America's allies had contributed 75–80 percent of the soldiers and money, the United States was going to have to man and pay for this operation largely on its own.

Arrival and Early Decisions

Bremer inherited the Office of Reconstruction and Humanitarian Assistance (ORHA), which had been structured in the belief that the Iraqi administration would remain in place, any American occupation would be short-lived, and the main challenge would be dealing with the consequences of the use of weapons of mass destruction and other war-related damage. The transition from ORHA, headed by Army Lieutenant General Jay Garner (Ret.), to the CPA did not go smoothly. Secretary of Defense Donald Rumsfeld informed Garner about the impending change in leadership only one day after Garner's arrival in Baghdad; Garner had expected to be superseded, but not so quickly. Rumsfeld encouraged Garner to stay on under Bremer, but Garner declined, as he did again a couple of weeks later when Bremer made the same request.

Almost immediately on his arrival in Baghdad, Bremer announced two major steps that would prove to be the most controversial of his tenure. The first was to purge some 30,000 senior Ba'ath party members from public employment, and the second was to disband the Iraqi army. Both decisions, the details of which are considered further below, had been briefed to the President and his principal cabinet advisors and approved by Secretary Rumsfeld. Garner had not been consulted, however, and he advised Bremer against both steps on learning of them, as did other members of the ORHA team. Bremer declined to reconsider either measure.

No one in Washington had kept Garner apprised of the major changes in approach to the occupation being considered there, in part because no one in Washington short of Secretary Rumsfeld had been charged with keeping Garner so informed. Garner was supposed to be operating under the direction of General Tommy Franks, commander of U.S. Central Command, but Franks was soon to retire and also somewhat divorced from the policy discussions then under way in the Pentagon. Bremer might have been wise to have informed and consulted Garner on these issues before his arrival in Baghdad, but Bremer was not yet in charge, had never met Garner, and was fully occupied with preparing for his assignment. The result was to leave a residue of bitterness and recrimination from the very start of Bremer's tenure.

Building the CPA

Bremer personally recruited a number of experienced and accomplished people to serve as his senior staff, although some of them showed up late and few stayed for the duration. Their successors were generally also of good, sometimes superior, quality, but rapid and frequent staff turnover had a decidedly negative effect on continuity. Even more debilitating was Washington's persistent inability to fill more than half the mid-level and junior positions in the CPA and these seldom for more than three to six months at a time. As a result, while intended to be a dominantly civilian organization, the CPA remained heavily military. Many personnel were reserve officers, although a number of them possessed relevant civilian experience. Many younger staff were recruited through the administration's political patronage machinery. There was a chronic shortage of experienced middle-level managers. In particular, there was a shortage of Arabic-speaking regional experts and officials who had worked in previous postconflict stabilization efforts. The result was an organization made up largely of senior supervisors and junior subordinates.

Bremer rapidly established the skeleton of an organization intended to serve as a "government within a government." Half a dozen offices supervised a larger number of Iraqi ministries. Alongside these line units was a staff that included a general council, a financial management office, a policy planning unit, and an executive secretariat. Bremer did not, at first, formally appoint a principal deputy, although Clayton McManaway, a former ambassador with extensive service in wartime Vietnam, served as his closest advisor and assumed charge when Bremer was out of the country. Neither Bremer nor his chief lieutenants ever had any precise idea how many people were working for them on any given day. The Green Zone mess halls were feeding up to 7,000 people, but most of them were either under the military command or contractors working for the military. At its peak, the CPA's notional staff was around 2,000, of whom perhaps half were in the country at any one time. Those who were present routinely worked 80-hour weeks. A significant minority of positions within the CPA were filled by non-American officials from allied countries.

Bremer's management style was very hands-on. He exhibited great energy and a quick grasp of complex issues. He was willing to take responsibility and make difficult decisions. He was able, through his own example, to secure the respect, loyalty, and affection of his numer-

ous staff. Despite these strengths, the CPA structure was overly centralized, and Bremer was excessively burdened by the number of subordinates reporting directly to him and the variety of issues requiring his
attention. The lack of any agreed-on plan, the improvised nature of the
organization, and the rapidity of staff turnover made a greater degree
of delegation difficult and, in the early days, possibly dangerous; but
Bremer would have been better served by formally empowering one or
two deputies, as he eventually did six months later.

The CPA was built from scratch, and every bureaucratic relationship had to be crafted from whole cloth. This went from determining
who paid for use of the motor pool or mess hall to Bremer's relationship with his military partner and Washington superiors. American
and coalition military forces came under Lieutenant General Ricardo
Sanchez, the commander of CJTF-7. Bremer and Sanchez, by their
own accounts, maintained cordial relations. Sanchez was under formal
orders from Secretary Rumsfeld to support Bremer, which he and his
command did quite extensively. This injunction, in Sanchez's view, did
not accord Bremer oversight of, or even necessarily visibility into, military operations at the tactical level. The two men differed on occasion
and their staffs did so more often, but they also collaborated closely.
The extensive overall level of CJTF-7 support for the CPA is notewor-
thy given that Sanchez's staff resources, although more numerous than
Bremer's, were almost as undermanned as were those of the CPA.

The CPA's relationship with Washington was also improvised
and unclear, as was Bremer's with his bosses. The CPA was, at one
and the same time, an element of the Defense Department, a multi-
national organization, and a foreign government. In their capacity as
the government of Iraq, CPA managers rejected efforts by Washington agencies, most notably the White House Office of Management
and Budget, to impose strictures on how the CPA spent Iraqi funds.
Bremer was subordinate to Secretary Rumsfeld but also a presidential
envoy. He communicated directly with the President and the White
House staff. This eventually led Rumsfeld to complain to both Bremer
and Condoleezza Rice, the President's 'national security advisor. These
complaints were unavailing, and from the fall of 2003, Bremer effectively worked under guidance from the White House. Rumsfeld seems

to have felt that Bremer was trying to circumvent him, but the main problem was that Rumsfeld had never established within the Defense Department an adequate mechanism to monitor, support, and guide the CPA's activities and to keep the White House and other relevant agencies informed on what was going on in Baghdad.

Putting both Bremer and Sanchez under the Secretary of Defense was intended to reduce tension between the civil and military components of the U.S. effort, but it probably had the opposite effect. Rumsfeld's management style was to nag his subordinates, peppering them with frequent suggestions but seldom issuing firm instructions or taking clear responsibility. As a result, disagreements between Bremer and Sanchez were rarely adjudicated in a timely fashion. Additionally, the sheer novelty of the arrangement made for difficulties. Friction between American ambassadors and local American military commanders is not infrequent, nor is it unheard of for American diplomats to deal directly with the White House. Such relationships are governed, however, by law, regulation, presidential directive, and decades of customary practice. As a consequence, it is well understood how all the players should behave, even if they do not always do so. With the CPA, a unique political experiment under Defense Department auspices in what became an active war zone, all such relationships had to be worked out anew.

Bremer and his staff were fond of complaining about Washington's "thousand-mile screw driver," and they were indeed the recipients of copious advice and a good deal of micromanagement on the use of U.S. funds. During the CPA's early months, however, Bremer was the victim not of too much policy oversight but of too little. Rumsfeld seems to have refused to allow non–Defense Department personnel in the CPA to communicate directly and formally with their own agencies. In addition, for the first few months the Defense Department failed to repeat Bremer's reports to the State Department, the White House, or the CIA, and was sometimes slow to do so thereafter. The White House, for its part, had decided to delegate responsibility for interagency coordination about Iraq to Bremer, a manifestly impossible task, given his limited staff, manifold other responsibilities, and the fact that non–Defense Department personnel in Baghdad had only limited capacity to communicate with their home agencies, particularly in the early months. In consequence, other agency personnel in Baghdad were not in a position to fully tap the expertise of their home offices or fully represent their agency's views, although a great deal of coordination was achieved through informal phone calls and unclassified, unofficial email. Eventually, Rice was reduced to sending her staffers on forays into the Pentagon to find out what was going on. As a result of these communication blockages, which persisted throughout the CPA's lifespan to some degree but were particularly acute in the spring and summer of 2003, senior Washington officials were less informed and more surprised by events and decisions emanating from Baghdad than they should have been.

Creating the Governing Council

Prior to Bremer's appointment, American planning for post-Saddam Iraq had proceeded along two ill-defined but divergent tracks, one moving toward the extended occupation, as finally eventuated, the other toward a swift handoff to a nonelected Iraqi successor regime, as had occurred in Afghanistan 15 months earlier. In Washington, the issue had been papered over in an interagency agreement to form an interim Iraqi administration. What was left undefined was whether this administration would have independent authority or would simply provide a vehicle through which the United States would govern Iraq— much as the Japanese government, which remained in existence after its 1945 surrender, was the medium through which General Douglas MacArthur had ruled Japan.

General Garner, the head of ORHA, and Zalmay Khalilzad, a National Security Council (NSC) staffer and presidential envoy to the Iraqi opposition, seemed to be proceeding on the assumption that the occupation would be short-lived. This is certainly the impression they conveyed to Iraqi leaders with whom they were consulting in an apparent effort to form an Iraqi government.

Before leaving for Baghdad, Bremer secured President Bush's agreement that there would be only one American envoy in Iraq. On his arrival, he began consultations leading to the formation of the Governing Council, a body of Iraqi émigré and internal leaders chosen by Bremer with the help of the UN and a team of American and British regional experts. This body was to be largely advisory, although its influence and prerogatives would grow over the succeeding months. Bremer observed, in defense of his decision not to accord this group executive or legislative power, that a body that could not agree on its own chairman (the Governing Council chose to rotate that position on a monthly basis) could hardly be ready to rule. Others have speculated that if given real authority, the council might have behaved more responsibly. As with any counterfactual, it is impossible to prove or disprove this hypothesis, but the behavior of these same politicians when they were accorded real power a year later does not suggest that their reformation would have been rapid.

Bremer regarded the decision to mount an extended occupation rather than immediately turn power over to an Iraqi interim government as having been made, in principle, prior to his appointment and embodied in the general guidance he received from the President and Rumsfeld. The record on this point is unclear. The continuing debate over when and by whom a decision was taken to mount an extended occupation reflects the general lack of clarity characteristic of the administration's planning for and early management of its intervention in Iraq. Given that neither the President nor any of his principal advisors had so much as met Bremer prior to his selection, something more than simple confidence in his judgment seems to have been in play in the leeway he was given. It seems likely, therefore, that the decision to supersede Garner almost immediately on his arrival in Baghdad was occasioned by the mounting chaos there and was accompanied by an inclination to assert a firmer American grip, one result of which was the selection and dispatch of Bremer.

What is certain is that this shift in policy left the Iraqi leaders feeling deceived, military commanders uninformed, and senior levels of the administration unconsulted. It was not inappropriate for the administration to have retained two options for governing Iraq, given uncertainties about what they would encounter once Saddam fell. In the event, finding Iraq descending into chaos and the Iraqi elites badly divided, it was not unreasonable to decide in favor of a more extended occupation. What was censurable was to have failed to distinguish between the two approaches, engage all the President's principal advisors on the decision, make a clear-cut choice in the end, arm Bremer and Sanchez with more than general and largely oral instructions, resource the operation commensurate with the expanded mission, and take a consistent line with the Iraqi political leadership.

For several months the Iraqi ministries were run, to the extent they functioned at all, by CPA senior advisors who directed the activity of their principal Iraqi subordinates. In August, Bremer allowed the Governing Council to appoint Iraqi ministers to head each agency. Thereafter, CPA advisors played a slightly less prominent role, although they retained veto authority over major decisions and controlled many of the purse strings.

Establishing Security

Bremer understood the preeminent importance of establishing security as the first task of any occupying force. While still in Washington, he was told that most American troops were to be withdrawn from Iraq within the next few months, leaving as few as 30,000 by the fall of 2003. He immediately raised the issue of troop levels with Rumsfeld and President Bush before leaving for Baghdad. On the day of his arrival, he told his senior staff that law and order would be their first priority. He repeated this in a message to the President ten days later.

Bremer made an early decision to retain the Iraqi police but to build an entirely new army from scratch. Neither approach produced positive results. The new Iraqi army eventually became a relatively competent and reliable force, but it took several years. The police force, which had not been disbanded, was even slower to develop; it became, indeed, a serious source of insecurity for the next several years. This experience indicates that the CPA's critical failure lay not so much in retaining police or in disbanding the army, as some have charged, but rather in failing to reform and rebuild either of these forces in a timely fashion. Yet it is not clear whether the capacity to raise and train foreign security forces on the scale needed then existed anywhere in the U.S. government. In early 2004, the U.S. military assumed responsibility for rebuilding both the army and police but initially did only marginally better. Numbers increased but quality was much slower to follow.

The decision to disband the army has become the single most-cited criticism of the CPA's 14-month reign. This step was not taken without considerable forethought. Walter Slocombe, who had served as Under Secretary of Defense for Policy throughout much of the Clinton administration, had been chosen to head the security-related components of the CPA well before Bremer's appointment. He had been consulting with various DoD officials in preparation for that task when Bremer was named. By then it had become clear that the Iraqi army had disintegrated under U.S. military pressure and that most of its facilities had subsequently been destroyed in the looting. Slocombe and other senior DoD officials decided that it would be better to disband the existing Iraqi army and raise a new one, employing many current army officers in the process but not building on the old foundation. This step would obviate the need to employ a bloated and politicized officer corps or to force the return of reluctant and ill-paid conscripts. It would also, in Bremer's view, help persuade the Iraqi population that the break with the former regime was final and irreversible.

An order to that event was drafted and cleared throughout the Pentagon shortly before Bremer's departure for Baghdad. It also seems to have been cleared with Central Command and the staff of the senior American military commander in the field, although Lieutenant General David McKiernan, Sanchez's predecessor, has since denied approving or even knowing of the decision in advance. Slocombe discussed the proposal with British officials in London on his way to Baghdad. They raised no objection. On May 15, Garner tried, unsuccessfully, to persuade Bremer to reconsider the measure. On May 19, Rumsfeld approved the order. On May 22, the President and the other members of the National Security Council were briefed. Again, no one raised any objection. The next day Bremer issued the order.

This decision process, while more orderly, inclusive, and clear-cut than some administration actions of the period, was far from perfect.

The order had not been discussed on an interagency basis until the President and his chief advisors were informed on the day before its announcement. Stephen Hadley, Rice's deputy on the NSC staff, and Air Force General Richard Meyers, Chairman of the Joint Chiefs of Staff, both complained later that they had not been consulted, although the draft order had been shown to more junior NSC and military officials. At least some of the military officers who acquiesced in this decision did so in the misapprehension that large elements of the army would quickly be recalled to form the basis for a new force, which was not what Bremer and Slocombe intended. Their failure to clarify their intention in this regard from the beginning led to considerable subsequent resentment and recrimination. Given that the Iraqi army had already dissolved, there was no immediate necessity to issue such an order, other than the desire to demonstrate to the Iraqi population that there would be no return of a Saddamist-style government.

The order was certainly remiss in one respect: It made no provision for payments to the separated soldiers or for their reintegration into civilian society. A month later, provision was made for stipends to be paid to former career personnel; a month later still, such payments actually began.

A fully thought-through program for disarming, demobilizing, and reintegrating the old army would undoubtedly have expedited the selective recall of individuals and perhaps elements into the new army. It might also have recovered at least some of the weapons the dispersing soldiers had taken with them. Given that disarmament, demobilization, and reintegration schemes had by 2003 become a standard part of postconflict reconstruction missions, there was no good reason not to have incorporated all aspects of such a program in the original order, even if it had been necessary to delay its promulgation to do so. Approaching the issue in this more comprehensive fashion could have attenuated the negative reaction among former soldiers and their families, recouped some of the weapons former soldiers had take taken with them, provided those separated from the service a constructive outlet for their continued activity, and facilitated recruiting some of them back into the new army in due course. ORHA's plans had called for such a program, but it assumed the army would be present for duty.

These plans were thus irrelevant. In the aftermath of Order Number 2, which dissolved the Iraqi army and other entities such as the Iraqi Intelligence Service, the CPA began work on a reintegration plan of its own but was unable to implement it.

In retrospect, it would have been better to put all Iraqi army personnel on inactive status, continue to pay them, and recall individuals incrementally and selectively. This is not too far from what eventually occurred. Most former soldiers were eventually paid and some were recalled to duty. But doing so without formally disbanding the army would have avoided the traumatic effect of abolishing a force and a national symbol that, unlike the Ba'ath party, was respected in parts of the Shi'ite and Sunni communities. It would also have allowed an accelerated recall of individuals and a selective recall of entire units, as the need emerged.

Efforts to rebuild both the army and police got off to a slow start. Initial CPA plans called for the gradual buildup of an Iraqi army that would concentrate on external defense. Despite pressure from CJTF-7, the CPA was slow to adjust the pace and refocus this training to meet the mounting internal threat. It was far from alone in this regard. Washington was even slower to appreciate what was happening in Iraq. Early on, Bremer agreed with Sanchez that the U.S. military should take over responsibility for training the new army, but lower-level disagreements and an absence of adequate follow through from the top seems to have blocked implementation of their decision until Rumsfeld ordered the shift in March 2004.

Bernard Kerik, former New York City police chief, had responsibility for directing, improving, and expanding the Iraqi police. He spent most of his energy on the first of these tasks, overseeing street operations in Baghdad but doing little to recruit and train the much larger and more professional force that was needed. His lack of federal and international experience was a serious handicap in this regard because all the required resources and expertise for such an effort would have to come from those sources.

After four months, Kerik left Baghdad, to the relief of many. Bremer had turned over responsibility for locating the site for a police training facility to Clayton McManaway, Bremer's *de facto* deputy and close advisor, who secured the agreement of Jordan's King Abdullah to establish such a center there. Large-scale police training did not begin until late in the fall of 2003, six months after Saddam's regime was overthrown and well after the insurgency had begun to take root.

As American casualties mounted, the CPA came under increasing pressure from Washington to boost the number of Iraqi security personnel, almost regardless of quality. The result was to generate additional numbers of incompetent, corrupt, and increasingly abusive police and the formation of numerous, minimally trained militia (labeled the Iraqi Civil Defense Corps) attached directly to American units. Initial Iraqi army contingents, having been told their mission was external defense, balked the first time they were thrown into the counterinsurgency effort. Bremer resisted persistent military efforts to take over police training. In the spring of 2004, Rumsfeld eventually transferred responsibilities for training both the police and the army to the U.S. military. By the second half of 2004, there was a significant increase in the quantity of security forces, but their quality rose much more slowly.

The State Department had originally proposed sending several thousand American and international civilian police to Iraq, based on experience in Haiti, Bosnia, Kosovo, and more than a dozen UN operations, where civilian police had provided a valuable supplement to military contingents in providing for public safety and professionalizing local security forces. The White House cut this number drastically and decreed that those police who were deployed should be unarmed. The State Department ultimately proved unable to deploy even the reduced number that had been authorized. Bremer persisted in pressing for more such police. The NSC staff advised against seeking UN police (on which NATO had relied in Bosnia and Kosovo) on the grounds that they had proven to be incompetent and corrupt. Eventually, the security situation in Iraq deteriorated beyond the point at which lightly armed civilian police would have been of much assistance, and efforts to secure their deployment were abandoned.

Governing Iraq

Bremer's first decision on reaching Baghdad was to dismiss from public service thousands of senior Ba'athist officials. This may have been the most popular step taken by Bremer during his entire 14-month stay, at least among the Shi'ite and Kurdish majority, but it further antagonized the Sunni community, from whence an insurgency soon arose.

Again, this was not a hastily conceived measure, nor was Bremer its originator. The idea of excluding senior Ba'athists from public office had been raised by the Iraqi émigrés working on the State Department's Future of Iraq Project. It had been briefed to the President and the rest of the NSC on March 10. An order to this effect had been prepared for Garner to issue. Learning of it, Bremer asked that its promulgation be postponed until his arrival. As a result, it became CPA Order Number 1, issued on May 1.

Like the decision to dissolve the army, the de-Ba'athification decree was approved by Bremer's superiors in the Pentagon. Unlike the army decree, the decision on de-Ba'athification was also discussed with other agencies, although its exact nature and extent may not have been thoroughly reviewed outside the Defense Department.

There now seems little doubt that the decrees on dissolving the army and on de-Ba'athification could have profited from further review. Arguing against the delay needed to conduct such a review was the expanding chaos in Iraq and the sense of drift occasioned by uncertainty over the governance of the country. Bremer had been recruited to show a firmer American hand. He was anxious to establish the CPA's authority, and felt that both measures would reassure the bulk of the population that Saddam's dictatorship was truly over and not destined to return. Although informed of both actions in advance, President Bush was nevertheless surprised by their extent and rapidity of execution, but he was inclined to defer to Bremer's judgment as the man on the spot. For his part, Bremer believed he was acting on the basis of clear Washington guidance.

Although consciously modeled on de-Nazification in post–World War II Germany, the de-Ba'athification decree was designed to be much less far-reaching. In Germany, 2.5 percent of the population was affected; in Iraq the intent was to cover only about .01 percent, or 25 times less. In Germany, senior Nazi party members were barred from all employment save manual labor; in Iraq, senior Ba'athists were barred only from government jobs.

Bremer intended that those purged who had not been guilty of personal abuses should be able to rehabilitate themselves through a process of review and exoneration. Unfortunately, encouraged by the Defense Department, Bremer turned implementation of this decree over to the Governing Council, which awarded it to Ahmad Chalabi. He and other Iraqi politicians exhibited little interest in restoring former Ba'athists to public office, no matter how free of personal guilt they might be.

Given the scale of Ba'athist abuses and the intense resentment toward the party among the Shi'ite and Sunni populations, some level of de-Ba'thification was justified and unavoidable. A rebalancing of government employment in favor of the Shi'ite and Kurdish majority was inevitable and inherent in the concept of representative government. Bremer's measure may, indeed, have been minimalist in this regard. Certainly, it is inconsistent to criticize the CPA for both delaying a return of sovereignty and purging the Iraqi bureaucracy too heavily, since a representative Iraqi government would probably have acted to free up even more jobs for its supporters.

Bremer soon regretted turning administration of this program over to Iraqi leaders, and he has since acknowledged that it was a mistake. Throughout his tenure, he came under continued pressure from Shi'ite and Kurdish politicians to extend de-Ba'atification further. Bremer largely resisted these entreaties, and, over Governing Council objection, eventually forced the rehiring of several thousand teachers who had been dismissed as a result of the decree.

U.S. officials were shocked at the state in which they found Iraq's electric, water, health, and education systems. Iraq's infrastructure had been relatively unaffected by the war, but it was badly run down by years of mismanagement and economic sanctions and further damaged by the widespread looting that followed Saddam's fall. Prewar American planning had called for fixing only what the invasion had broken. It soon became evident, however, that a much vaster program of reconstruction was called for.

Electric generation is one of the metrics by which the CPA is often judged a failure. To some extent, the CPA has itself to blame for that, since Bremer publicly promised a large-scale increase in electrical generation. The judgment of failure is largely unjustified, however. By October 1, 2003, the CPA had brought electric power generation to a higher-than-prewar level. (It fell to slightly below prewar levels over subsequent months in consequence of both the antiquated state of the electrical grid and insurgent attacks on it.) Bremer also allocated available electricity more fairly throughout the country. Under Saddam, Baghdad had enjoyed more-or-less continuous service, while less favored areas of the country experienced frequent blackouts. Now these shortages were more evenly distributed. Unfortunately, most political leaders and nearly all foreign journalists lived in Baghdad, so the impression of an overall degradation in service gained currency.

Nor was Iraq's electricity production under the CPA substandard for a country at Iraq's overall level of development. On a per-capita basis, under Saddam and then under the CPA, Iraq generated electricity at levels equivalent to that of other countries, such as Jordan, that were more prosperous and more industrialized. But those countries did not experience chronic shortages. Iraq's problem was not principally an inadequate supply of electricity but excessive demand brought on by a failure to charge customers for what they consumed. Fees charged consumers for electricity under Saddam had fallen to negligible levels due to inflation, and the CPA chose to stop collecting them altogether. Unconstrained demand for electricity was further stimulated by the sharp rise in imports of white goods occasioned by the CPA's decision to largely eliminate external tariffs, an otherwise quite beneficial move. With much lower prices, ordinary Iraqis stocked up on refrigerators and air conditioners, giving no thought to the cost of their operation because, from the consumer's standpoint, there was none.

The CPA committed large sums, ultimately over $5 billion, to increase electricity production over prewar levels. Some of this money was necessary because insurgent attacks on the grid required continued investments simply to keep it at existing levels. Much of it was ultimately wasted, however, in the years following the CPA's lifespan, as a result of rising violence and lack of Iraqi maintenance.

In contrast, the CPA was able to raise the delivery of health and education services well beyond prewar levels. Spending on public health increased under the CPA by 3,200 percent. Thousands of schools were refurbished, textbooks rewritten to eliminate Ba'athist content, and millions distributed. Higher education continued to lag, however.

The CPA charted an uncertain course with respect to local government. This may have been unfortunate, but it was hardly unusual. Most democratization experts recommend beginning at the grass roots, holding local elections, allowing a new generation of leaders to emerge thereby, and proceeding to national elections only when civil society, free media, and nonsectarian political parties have had time to get organized. In practice, this almost never happens. The international community often has little presence beyond the capital, there is frequently great urgency attached to forming a national government, the powers of local governments are seldom well established, and postconflict societies are often prey to serious centrifugal forces which make empowering such governments dangerous in the absence of an established and functioning central authority.

All of these conditions applied in Iraq. Only with great difficulty was the CPA able, by the end of its lifespan, to deploy a handful of its personnel—often no more than one or two—to Iraqi provinces. Iraqis had no modern experience with federalism, and considerable skepticism regarding it. By contrast the CPA was under great pressure, from the moment of its creation, to hold national elections and restore sovereignty to the resultant government. And finally, the fragmentation of Iraq into three or more warring states was an ever-present danger that might have been advanced by empowering local governments before establishing a national one.

As a result, the CPA proceeded cautiously in this sphere. Bremer instructed U.S. military commanders to stop holding elections for local councils. The CPA was slow to provide funding for the councils that had been created. In the early fall of 2003, Bremer's staff was recommending that the CPA authorize caucuses to refresh local councils that had previously been formed by appointment. Once the mid-November decision to accelerate the handover of sovereignty was made, however, Bremer pulled back, feeling that the interim constitution, then under negotiation among Iraqi leaders, should deal with and establish the role of such governments. Only in April 2004 did Bremer finally sign an order establishing the authorities and responsibilities of provincial and municipal councils, governors, and mayors.

In retrospect, the CPA's failure to do more to foster local government was a lost opportunity, but it was understandable under the circumstances. Elections at the municipal and provincial level might have been particularly useful in the Sunni regions, providing an overt and democratically legitimated alternative leadership to that of the mounting insurgency.

Promoting the Rule of Law

In no sector were staffing shortages more keenly felt than in those CPA units overseeing the Ministry of Justice, the courts, and the prisons. The U.S. Department of Justice was not prepared to draw significantly on its Washington staff or that of U.S. Attorneys' offices throughout the country to staff positions in Iraq. As a result, the Justice Department had the worst record of any U.S. agency in meeting the CPA's staffing needs. Nevertheless, the CPA was able to purge Iraqi laws of Ba'athist influence, reopen the courts, and begin to build an independent judiciary. By September 2003, the CPA had reported 90 percent of the courthouses open, although the justice system was far from fully functional. In April 2004, Iraqi judges adjudicated more than 3,000 cases, an all-time record for the country.

In June 2003, the CPA created a Central Criminal Court in Baghdad to handle major cases of national interest. A number of high-profile trials ensued, involving corruption, arms smuggling, and abuse of office. The CPA also laid the groundwork for prosecution of war crimes and crimes against humanity. It trained Iraqi judges, investigators, and prosecutors; oversaw the collection of forensic evidence; registered and examined mass grave sites; and established a mass grave database. It also resisted pressure from Washington to introduce international judges and prosecutors into the Iraqi process. The resultant trials of Saddam and other of his henchmen, which came after the CPA's demise, built on this work. The result was an Iraqi process that exhibited some imperfections but was vastly more expeditious and inexpensive than any international tribunal would have been.

Bremer gave high priority to anticorruption measures, and the CPA introduced a number of reforms designed to reduce the incidence of corruption. Among the most important was the assignment of independent inspectors general to each of the ministries. When one minister exceeded his authority and fired an inspector general, Bremer forced a reinstatement.

Bremer deflected pressure from Iraqi and American political figures to open a CPA investigation into corruption associated with the UN-run Oil-for-Food Program, supporting instead the UN's inquiry and turning the Iraqi investigation over to the independent and apolitical Board of Supreme Audit rather than the Governing Council, as Chalabi and others were demanding. Bremer also resisted pressures from both Washington and the Governing Council to close down Al Jazeera's broadcast operations in Iraq. The CPA did proceed with the prosecution of several Al Jazeera journalists who were accused of profiting from advance knowledge of insurgent attacks to secure exclusive film footage, rather than warning the authorities, but it refrained from acts of censorship, despite the station's sometime incendiary content.

Bremer decided to reopen Abu Ghraib prison after the invasion, determining that there were no short-term alternatives for housing the growing number of detainees. But the CPA had no authority over the handling of detainees, and the eventual prisoner abuses at Abu Ghraib, once they became known, were a major blow to the credibility of U.S. efforts in the rule of law area, as they were more generally.

Growing the Economy

Economic growth in Iraq for 2004, the first year after the CPA's arrival, was 46.5 percent. This is the second-highest figure in any of the 22 postconflict environments studied in previous RAND publications. It was exceeded only by Bosnia and is much higher than growth registered in post–World War II Germany or Japan or any of the many UN-led post–Cold War nation-building endeavors.

The CPA achieved these results by curbing inflation, issuing a new currency, working with the Central Bank to stabilize the dinar through transparent daily auctions, reducing external tariffs, reforming the banking system, expanding liquidity, and stimulating consumer demand. This growth was achieved without a large influx of U.S. or other external assistance. Substantial U.S. government aid began to flow into Iraq only after the end of the CPA. The CPA also promoted, supported, and helped broker what became the largest debt relief package in history, one that will ultimately free Iraq of some $100 billion in public and privately held debt.

Iraqis were nevertheless disappointed with the state of their economy under the CPA. This was the product both of unrealistic Iraqi expectations and the Bush administration's own rhetoric, which had emphasized the material improvements in Iraqi well-being that would flow from the occupation.

The CPA failed to make significant cuts in Iraq's comprehensive and vastly counterproductive system of subsidies for electricity, fuel, food, and state-owned enterprises. Given the deteriorating security situation and the distinct possibility that making such cuts would generate further unrest, this may have been a prudent choice and was understandable in any case. The CPA economic policy has been criticized as being naively ideological in its devotion to free-market principles. Some of its programs fit this mold. Certainly the CPA's effort to create a Baghdad stock exchange was premature, given the state of the Iraqi private sector. On the whole, however, the CPA's economic policies were consistent with established best practices in postconflict environments and, if anything, were too cautious when it came to cutting subsidies.

Temporary employment-generating schemes are seldom a good choice for scarce public resources in postconflict environments. Such efforts almost invariably produce only a limited and very short-term impact; the CPA's efforts in this regard were no exception. Combined with a well-considered counterinsurgency strategy, job schemes might have made some sense, but the U.S. military was still several years away from adopting such a strategy. If the CPA is to be criticized in this area, it would be for putting too much money into temporary job schemes rather than too little.

Problems in the allocation of the $18.2 billion in supplemental funding voted by Congress in November 2003 became evident after the demise of the CPA and were due, in part at least, to choices made under its authority. In particular, the large proportion of those funds devoted to electricity generation and other forms of heavy infrastructure was unwise. As noted earlier, by the end of the CPA, Iraq was generating per-capita kilowatt hours at a level comparable to those in Jordan and other countries at Iraq's level of development. It was experiencing chronic blackouts primarily because of excess demand arising from the fact that it was not charging consumers for the power they used, not just because of deficiencies in the electric power system. Assistance in this sector, beyond the emergency repairs that the CPA had successfully implemented, should have been conditioned on the elimination of this subsidy and the implementation of plans to maintain and eventually amortize the costs of new power plants. Some of the money originally designated for the heavy infrastructure sector was eventually reprogrammed for capacity-building within the Iraqi government, which should have had a higher priority from the beginning.

Although the Iraqi economy rebounded dramatically in 2004, economic output fell in 2005 as a result of the rising civil war. Assuming the gains in security of 2007–2008 can be sustained, the reforms introduced by the CPA should provide the basis for a growing economy less entirely dependent on oil.

It nevertheless proved a serious mistake for the United States to have premised so much of its appeal to the Iraqi people on an improvement in their economic circumstances, particularly when it proved impossible to deliver on these promises, due to rising violence. That is not to say that the United States should have withheld economic assistance. Rather, it should have deflated expectations of a rapid rise in the Iraqi standard of living. It would have been better to have confined American promises to (1) liberating the Iraqi people, (2) protecting them, and (3) allowing them to choose their own government, while stressing that eventual prosperity would depend on hard work and the policy choices that their government made. Had these three promises been made and kept, the substantial, and for the most part well-considered, economic reforms put in place by the CPA would have paid larger and more enduring dividends than did the massive American aid package introduced at the end of the CPA's tenure—much of which was dissipated in security costs and ill-conceived, often uncompleted, projects.

Running and Reorganizing the CPA

As time wore on, Bremer made a number of changes in the CPA's structure. He opened a CPA office in the Pentagon to improve backstopping, recruitment, and other forms of support. President Bush himself had noted to Bremer that the latter had far too many subordinates reporting directly to him. In November 2003, Bremer appointed two formal deputies, the senior responsible for policy and filling in for him when away, the second for operations, principally those related to reconstruction. Bremer also bolstered the strategic planning function. The CPA was eventually able to establish small teams in each of Iraq's 18 provinces. Beginning in November, when the decision was made to speed the return of sovereignty, Bremer refocused the work of the CPA on improving the capacity of the various Iraqi ministries to take on these responsibilities, and began to graduate individual ministries from CPA oversight.

Throughout the CPA's lifespan, staffing remained an acute problem. In fact, it became even more difficult as the end of the CPA's mandate neared, and agencies other than the State Department became even less inclined to send people forward. The CJTF-7 staff was similarly short of personnel. Although blame for the failure to fully man the CPA and CJTF-7 can be widely distributed, in the end, it was the responsibility of the Secretary of Defense to ensure that both organizations were staffed and the President's job to see that he did so.

Promoting Democratization

In early September, Bremer published a seven-step plan for the restoration of sovereignty. This plan required the drafting and ratification of a constitution and the holding of national elections to precede the formation of an Iraqi government, a sequence likely to take a couple of years. This timetable proved too slow for the Iraqi political leadership, and, as it turned out, for Washington as well.

By the fall of 2003, the original American project for Iraq was clearly faltering. Violent resistance was rising, and most Iraqis, however unfairly, blamed Americans for the damage and wanted them gone. The President and his advisors concluded that it was important to end the occupation as soon as possible. In mid-November, Bremer secured Governing Council agreement to an expedited timetable that called for negotiation of an interim constitution, known as the transitional administrative law, a transitional national assembly that would be chosen by provincial caucuses rather than by ballot, and a transitional government to be chosen by this assembly. The entire process was to be completed by mid-2004.

The new plan ran into immediate opposition from Grand Ayatollah Ali Husseini al-Sistani, Iraq's leading Shi'ite cleric. As a result, it had to be amended once again to eliminate both the caucuses and the transitional assembly. Instead, a UN envoy, Lakhdar Brahimi, was invited by Washington and the Governing Council to help select the members of the transitional government. One was eventually formed with the secular Shi'ite Ayad Allawi at its head.

During this period, Bremer and his staff focused heavily on supporting and influencing the negotiation of the transitional administrative law, which Bremer correctly believed would largely determine the contents of the permanent constitution that was to be drafted and ratified a year later.

The erratic nature of U.S. policy—first signaling in early 2003 that the formation of an Iraqi government was immnent, then shifting in May to a much slower timetable, then shifting in November back toward a more expedited process, and finally in early 2004 abandoning the caucus system in favor of a UN-conducted selection process—

undoubtedly created confusion and irritation among Iraqi leaders and their constituents. In retrospect, it would probably have been better to have begun in the spring of 2003 with the more expedited process that was finally adopted, thereby anticipating that the United States was not going to deploy enough of the assets needed, in terms of troops, civilian officials, and money, to effectively secure and govern Iraq for the extended period needed to first write a constitution and then hold elections. But that conclusion was not as evident at the time, and Bremer's more deliberate plan was consistent with the best available expert advice. Indeed, to the extent Bremer was criticized by democratization and nation-building experts throughout the first half of 2003, it was for moving too quickly toward the transfer of sovereignty, rather than too slowly.

Of the three speeds toward sovereignty—slow, fast, and immediate—only the last was not tried. Some believe this might have yielded the best results, citing the Afghan example of late 2001. There is little reason, however, to think that an Iraqi government formed in the spring of 2003 would have performed any better than the one that was finally empowered in the summer of 2004. The result of such an attempt might well have been to simply accelerate the descent into civil war.

In the event, the CPA adjusted to the new and much accelerated November timetable, and set in train the various steps needed to effectuate the transition by the mid-2004 deadline, including the elaboration of a liberal interim constitution, a strengthened bureaucracy, and the beginnings of a more coherent interagency structure for managing Iraq's national security affairs. These contributions endured through the civil war that raged in the aftermath of the occupation, and they provide what hope there is that Iraq will neither fragment nor return to the savage dictatorship it experienced under Saddam.

Disarming Militias and Countering Insurgents

The CPA's closing months were dominated by mounting war on two fronts with both Sunni insurgents and Shi'ite militia. In the spring of

2004, these threats came together in a manner that almost derailed the approaching transfer of power. For months, the CPA and CJTF-7 had been steeling themselves to neutralize the most militant of Shi'ite militia leaders, Muqtada al-Sadr and his Mahdi army. On March 31, Sunni gunmen in Fallujah ambushed and killed four armed American contractors, after which a frenzied mob mutilated and exhibited the Americans' charred bodies. Washington wanted quick retribution, insisting that Fallujah be secured forthwith.

The decision to respond to these Sunni and Shi'ite provocations by launching simultaneous offensives against both the Fallujah insurgents and Muqtada al-Sadr's militia was ill conceived. The effort to retake Fallujah was pushed by Washington despite reservations on the part of the American civilian and military leadership in Iraq. The resultant fighting enraged both communities. Members of the Governing Council threatened to resign. The UN threatened to abandon its efforts to form a transitional government. In the end, Bremer and Sanchez were instructed by Washington to discontinue both efforts.

The failure to provide timely and consistent guidance to both the CPA and CJTF-7 on how to respond to months of repeated provocations from Muqtada al-Sadr was also primarily a Washington problem. Bremer and Sanchez had both sought a go-ahead from Washington to arrest al-Sadr on a number of occasions, but never received a clear decision from Rumsfeld or the President. Bremer and Sanchez were not always in synch and most agencies in Washington were opposed to taking on al-Sadr; but ultimately it was the President and Secretary of Defense who failed to make a clear-cut decision, and the NSC staff that failed to bring the issue to a head.

In its closing months, the CPA developed a pathway toward disarming several of the Shi'ite militias and even secured the agreement of their leadership, al-Sadr excepted, to do so. These plans could only have been implemented, however, if backed by more money and military muscle than Washington was prepared to deploy and employ for the purpose. In the end, the problem was passed on to the new Iraqi government, which did not pursue the effort further for the next several years.

Mission Accomplished or Mission Impossible?

On June 28, two days earlier than expected, Bremer formally transferred sovereignty to the Iraqi people and their interim government. He
left Iraq later that same day. Following Bremer's departure, the CPA
was quietly dismantled. Neither the Defense Department nor the State
Department was eager to claim its legacy. The Department of Defense
had no desire to repeat a foray into the political and economic aspects
of reconstruction, nor was State inclined to look to the CPA for positive
lessons. For both agencies, as for many Americans and Iraqis, this now
defunct organization became a convenient repository for blame about
everything that had gone wrong over the preceding 14 months.

Yet in the course of that relatively brief period, the CPA had
restored Iraq's essential public services to near or beyond their prewar
level, instituted reforms in the Iraqi judiciary and penal systems, dramatically reduced inflation, promoted rapid economic growth, put in
place barriers to corruption, began reform of the civil service, promoted
the development of the most liberal constitution in the Middle East,
and set the stage for a series of free elections. All this was accomplished
without the benefit of prior planning or major infusions of U.S. aid.
Measured against progress registered over a similar period in more than
20 other American-, NATO-, and UN-led postconflict reconstruction
missions, these accomplishments rank quite high.

But the CPA could not halt Iraq's descent into civil war. With
the return of sovereignty, violent resistance to an occupation devolved
into an even more violent conflict between Sunni and Shi'ite extremist
groups. With respect to security, arguably the most important aspect of
any postconflict mission, Iraq comes near the bottom in any ranking of
postwar reconstruction efforts.

The CPA thus largely succeeded in the areas where it had the lead
responsibility but failed in the most important task, for which it did
not. The degree to which one judges the CPA's overall performance,
therefore, must depend heavily on how one assesses its contribution to
the deteriorating security situation.

Principal responsibility for rising insecurity must be attributed to
the U.S. administration's failure to prepare its forces to assume respon-

sibility for public safety after the collapse of Saddam's regime, to deploy an adequate number of troops for that purpose, and to institute appropriate counterinsurgency measures when widespread and violent resistance emerged. These omissions cannot be laid solely, or even principally, at the CPA's door.

The United States went into Iraq with a maximalist agenda—standing up a model democracy that would serve as a beacon to the entire region—and a minimalist application of money and manpower. In particular, it deployed only enough troops to topple the old regime, but not enough to deter the emergence of violent resistance or to counter and defeat the resultant insurgency. The subsequent difficulties encountered owe much to this disjunction between the scope of America's ambitions and the scale of its initial commitment. Given what the CPA had to work with, it now seems apparent that its mission could never have been fully executed with the manpower, money, and time available to it.

It will be endlessly debated whether disbanding a then-absent army and purging an abusive (and incompetent) bureaucracy contributed to the disorder. These were certainly not the proximate causes of Iraq's decent into chaos, since that had started before the CPA was created and these steps were taken. Those two actions did antagonize the Sunni community from which the insurgency soon arose, but a high level of Sunni dissatisfaction was inherent in any effort to provide Iraq a representative government responsive to the bulk of its population. The two steps also further limited the ability of the Iraqi state apparatus to respond to the CPA's needs. Given the performance of the Iraqi police, which notably were not purged, this further degradation may or may not have made much of a difference. Formally dissolving the army was probably an unnecessary and counterproductive gesture; failing to recall a larger number of former soldiers more quickly was the more costly mistake.

Having been set an impossible task, Bremer and his team were then left bereft of adequate support, backstopping, and oversight. Throughout the CPA period, the administration never committed manpower or money commensurate with its rhetoric or its ambitions. Both the CPA and CJTF-7 staffs were grossly undermanned. Bremer governed and reconstructed Iraq with largely Iraqi funds. By the time large-scale U.S. financing became available in the waning days of the CPA, some key programs, including those for building up local security forces, were running on fumes.

The White House was mistaken in thinking that its responsibilities for interagency management could be delegated first to the Department of Defense and then to Baghdad. Bremer's two most controversial decisions, disbanding the army and firing thousands of Ba'athist officials, had been thoroughly discussed in the Department of Defense and approved by his superiors, but they had not been adequately debated or fully considered by the rest of the national security establishment, something that Condoleezza Rice and her staff should have insisted on. President Bush himself was sometimes surprised by Bremer's decisions, although inclined to back him up and defer to his judgment. If senior people at the Departments of State and Defense and at the White House were often surprised and sometimes displeased by decisions made by the CPA, the failure lay principally with Washington for not establishing a clear and transparent channel for reporting and instruction.

The decision to give oversight for all nonmilitary tasks in Iraq to the Defense Department was an important contributing factor to this lack of support and oversight. By doing so, the President took himself and his staff out of the daily decision loop. Whatever sense it may have made in the abstract to transfer nonmilitary responsibilities to the Department of Defense, doing so only a few weeks before the invasion imposed immense start-up costs on the operation. That department, despite its wealth of resources, had no experience in setting up, supporting, and running a branch office of the U.S. government half a world away, a core mission of the State Department. The result was a series of heroic, but in many cases unnecessary, improvisations as new arrangements had to be established to handle tasks familiar to the State Department but new to the Department of Defense.

One of the administration's most serious conceptual and rhetorical errors was to model its efforts in Iraq on those of the post–World War II occupation of Germany and Japan. Those two countries were both highly homogenous societies, with no proclivity toward sectarian conflict. They were first-world economies whose populations did not need to be taught how to run a successful free-market system. And they had surrendered unconditionally. By contrast, Iraq in 2003 looked a lot more like Yugoslavia in 1995—ethnically and religiously divided, with an economy wrecked by war and sanctions and a pattern of historic sectarian grievances. The deceptive ease with which a democratic transition had been arranged in Afghanistan 15 months earlier encouraged an underestimation of the costs and risks of nation-building on this scale. Had the administration recognized that it was taking on tasks comparable to those NATO had assumed only a few years earlier in Bosnia and Kosovo, but in a society ten times bigger, it might have scaled up its initial military and monetary commitments and scaled back its soaring rhetoric. (Alternatively, of course, such a realization might have caused the administration to reconsider the entire enterprise.)

By the same token, it was, as has been noted, a mistake for the United States to have premised so much of its appeal to the Iraqi people on an improvement in their material circumstances. This emphasis on the economic aspects of reconstruction derived, in some measure, from an inaccurate reading of history, in particular, of the post–World War II German and Japanese occupations. Germany did not receive reconstruction aid until 1948, and Japan never did. In both cases, democratic political reforms had been put in place well before their subsequent economic take-offs. Experience in these and many other cases has dictated a prioritization of postwar tasks: beginning with security, then restoring basic public services, stabilizing the economy, and finally reforming the political system. Some level of growth will automatically resume when and if the fighting stops and people stop killing each other and go back to work. Further economic growth is helpful to consolidate political reforms and sustain peace, but it is not a prerequisite for the initial application of those reforms and cannot, in any case, be sustained in the absence of security. The United States should have aided Iraq's economic development, as it did, but it should also have depressed rather than stimulated Iraqi expectations for rapid improvement in their standard of living and directed a larger share of its assistance to rebuilding the Iraqi army, police, and government as a whole.

Given the circumstances in which they found themselves, Bremer and his team performed credibly. Senior levels of the CPA staff were generally competent and experienced. Everyone worked very hard. Not every decision was optimal, but choices were made in an orderly fashion on the basis of professional advice despite the hectic pace of events. Bremer was restrained and judicious in the use of his extraordinary powers, sometimes resisting or ignoring ill-considered advice from Washington. Most CPA policies were consistent with best practices that had emerged in the conduct of postconflict reconstruction missions over previous decades. The results in most spheres, security excepted, bear comparison—in some cases quite favorable comparison—with the record of earlier such operations.

On the negative side, the CPA structure was overly centralized, particularly during the first six months. Staff turbulence exacerbated this problem, leading Bremer, not unnaturally, to rely increasingly on those few key staffers who stayed for the duration. The frustration of CPA officers assigned to the provinces was particularly acute, as their capacity to communicate with and influence the center was limited—both practically, as a result of inadequate communications, and organizationally, as a result of this centralization of decisionmaking.

Planning for the occupation of Iraq has been rightly criticized, but it is also important to stress the lack of preparation. There was a good deal of planning in the State and Defense Departments and in several military commands. But these disparate activities were never fully integrated into a national plan that could have been given to the ORHA and then CPA leadership when they deployed. This meant that the CPA, chronically understaffed, had to create a strategic plan "on the fly."

Contrast this to the planning and preparation for the conventional battle that toppled Saddam. Those plans represented more than a year of intellectual work on the part of the administration's top military and civilian leadership. Even more important, that planning process was accompanied by the movement of hundreds of thousands of men and tens of thousands of machines into position for battle and by the allocation of tens of billions of dollars for its execution. By contrast, the CPA was initially bereft not just of a plan, but of the manpower and money needed to carry one out.

It has been rightly said that no war plan survives first contact with the enemy. It is also true that no postwar plan is likely to survive first contact with the former enemy. In any postconflict situation, some degree of improvisation is therefore inevitable, no matter how good the prewar preparations. The true test of any planning process is not whether it accurately predicts each successive turn in an operation, but whether it provides the operators with the resources and flexibility to carry out their assigned tasks. This the planning process for postwar Iraq signally failed to do.

It is unlikely that American officials will again face decisions exactly like those required of the CPA in the spring of 2003. Second-guessing those decisions will take one only so far in preparing for future challenges. But it is certain that the United States will again find itself assisting a society emerging from conflict to build an enduring peace and establish a representative government. Learning how best to prepare for such a challenge is the key to more-successful future operations. In this regard, Iraq provides an object lesson of the costs and consequences of unprepared nation-building.

Abbreviations

CJTF-7	Combined Joint Task Force-7
CMATT	Coalition Military Assistance Training Team
CPA	Coalition Provisional Authority
CPI	Commission on Public Integrity
DFI	Development Fund for Iraq
DOJ	Department of Justice
FDI	foreign direct investment
GC	Governing Council
ICITAP	International Criminal Investigative Training Assistance Program
IFES	International Foundation for Electoral Systems
IGC	Iraqi Governing Council
IIA	Iraqi Interim Authority
INL	Bureau of International Narcotics and Law Enforcement Affairs
IRGC	Islamic Revolution Guard Corps
IRMO	Iraqi Reconstruction Management Office
KADEK	Kurdistan Freedom and Democracy Congress
KBR	Kellogg Brown & Root
MOIS	Ministry of Intelligence and Security (Iran)
NAC	Neighborhood Advisory Council

NIA	New Iraqi Army
NSC	National Security Council
OFF	Oil-for-Food (program)
ORHA	Office of Humanitarian and Reconstruction Assistance
PKK	Kurdistan Workers Party
PMO	project management office
POTUS	President of the United States
SCIRI	Supreme Council of the Islamic Revolution in Iraq
SIGIR	Special Inspector General for Iraqi Reconstruction
SOE	state-owned enterprise
TAL	Transitional Administrative Law
TNA	Transitional National Assembly
USAID	U.S. Agency for International Development
WHLO	White House Liaison Office

CHAPTER ONE

The Origin of the CPA

As U.S. forces began their invasion of Iraq in late March of 2003, L. Paul Bremer received a phone call from Lewis ("Scooter") Libby, Vice President Dick Cheney's chief of staff. "He asked whether I would be interested in coming back into the government to serve in Iraq," recalled Bremer. "I couldn't serve for long," Bremer responded, "since I am busy running my company." At the time, Bremer was chairman and chief executive officer of Marsh Crisis Consulting, a crisis management firm owned by the financial services firm Marsh & McLennan. Libby replied that "it wouldn't be a full-time job. Perhaps 90 days." Interest in Bremer then began to percolate among senior U.S. officials in Washington. Paul Wolfowitz, the deputy secretary of defense, asked Bremer to come to his office shortly after the conversation with Libby. He began with a rather pointed question: "Do you believe in democracy with the Arabs?" Bremer responded that, in his view, democracy was certainly possible within the Arab world. Wolfowitz then asked, "Would you be willing to have your name considered in running the occupation of Iraq?" The list, he said, included a range of experienced U.S. diplomats, such as Thomas Pickering, a career foreign service officer who had been U.S. ambassador to the UN and half a dozen countries. After consulting with his wife, Bremer said he was willing to be considered.[1]

Following a meeting with Secretary of State Donald Rumsfeld in the Pentagon in late April, Bremer was invited by President George W.

---

[1] Author interview with L. Paul Bremer III, August 12, 2008.

Bush to come to the Oval Office the following day. "Why would you want this impossible job?" the President asked.

"Because I believe America has done something great in liberating the Iraqis, sir, and because I think I can help," Bremer replied.[2]

A few days later Bremer left for Baghdad, arriving on May 12. A week later he penned a note to President Bush, conveying his initial impressions.

May 20, 2003

Mr. President:

After a week on the ground, I thought it might be useful to give you my first impressions of the situation here.

We have two important goals in this immediate period. We must make it clear to everyone that we mean business: that Saddam and the Ba'athists are finished. And we must show the average Iraqi that his life will be better.

I have now visited cities in the North and South and have traveled around Baghdad every day, speaking often to Iraqis on the streets or in stores. As I have moved around, there has been an almost universal expression of thanks to the US and to you in particular for freeing Iraq from Saddam's tyranny. In the northern town of Mosul yesterday, an old man, under the impression that I was President Bush (he apparently has poor TV reception), rushed up and planted two very wet and hairy kisses on my cheeks. (Such events confirm the wisdom of the ancient custom of sending emissaries to far away lands). . . .

Our immediate goal will be to arrange a National Conference this summer, which will set in motion the writing of a constitution, and reform of the judicial, legal and economic systems. As the Iraqis are

---

[2]   L. Paul Bremer, *My Year in Iraq: The Struggle to Build a Future of Hope* (New York: Simon and Schuster, 2006), pp. 6–8.

*progressively more prepared to assume responsibility, we would be
prepared to give it to them. But we must be firm and clear: a legitimate sovereign Iraqi government must be built on a well-prepared
base.*

Respectfully,

Jerry Bremer

Baghdad [3]

Preparation for the war in Iraq had begun only two months
into the Afghan campaign. On November 27, 2001, the Secretary of
Defense directed U.S. Central Command (CENTCOM) to develop
a plan to remove Saddam Hussein from power. What was later designated OPLAN 1003V laid out four phases of American engagement:
securing foreign support and preparing for deployment; shaping the
battle space; conducting combat operations; and engaging in limited
postcombat operations. The last component was accordingly referred
to as Phase IV.[4]

It was only on January 20, 2003, however, that President George
W. Bush issued National Security Presidential Directive 24, which
gave the Department of Defense lead responsibility for postwar Iraq
and directed it to form a new office to take charge of planning and subsequent implementation of the nonmilitary tasks involved. Secretary of
Defense Donald Rumsfeld asked retired Army Lieutenant General Jay
Garner to lead what became known as the Office of Reconstruction
and Humanitarian Assistance (ORHA), asking Garner to "horizon-
tally connect the plans" for postwar Iraq across a range of U.S. government agencies and "find out what the problems are and work on those
problems and anything else you find."[5] Under Secretary of Defense for
Policy Douglas Feith explained to Garner in a phone conversation that

---

[3] Letter from Jerry Bremer to President George W. Bush, May 22, 2003.

[4] Nora Bensahel et al., Olga Oliker, Keith Crane, Richard R. Brennan, Jr., Heather S.
Gregg, Thomas Sullivan, and Andrew Rathmell, *After Saddam: Prewar Planning and the
Occupation of Iraq* (Santa Monica, Calif.: RAND, 2008), p. 54.

[5] Bob Woodward, *State of Denial: Bush at War, Part III* (New York: Simon and Schuster,
2006), p. 108.

the appointment was short-term and he "could count on being relieved fairly soon, once the President appointed a formal political leader or diplomat." Garner responded wryly that the President ultimately wanted a real "person of stature" to run the occupation of Iraq.[6]

ORHA was tasked with overseeing repairs to the war-damaged Iraqi infrastructure, such as oil fields, hospitals, roads, and telecommunications networks, and averting a humanitarian disaster. In Feith's view, ORHA was intended "to help U.S. Central Command temporarily fulfill Phase IV responsibilities," not to assume those responsibilities itself.[7] Garner's background seemed well suited for this assignment. He understood the Department of Defense bureaucracy as a retired Army lieutenant general, and he had been involved in U.S. reconstruction efforts in the Kurdish region of Iraq after Operation Desert Storm. After agreeing to take the job, Garner started assembling his staff. This process proved unusually divisive. Rumsfeld personally vetoed two of Garner's choices, Thomas Warrick and Meghan O'Sullivan, both State Department employees, explaining that he had instructions to do so by a higher level, which Garner interpreted as Vice President Cheney.[8]

---

[6] Author interview with Douglas Feith, November 4, 2008. Douglas J. Feith, *War and Decision: Inside the Pentagon at the Dawn of the War on Terrorism* (New York: HarperCollins, 2008), p. 348.

[7] Author interview with Douglas Feith, November 4, 2008.

[8] Woodward, *State of Denial*, p 127. Other observers attributed Rumsfeld's objection to Warrick to the latter's earlier efforts to woo Iraqi exile leaders and steer them away from contacts with the Pentagon. Douglas Feith writes: "When Paul Wolfowitz went to visit with a group of Iraqi Americans in Dearborn, Michigan [in Feb. 2003] he learned that Warrick had urged them to stay away from the Wolfowitz meeting, threatening to exclude them from the Future of Iraq Project. . . . I heard Wolfowitz inform Rumsfeld about Warrick." Similarly, Michael Rubin, a Department of Defense and CPA official who lived in northern Iraq during Saddam's rule and had extensive contacts in the Iraqi exile community, noted: "The State Department sanctioned Warrick for professional misconduct upon determining the credibility of complaints leveled by Iraqis who resented both assertions that his Rolodex would be the future Iraqi government and threatened to blackball them unless they altered their positions." David Phillips, who worked with Warrick on the Future of Iraq project, writes that one exile recalled that Warrick told Iraqi exiles that "if you work with Paul Wolfowitz, the State Department will not give you anything." See Feith, *War and Decision*, p. 377; Michael Rubin, "Iraq in Books: Review Essay," *Middle East Quarterly*, Vol. 14, No. 2, Spring 2007, p. 25; and David Phillips, *Losing Iraq* (New York: Westview Press, 2005), p. 127.

Rumsfeld eventually relented concerning O'Sullivan, who remained in Iraq throughout the CPA's lifespan and became one of Bremer's most influential advisors. Garner had selected most of his senior staff intended to oversee the various Iraqi ministries from the State Department, including several former ambassadors. At a Pentagon meeting in late February 2003, Rumsfeld vetoed many of these choices, infuriating Secretary of State Colin Powell and other senior State Department officials when they learned of it. Deputy National Security Advisor Stephen Hadley called Feith to say, "Powell was enraged beyond anything that anyone had seen before." Recalling this episode, Feith concluded, "Of all the Iraq-related quarrels between State and Defense, none produced more toxic antagonism at the top levels of State."[9]

Staffing ORHA remained a challenge, presaging the even greater difficulties that Bremer would ultimately encounter in populating the much larger CPA with qualified people. In late January, Garner met with National Security Advisor Condoleezza Rice and other members of the National Security Council (NSC) to discuss which federal agencies would provide personnel for specific positions within ORHA. The NSC staff verbally tasked the agencies to support ORHA, but few moved quickly to provide personnel. In February, senior ORHA officials again requested that the NSC staff press the agencies to send personnel as quickly as possible. But many interagency representatives did not join ORHA until after the organization moved to Baghdad, and others did not arrive until ORHA had been folded into the CPA.[10]

By late January 2003, ORHA consisted of several components, as shown in Figure 1.1: a reconstruction coordinator, who covered such sectors as oil and power; a civil administration coordinator, responsible for a range of somewhat disconnected sectors from law enforcement to agriculture; a humanitarian assistance coordinator; and an expeditionary support staff. The organizational chart also notes a separate security coordinator, who was responsible for all military, public order,

---

[9]   Feith, *War and Decision*, pp. 386, 387.

[10] Bensahel et al., *After Saddam*, p. 54.

Figure 1.1
ORHA Organizational Chart, January 2003

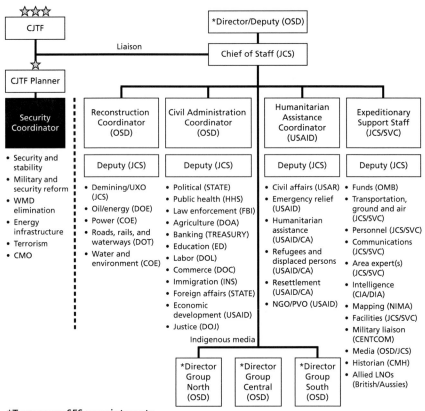

*Temporary SES appointments.
SOURCE: Chart provided to ORHA, late January 2003.
RAND *MG847-1.1*

weapons of mass destruction, and terrorism issues within Iraq. The security coordinator was to report to the three-star commander of Combined Joint Task Force–Iraq, while Garner himself was to report directly to Rumsfeld, although this was later modified to have him report via the CENTCOM commander, General Tommy Franks.

ORHA officials discovered that the administrative hurdles necessary to set up the organization left little time for serious planning. The initial staff members were crowded into a small space in the Pentagon, which had few desks, phones, or computers. New staff members arrived almost daily, which posed ever-increasing requirements for office space and supplies, and also required time for orientation and training. In addition, ORHA had to prepare for deployment to the theater on short notice, which involved medical exams, weapons training and certification, and personal arrangements. The staff was able to accomplish some work while juggling its administrative demands, but they lacked the time and senior-level attention required for real strategic planning.[11] A U.S. Army review later observed that Garner

> had 61 days between the announcement of ORHA's creation and the start of the war to build an organization, develop interagency plans across the administration, coordinate them with CENT-COM and the still undetermined military headquarters that would assume the military lead in post-Saddam Iraq, and deploy his team to the theater. It proved to be an almost impossible set of tasks.[12]

Even though new personnel showed up regularly through February and March 2003, staffing remained an ongoing challenge for ORHA. Even the Department of Defense proved slow to provide the necessary personnel.

ORHA moved to Kuwait in March 2003 and deployed to Iraq at the end of April. Garner himself entered Iraq for the first time on April 11, 2003—two days after the toppling of the statue of Saddam Hussein—on a day-long visit to the southern city of Umm Qasr. He returned to Iraq in late April and began, during his few remaining weeks, to deal with humanitarian and other reconstruction tasks in Iraq. Garner expected to hold elections within as few as three months. He intended to sell Iraqi oil to raise revenue, use lower-level Ba'athists to sustain governmental functions, and utilize between 200,000 and 300,000 former members of the Iraqi army to help in postwar reconstruction. "I had briefed the President on bringing back the Army and he agreed with me," Garner later recalled. "I briefed Condoleezza Rice

---

[11] Bensahel et al., *After Saddam*, pp. 56–58.

[12] Donald P. Wright and Colonel Timothy R. Reese, *On Point II: Transition to the New Campaign* (Fort Leavenworth, Kan.: Combat Studies Institute Press, 2008), p. 71.

every week on it and she agreed. I would bring it up with Rumsfeld every time I talked with him and he agreed with it. Wolfowitz and Feith agreed with it, and the President agreed with it."[13] Garner also set up an interim Iraqi advisory group made of key Sunnis, Shi'ites, and Kurds to put a local face on the occupation government.

Garner and ORHA faced several obstacles. First, the insufficient breadth and depth of prewar planning for the postcombat Phase IV made it extremely difficult to begin restoring vital public services. Virtually no analysis had been done on Iraq's government sectors, from transportation to education. The relatively small number of U.S. forces on the ground and their slowness in assuming responsibility for public security opened a power vacuum when the old regime collapsed and resulted in widespread and largely unchecked looting. The consequent physical destruction of key Iraqi public buildings made it difficult for ORHA to identify ministry personnel, since they had nowhere left to work. American advisors to Iraqi ministries were forced to rely on word-of-mouth to locate ministry staff.[14] ORHA had far too few people to carry out its many responsibilities. It numbered only 151 staff in Kuwait by March 16. This number grew to nearly 300 over the next few weeks, most of whom were active-duty or retired military personnel.[15] "ORHA's senior ranks lacked significant depth in diplomatic experience and had limited understanding of the Middle East," one U.S. Army study subsequently concluded.[16] This was not by intent. Garner's preference was for a largely civilian staff, but the reluctance of other agencies to provide personnel in a timely fashion, in State's case perhaps exacerbated by Rumsfeld's rejection of many State Department candidates, made this impossible.

On Garner's first night in Baghdad, Rumsfeld called to inform him that the President would be appointing L. Paul Bremer to form and take over the Coalition Provisional Authority. Garner knew that

---

[13] Wright and Reese, *On Point II,* p. 85. On support for keeping the Iraqi army intact, also see Feith, *War and Decision,* pp. 366–368, 428–433.

[14] Bensahel et al., *After Saddam,* pp. 68–69.

[15] Wright and Reese, *On Point II,* p. 71; Bensahel et al., *After Saddam,* p. 58.

[16] Wright and Reese, *On Point II,* p. 14.

he was eventually to be superseded by a more senior figure, but was surprised that it had come so quickly. Rumsfeld encouraged Garner to stay on under Bremer, but Garner declined, stating that he would come home once Bremer arrived.[17]

---

[17] Interview with Jay Garner; Charles H. Ferguson, *No End in Sight: Iraq's Descent into Chaos* (New York: PublicAffairs Books, 2008), p 144.

CHAPTER TWO

Building the CPA

On April 16, 2003, General Tommy Franks, the commander of U.S. Central Command, issued a "Freedom Message to the Iraqi People," in which he noted that "I am creating the Coalition Provisional Authority to exercise powers of government temporarily."[1] He also, in this message, outlawed the Ba'ath party. Three weeks later, on May 6, President Bush announced the appointment of L. Paul Bremer III to head that organization. President Bush said that the CPA would establish "an orderly country in Iraq that is free and at peace, where the average citizen has a chance to achieve his or her dreams."[2] Technically, Bremer replaced Franks as the head of the CPA. As a practical matter, Franks had never exercised those responsibilities. In most people's eyes, including Garner's, it was Garner who was being superseded.

During his 23-year State Department career, Bremer had served as special assistant or executive assistant to six secretaries of state. He had been ambassador to the Netherlands from 1983 to 1986 and then the State Department's Ambassador at Large for Counter-Terrorism. He had left government in the early 1990s, worked as an executive in Henry Kissinger's consulting firm, and then become chairman and chief executive officer of Marsh Crisis Consulting. Bremer arrived in Baghdad on May 12, 2003.

---

[1] Feith, *War and Decision*, p. 418.

[2] *President Names Envoy to Iraq: Remarks by the President in Photo Opportunity After Meeting with the Secretary of Defense* (Washington, D.C.: Office of the Press Secretary, May 6, 2003).

The administration never issued a formal order dissolving ORHA. A briefing to Bremer on May 25, 2003, noted that the ORHA staff "is not designed to separately support the Coalition Provisional Authority" and was too "military heavy."[3] Bremer fundamentally restructured and reorganized U.S. reconstruction efforts in Iraq. Some ORHA staff were integrated into the new organization, but others felt unwelcome and decided to leave. Bremer asked Garner to stay on in a senior capacity, but the latter agreed to remain only briefly.

Legal Basis

The CPA's authority derived formally from the status the United States and Great Britain assumed as occupying powers under the laws of armed conflict, as acknowledged in UN Security Council Resolution 1483, of May 22, 2003. This resolution recognized "the specific authorities, responsibilities, and obligations under applicable international law of [the United States and United Kingdom] as occupying powers under unified command (the "Authority")."[4]

The decision to treat Iraq as a conquered country freed the United States from the constraints usually associated with UN-mandated multilateral peace operations. The UN Security Council recognized American authority over Iraq but did not endorse it, nor was the United States under any obligation to report back to the Security Council or seek periodic renewal of its mandate. But while this arrangement left the U.S. government legally unbound, the lack of a UN endorsement also left it bereft of substantial external support. Only the United Kingdom had contributed significant forces to the invasion, and even the British troop commitment was soon cut dramatically. Even more crippling, over the longer term, was the hostility of most of Iraq's neighbors

---

[3]   Coalition Provisional Authority, Proposed ORHA Reorganization, May 25, 2003.

[4]   United Nations Security Council Resolution 1483, S/RES/1483, May 22, 2003. Also see, for example, Sir Jeremy Greenstock, Permanent Representative of the United Kingdom, and John D. Negroponte, Permanent Representative of the United States, letter to the President of the Security Council, May 8, 2003.

to an enterprise about which they had not been consulted and to which they were mostly opposed.[5]

By formally accepting the designation as an occupying power, the United States further enhanced the controversy associated with its presence in Iraq. Since the end of the Cold War, nearly all military interventions and subsequent efforts at postconflict reconstruction had been mandated by the UN Security Council. These had been characterized not as occupations but as peace enforcement or peacekeeping operations. This alternative was theoretically open to the United States, but in light of Washington's failure to secure UN Security Council sanction for the invasion, the U.S. administration was not open to accepting UN oversight for the reconstruction phase, nor was the UN eager to assume such a responsibility

Among the American public this reversion to an earlier terminology and legal form passed largely unnoticed. Nor was the administration shy about comparing the American role in Iraq to that in post–World War II Germany and Japan half a century earlier. For Americans, this historical reference was a positive one, given the benign results of those efforts. For the population of Iraq and its region, however, the occupations with which they were most familiar were Britain's of their own country after World War I and Israel's then 36-year hold on the West Bank and Gaza. These antecedents were not likely to commend the CPA to its new charges.

The CPA was a multinational organization with a significant minority of personnel from countries represented in the military coalition. As a practical matter, however, the CPA was run by the United States and took instructions only from Washington. The United Kingdom had provided a substantial proportion of the initial invading force, but its troop numbers soon dropped to less than 10 percent of the U.S. total. Consequently, London never pushed its case for parity in running the CPA. The British Foreign Office sent a series of senior diplomats, initially John Sawers, its Ambassador to Egypt, and later Jeremy Greenstock, its former ambassador to the United Nations, to represent

---

[5]   Kuwait, and Jordan to a lesser extent, supported the U.S. intervention; Syria, Iran, Turkey, and Saudi Arabia were opposed to one degree or another.

it within the CPA, but these individuals exercised no line responsibility within the organization. In deference to British sensibilities, Bremer initially refrained from formally appointing an American deputy, but he did not use his UK colleague in this capacity, nor, given that 90 percent of the money and manpower were coming from the United States, would such an arrangement have been very workable. Nevertheless, the absence of a deputy or deputies resulted in an overcentralization of CPA decisionmaking, because a dozen or more units responded to Bremer directly.

The Chain of Command

If Bremer was Washington's senior representative in Baghdad, he was also Iraq's top official. Within the U.S. government, this led to varying interpretations regarding the sources and extent of his authority. Bremer felt that "it is not entirely clear that the CPA was a U.S. government entity."[6] Clayton McManaway, Bremer's closest associate during the first half of his year in Iraq, similarly felt that the "CPA was the Iraqi government; it was not an American entity. Many American policymakers, including at the Pentagon and Office of Management and Budget, didn't see it this way." One of the most vocal critics of Bremer and McManaway's interpretation was Robin Cleveland, the Associate Director for National Security Programs at the Office of Management and Budget, who argued that the CPA was subject to the same limitations as any other U.S. government agency, even regarding its handling of Iraqi public funds. "I was on the phone with senior Pentagon officials in Washington," noted McManaway, "and argued that the money the UN owed (to Iraq under the Oil-for-Food Program) was Iraqi money, not U.S. money. Jerry Bremer was the custodian of the Iraqi people."[7]

As the administrator of Iraq, Bremer exercised supreme executive, legislative, and judicial powers. He could issue decrees, of which there were four main types. *Regulations* defined the institutions and authori-

---

[6] Author interview with L. Paul Bremer, August 12, 2008.

[7] Author interview with Clayton McManaway, July 22, 2008.

ties of the CPA. In Regulation Number 1, for example, Bremer stated that "the CPA shall exercise powers of government temporarily in order to provide for the effective administration of Iraq during the period of transitional administration."[8] This document also noted that the laws in force in Iraq as of April 16, 2003, would continue to apply in Iraq as long as they did not conflict with any other regulations or orders issued by the CPA. *Orders* were directives to the Iraqi people that created penal consequences and altered Iraqi law. These included some of the most controversial decisions made by the CPA. Order Number 1, for example, laid out the CPA's de-Ba'athification policy, and Order Number 2 dissolved the Iraqi army, along with a number of other institutions.[9] *Memoranda* expanded on orders and regulations by creating or adjusting procedures. *Public Notices* communicated the intentions of Bremer to the public or reinforced aspects of existing law that the CPA intended to enforce.

There was consistent pushback from some Iraqis on enforcement of CPA's decrees. Ali Allawi, who served as Minister of Defense and Minister of Finance, later wrote that "the 'legality' of Bremer's Orders was always a contentious issue. The Iraqi judiciary was loath to implement the more controversial aspects of Bremer's decrees, and a number of them were left to gather dust." By the end of the CPA's tenure in 2004, Iraqi lawyers and judges increasingly procrastinated in implementing or interpreting CPA's laws, preferring the established Iraqi version if it contravened CPA's orders. "Enforcement of the Orders," Allawi thus concluded, "was an ongoing problem."[10]

---

[8] Coalition Provisional Authority Regulation Number 1, CPA/REG/16 May 2003/01, May 16, 2003.

[9] Coalition Provisional Authority Order Number 1, "De-Ba'athification of Iraqi Society," May 16, 2003; Coalition Provisional Authority Order Number 2, "Dissolution of Entities," August 23, 2003.

[10] Ali A. Allawi, *The Occupation of Iraq: Winning the War: Losing the Peace* (New Haven, Conn.: Yale University Press, 2007), p. 160.

Bremer was designated as President Bush's special envoy to Iraq, but he was supposed to report to and through the secretary of defense.[11] Bremer nevertheless sometimes talked directly to President Bush and the White House, and this eventually became a sore spot for Rumsfeld. In December 2003 in a meeting in Baghdad, for example, Rumsfeld pulled Bremer aside. "Look," he said, "it's clear to me that your reporting channel is now direct to the president and not through me."[12] Bremer, for his part, found that his reports to the Pentagon were initially not getting to other U.S. agencies. "It became a serious problem that reports I was sending to Secretary Rumsfeld and the Pentagon leadership were not being shared outside of the Pentagon."[13] In Washington, Condoleezza Rice, Bush's national security advisor, asked her staff to use their informal contacts in the Pentagon to find out what was going on in Baghdad, since she was receiving so little information through formal channels.[14]

An initial point of contention related to the role of Zalmay Khalilzad, a National Security Council staffer who had been appointed a presidential envoy to the Iraqi opposition in December 2002 and had visited the country in mid-April to help Jay Garner deal with political leaders. Khalilzad and the State Department believed he would continue in this capacity, assisting Bremer in the same way. At a lunch meeting with President Bush before leaving for Iraq, Bremer raised this issue. "Mr. President," he said, "I must have full authority to bring all the resources of the American government to bear on Iraq's reconstruction." Bush responded: "I understand and agree."[15] Khalilzad's role was terminated as a result. Half a year later, Robert Blackwill would be sent to Baghdad from the NSC staff to play a role somewhat similar to the

---

[11] Memo from Secretary of Defense for Presidential Envoy to Iraq, "Designation as Administrator of the Coalition Provisional Authority," May 13, 2003. See also Letter from President George W. Bush to Bremer, May 9, 2003.

[12] Bremer, *My Year in Iraq*, p. 245.

[13] Author interview with L. Paul Bremer, III, November 15, 2007.

[14] Author interview with Frank Miller, June 6, 2008.

[15] Bremer, *My Year in Iraq*, p. 11.

one Khalilzad had abandoned, although without the formal designation as presidential envoy.

Bremer's opposite military number was Lieutenant General Ricardo Sanchez, the commander of Combined Joint Task Force 7 (CJTF -7).[16] The two maintained reasonably cordial relations, but their staffs often clashed. In his memoirs, Sanchez says that the "details of the command relationship between CPA and the military were never clearly defined by any level of command, all the way up to the Department of Defense."[17] Bremer agreed that "it was a vague and awkward relationship. I was not in the military chain of command. But there was an inherent need to coordinate between the military and CPA. I worked closely with the Commander of CJTF-7."[18] Rumsfeld had formally instructed Sanchez to support Bremer and the CPA. Bremer recognized that he was not in a position to instruct Sanchez, but felt that his guidance should be treated analogous to "commander's intent," by which he meant that the military should act on it in a general way. Sanchez was deferential in his personal relations with Bremer,[19] and his command afforded very substantial support to the CPA; but he did not interpret this to include CPA oversight, or even necessarily visibility into military operations at the tactical level. "Initially, Ambassador Bremer believed that the military was going to work for him," wrote Sanchez later. "No one in Combined Joint Task Force 7 thought that was a good idea. It was civilian *command* of the military, and that was not acceptable."[20] Bremer denies having any such expectation, noting that Rumsfeld's letter of instruction to him and his military colleagues

---

[16] Formally, Bremer and the CENTCOM commander—first General Franks and then General Abizaid—both reported to Rumsfeld, whereas Sanchez reported to CENTCOM, putting him one step lower in the command chain than Bremer. CENTCOM headquarters were in Tampa, Fla., however, and its commander had many other responsibilities, including the war in Afghanistan. Sanchez was thus Bremer's principal military partner.

[17] Ricardo S. Sanchez with Donald T. Phillips, *Wiser in Battle: A Soldier's Story* (New York: HarperCollins, 2008), p. 179.

[18] Author interview with L. Paul Bremer, III, November 15, 2007.

[19] Author interview with Clayton McManaway, Oct 27, 2008.

[20] Sanchez, *Wiser in Battle*, p. 179.

was based on the standard letter outlining the authority of American ambassadors abroad, who do not exercise command over military forces in their countries of residence.[21] It is clear from their respective recollections that Sanchez feared greater encroachment on his authority by Bremer than Bremer thought he was exercising.

While Bremer and Sanchez seemed to have gotten along fairly well, their staffs were often at odds. Sanchez later characterized this friction as "devastating," recalling being "continuously involved in sorting out differences between the ambassador's intent as we had discussed it and as I understood it, and the guidance and implementing approaches of his subordinates."[22]

On returning to Washington in the summer of 2003, Bremer was surprised, as noted above, to find that his reporting was not getting to State or the White House. He raised the issue with Pentagon officials and was assured that his messages would be henceforth shared.[23] This produced some improvement. The CPA started preparing a weekly summary of its activities for broad interagency distribution. Patrick Kennedy, Bremer's chief of staff, had sought permission to employ State Department communications personnel in Baghdad to send message traffic directly to State and other addressees in Washington. According to Kennedy, Rumsfeld refused. The State Department communications team in Baghdad consequently remained largely idle throughout the CPA's lifespan. Phone calls and unofficial email traffic circumvented this ban, and non-DoD personnel in Baghdad thereby remained in touch with their agencies, but lateral communication among multiple agencies, in Washington and in Baghdad, was inhibited by this heavy reliance on informal channels.[24]

Tensions in Bremer's relationship with the U.S. military were inherent in his anomalous and certainly unprecedented position as a presidential envoy governing an entire country in the midst of an

---

[21] Author interview with L. Paul Bremer, November 15, 2007.

[22] Author interview with Lieutenant General Ricardo Sanchez, January 27, 2009.

[23] Author interview with L. Paul Bremer, November 15, 2007.

[24] Author interview with Patrick Kennedy, September 5, 2008.

active conflict. Putting both Bremer and Sanchez under the Secretary of Defense was intended to ameliorate this problem but probably had the opposite effect. Rumsfeld's management style was to nag his subordinates, peppering them with frequent suggestions but seldom issuing firm instructions or taking clear responsibility. As a result, disagreements between Bremer and Sanchez were rarely adjudicated in a timely fashion. Additionally, the sheer novelty of the arrangement made for difficulties. Disagreements between American ambassadors and local American military commanders are not infrequent, nor is it unheard of for American diplomats to deal directly with the White House. However, such relationships are governed by law, regulation, presidential directive, and decades of customary practice. As a consequence, it is well understood how all the players should behave, even if they do not always do so. With the CPA, a unique experiment in civilian exercise of political responsibilities abroad under Defense Department authority, all such guidelines had to be worked out anew.

The difficulties encountered in getting CPA message traffic passed to the White House and other interested agencies is a case in point. The State Department routinely shares its diplomatic traffic with other agencies. State messages are withheld from the Department of Defense, CIA, or the White House only by exception, when the originator or main recipient makes a specific, one-time decision to do so. In contrast, Department of Defense communications are not routinely shared and go to other agencies only if the originator specifically directs that they do so. In this case, the standard Department of Defense protocol was followed. In U.S. embassies, non–State Department elements, including the Defense Department, are usually free to maintain their own communications as long as they keep the ambassador generally apprised of their activities. This was not permitted in Baghdad, except regarding the CIA. In effect, the CPA's reporting was thus being handled as if it was operational military traffic rather than the diplomatic reporting that it more closely resembled. The result was to keep most of Washington in the dark over what was going on in Baghdad for the first few months of the CPA's existence and to impede such communication throughout its lifespan.

Staffing and Organization

Despite the Defense Department's substantial personnel resources, the CPA was never adequately manned. One early CPA document asserted that the "CPA is best supported by an experienced, largely civilian interagency team," although a "military liaison cell is required to bridge between CPA and military." This memo went on to observe that the key goal in organizing the CPA was to build a structure to "support the long term organization and objectives; develop staff which enhances effectiveness and efficiency; ensure orderly, professional transition to the new team—set them up for success; internationalize; and use interagency and private contract where prudent to source requirements."[25] Yet the desire to establish a mostly civilian staff was never realized, and the CPA remained dependent on military personnel to fill many billets throughout its duration.

Bremer's senior aide and closest advisor, Clayton McManaway, was a former ambassador to Haiti with more than five years' experience in wartime Vietnam who had also served as a deputy to Bremer on two previous occasions. McManaway assumed charge in Bremer's absence from the country but did not formally function as a principal deputy, running the routine aspects of the CPA's work and leaving Bremer free to focus on the most important issues. Rather, McManaway acted as a counselor and high-level troubleshooter, offering private advice and focusing on important and urgent tasks.[26]

Figure 2.1 illustrates the CPA's organizational structure in mid-July 2003. The executive office supporting Bremer was run by Patrick Kennedy, his chief of staff. Kennedy was a senior Foreign Service officer with extensive administrative experience, having served as U.S. ambassador to the United Nations for Management and Reform and Assistant Secretary of State for Administration. After leaving Iraq, Kennedy became the Under Secretary of State for Management.

Under Kennedy were a number of staff positions. One was the office of the general counsel, headed by Scott Castle. Seconded from

---

[25] Coalition Provisional Authority, "Proposed ORHA Reorganization," May 25, 2003.

[26] Author interview with Clayton McManaway, July 22, 2008.

Figure 2.1
CPA Organizational Chart, July 2003

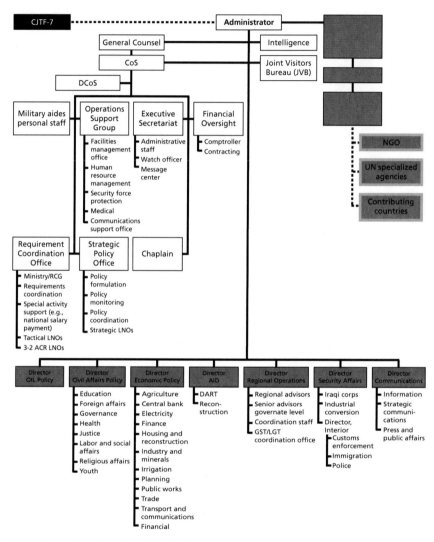

SOURCE: Coalition Provisional Authority.

RAND *MG847-2.1*

Rumsfeld's own legal staff, Castle had practiced as both a military and civilian attorney with a focus on administrative and federal appropriations law.[27] There was also an operations support group, an executive secretariat, a financial oversight office, and a small policy planning element led initially by Dayton Maxwell. Maxwell had a long career at the U.S. Agency for International Development (USAID), and had opened the USAID office in Bosnia in 1994.

Bremer asked Tom Korologos to take charge of congressional and media relations. Korologos had been a journalist with the *New York Herald Tribune, Long Island Press, Salt Lake Tribune*, and then the Associated Press. He had served as a senior staff member in the U.S. Congress, as an assistant to two presidents in the White House, and was president of Timmons & Company, one of Washington's premier lobbying and consulting firms.

The CIA's chief of station also reported, at least nominally, to Bremer, although most CIA activities in the country were in support of the military, or in search of weapons of mass destruction (WMD). Bremer appointed Marek Belka, a former Polish deputy prime minister, to chair the Council for International Coordination, on which sat representatives of the other coalition partners. Its purpose was to serve as "CPA's focal point and coordinating body for international assistance in the relief, recovery and development of Iraq," and to "encourage the international community to provide funds and other forms of assistance to Iraq."[28] Belka went on to become prime minister of Poland following his tour with the CPA.

Alongside these numerous staff functions, the CPA included seven line offices through which it sought to govern Iraq. The Office of Secu-

---

[27] Castle was an officer in the U.S. Army Reserve but served in Baghdad as a civilian. He was promoted to brigadier general during this period. Bizarrely, the Army required him to return to the United States for his two weeks of active duty, refusing to let him serve them in Baghdad.

[28] Coalition Provisional Authority, Regulation Number 5: "Regarding the Council for International Coordination," June 18, 2003. The regulation was slightly modified in August 2003 to include the following amendment: "Section 3, paragraph 1 of CPA Order No. 5, Council for International Coordination (18 June 2003), is hereby amended to read as follows: 1) The Administrator shall designate a Chairman and a Vice Chairman of the Council."

rity Affairs, initially responsible for both the Iraqi army and police, was headed by Walter Slocombe, who had been Under Secretary of Defense for Policy in the Clinton administration. Bernard Kerik, a former chief of the New York Police Department under Mayor Rudolph Giuliani, was named by Washington to serve under Slocombe to run the police and oversee the Interior Ministry. Other strategic units included offices for oil, civil affairs, economic development, regional operations, and communications. Most offices had jurisdiction over several ministries. The Office of Economic Policy, for example, oversaw the ministries of agriculture, electricity, transport, and trade, as well as the central bank. This office was headed by Peter McPherson, a former head of USAID, and Deputy Secretary of the Treasury. The Office of Civil Affairs oversaw the ministries of education, health, religious affairs, and justice. The Justice Unit of this office supervised senior advisors attached to the ministry of justice and the central criminal court.[29]

Under each of these line offices, senior advisors were assigned to every Iraqi ministry. Until August of 2003, these individuals were charged with actually running their respective ministries. Thereafter, on the appointment of Iraqi ministers, they were to act nominally as counselors, but in practice as coadjutors, since Bremer retained veto authority over ministerial decisions and the senior advisors retained considerable authority over spending.[30] Although most activities were naturally conducted by Iraqi employees of the ministries, the senior advisor and his support staff controlled funds and held contracting authority. These small ministerial advisory teams often consisted of a handful of people and tended to concentrate on immediate day-to-day tasks. Any longer-term planning was performed at higher levels of the CPA, including its planning staff.

The CPA was headquartered at a former royal palace nestled along the Euphrates River.[31] This secure area, encompassing over a square

---

[29] Action Memo from Clint Williamson to Presidential Envoy Ambassador L. Paul Bremer, Subject: CPA Justice Presence, June 16, 2003.

[30] Bensahel et al., *After Saddam*, p. 109.

[31] The CPA's location in one of Saddam's palaces rubbed many Iraqis the wrong way, necessary as this may have been for logistical and security reasons.

mile, became known as the "Green Zone." Most CPA staff worked in the palace, traveling to ministries and other destinations with security guards or in military convoys. For a short period of time in 2003 some staff secured their own vehicles and traveled to their destinations without escorts. This freedom was short-lived. As violence increased, providing security for CPA personnel traveling to and from downtown Baghdad and to points outside the city became a major task for CJTF-7. There were always more requests than available convoys. In the fall of 2003, routine trips had to be requested at least one week ahead of time. A few brave or foolhardy staffers continued to venture out of the Green Zone on their own, but this became increasingly rare.

Staff members were fed in the ballroom in the palace by employees of Kellogg Brown & Root (KBR), an international conglomerate based in Texas and a major service provider for the U.S. Department of Defense.[32] Although KBR served meals to as many as 7,000 people every day, only a minority of them worked for the CPA. Fred Smith, who served in the CPA office in Washington and later in Baghdad, noted, "We never got a good grip of how many people were in CPA. Sometimes people simply showed up in Baghdad. And sometimes people just left."[33] The challenge of keeping track of who was coming and going was acute from the beginning. "I tried to set up a country clearance system when Jerry and I first arrived in Baghdad," noted McManaway. "We needed to formalize a way for people to get into— and out of—Iraq. But we didn't even have the capability to send a cable when we arrived."[34] Patrick Kennedy, who oversaw the CPA housekeeping functions among other duties, said that he never had a clear notion of how many people were present for duty at any particular moment.[35] Rumsfeld found the inability of the CPA to keep an accurate account

---

[32] Bensahel et al., *After Saddam*, pp. 115–116.

[33] Author interview with Fred Smith, July 23, 2008.

[34] Author interview with Clayton McManaway, July 22, 2008.

[35] Author interview with Patrick Kennedy, September 5, 2008.

of its staff very frustrating, and this absence of hard figures undercut Bremer's repeated pleas for more personnel.[36]

About half the people working in the palace were military officers assigned to either CJTF-7 or CENTCOM. The CPA itself benefited from the assignment of a large number of U.S. Army civil affairs officers, almost exclusively reservists, who worked under the senior advisors. Some USAID contractors were also located in the palace. There were also contractors to run telecommunications systems, handle the motor pool, provide laundry services, and give haircuts. Many of CPA's staff slept in the Al Rasheed Hotel until it was closed following a rocket attack in late 2003. The rest were in trailers provided by KBR located on the palace grounds or on bunks in the palace ballroom or chapel. "I slept in a plastic trailer," noted Keith Crane, a CPA advisor on economic policy. "The Iraqi staff cleaned the bathrooms, but our rooms got pretty dusty over time."[37] In a private memo to Bremer, Jalal Talabani had warned that "it is my duty to inform you of the dangers that could result from using the Al-Rasheed Hotel. This is because all of its staff were either from the intelligence service (Al-Mukhabarat) or were Ba'athists. The danger could be concealed by poisoning those who are present, or by placing a time-bomb."[38]

CPA personnel received both hazardous duty and hardship pay, and all government departments were asked to strongly encourage staff to volunteer for duty in Iraq. A number of staff were retirees from both the private sector and government, including both Bremer and McManaway. At the other end of the age spectrum, the Defense Department hired a number of young people, some in their twenties, to help staff the ministerial advisory positions. Department of Defense had much more difficulty recruiting middle-level managers with relevant experience. In consequence, the CPA was characterized by an

---

[36] Author interview with former Department of Defense Comptroller Dov Zakheim. The CPA did submit a number of reports on the numbers and status of personnel assigned, but as remarks by Smith, McManaway, and Kennedy suggest, these were not regarded as complete or entirely reliable.

[37] Author interview with Keith Crane, July 16, 2008.

[38] Letter from Jalal Talabani to Paul Bremer, September 22, 2003.

unusual combination of old and young, senior and junior. It was also heavily male, on the order of ten to one, although women did hold a few senior positions.[39]

There was a significant minority of allied officials. "The British made the biggest contribution," noted Fred Smith. "Given British history, they know more about nation-building than we ever will. A range of other countries did as well, though we never had a clear sense of how many international personnel worked for CPA."[40] (A GAO report of June 2004 put the allied staff contribution at 13 percent of the total.) There were a number of Italians and Central and East Europeans from the Czech Republic, Estonia, Poland, Romania, and Ukraine. The palace also housed the headquarters of the Iraq Reconstruction and Development Council, which the CPA had inherited from ORHA. This organization of about 150 Iraqi expatriates, primarily U.S. citizens, was not formally part of the CPA. Its members provided links to Iraqi society, and some worked for the CPA as advisors at ministries and other government institutions.

The CPA sought, with only gradual and limited success, to establish a presence outside of Baghdad. It set up two major branch offices: one in Erbil, the capital of the Kurdish provinces, and one in the south, as well as smaller offices in each governorate, or province. The CPA's objective was to have a team in each of Iraq's 18 governorates, to be made up of a coordinator, a USAID local governance advisor, Iraqi locals, and a military component.[41] The CPA's chronic staffing problems hit these governorate teams particularly hard. Designed to have no more than seven to nine people, these teams sometimes consisted of one or two deployed officials during the early months of the occupation.[42]

On more than one occasion, Bremer stressed the CPA's staff shortfalls to Rumsfeld and asked that his requests be treated "with a sense

---

[39] Bensahel et al., *After Saddam*, p. 117.

[40] Author interview with Fred Smith, July 23, 2008.

[41] Andrew Bearpark to Paul Bremer, Governorate Team Concept, July 1, 2003.

[42] Author interview with Rory Stewart, June 21, 2007.

of urgency and high priority."[43] Despite these entreaties, it took the CPA six months to get civilian officials into each of Iraq's 18 provinces. Once there, these small staffs were often overwhelmed by the scale of their responsibilities.[44] The teams engaged in weekly political reporting. Governorate coordinators often felt they had little guidance from Baghdad, and the CPA's political staff sometimes had limited knowledge of what was taking place in the provinces, particularly in the first months of the occupation.[45] As a consequence, progress or the lack of it at the provincial and local level depended largely on the initiative and improvisation of individual governorate coordinators—even more so on that of the local coalition military commanders.[46]

A notable aspect of life in the CPA was the grueling workload. Most of the staff worked 16-hour days, seven days a week, in third-world conditions. Some joked that they had three days in the week: yesterday, today, and tomorrow. "I received a phone call from Jerry Bremer, who was in Washington," recalled Clayton McManaway. "It was a three-minute phone call, and Jerry had something like 48 items that he wanted me to take care of. Each of them would have taken a month to accomplish, and Jerry needed them done in two days. This was the environment we were operating in. It was very intense."[47] Matt Sherman, who served as deputy senior advisor to Iraq's Ministry of Interior, recalled, "I would usually take off either Friday or Sunday mornings, but there wasn't much else to do but work."[48] For recreation CPA staff sometimes congregated in a few bars on the palace grounds and the basement of the Al Rasheed Hotel, or by the pool.

---

[43] Memo from Paul Bremer to Donald Rumsfeld, "Moving Faster; A Problem or Two," July 7, 2003.

[44] Celeste J. Ward, *The Coalition Provisional Authority's Experience with Governance in Iraq*, United States Institute of Peace, Special Report 139, May 2005, p. 10.

[45] Author interview with Rory Stewart, June 21, 2007; author interview with Roman Martinez, July 1, 2007; author interview with Scott Carpenter, August 2, 2007.

[46] Author interview with Andrew Rathmell, July 13, 2007.

[47] Author interview with Clayton McManaway, July 22, 2008.

[48] Author interview with Matt Sherman, July 8, 2007.

Early Decisions

The transition from ORHA to CPA did not go smoothly. Garner expected to be superseded, but not so quickly. Rumsfeld informed him about the change in leadership, but no one in Washington kept him apprised of the changes being considered on the duration of the occupation, the treatment of the Ba'ath party, and the fate of the Iraqi army, in part because no one in Washington short of Secretary Rumsfeld had been charged with keeping Garner so informed. Bremer might have been wise to do so, but he was not yet in charge, had never met Garner, and was fully occupied with preparing for his assignment. The result was to leave a residue of bitterness and recrimination from the very start of Bremer's tenure.

Almost immediately on his arrival in Baghdad, Bremer announced two major decisions that would prove to be the most controversial of his tenure. The first was to purge some 30,000 senior Ba'ath party members from public employment, and the second was to disband the Iraqi army. Both decisions had been briefed to the President and his principal cabinet advisors and approved by Secretary Rumsfeld. Garner had not been consulted, however, and he advised Bremer against both steps on learning of them, as did other members of the ORHA team. Bremer declined to reconsider either measure. The content, effect, and decision path to each of these decisions is examined in more detail in Chapters Four and Five.

Conclusion

In a relatively short time, given the lack of preparation, Bremer was able to recruit a senior staff made up of accomplished and talented people, many with a good deal of relevant experience. Some arrived late, however, and others left early. Bremer had a hands-on managerial style. He exhibited great energy and a quick grasp of complex issues. He was willing to take responsibility and make difficult decisions. He was able, largely by personal example, to secure the respect, loyalty, and affecion of his numerous staff. Despite these strengths, Bremer was often over-burdened by the number of subordinates reporting directly to him and the variety of issues requiring his attention. The result was overcentralization and a tendency toward micromanagement.

All of Bremer's original senior staff left in the course of 2003, and while their successors were often of comparable or even superior quality, there was a serious loss of continuity. The CPA was never able to secure adequate middle-level staffing. As many as half the CPA positions were unfilled at any time, and the rest were often occupied by people who had just arrived or were just about to leave. When the CPA closed shop in June of 2004, there were, by one count, only seven people left from its initial cadre.

A number of CPA staffers were young and new to government, and their enthusiasm and appetite for hard work greatly exceeded their other qualifications. All staffers put in long hours, but their efforts were not always well coordinated or closely directed. Hierarchical and lateral relationships were uncertain in this scratch-built organization, and there was only limited accountability because most people's longer-term career prospects would not depend on how well or poorly they did in Baghdad.

The CPA's reporting went via the Pentagon, and the Department of Defense was not efficient in disseminating these reports further. Staff used cell phones and personal Internet accounts to maintain contact with their home offices, and indeed with each other, as the CPA's internal communications network was also quite limited. The Defense Department was not set up to oversee, manage, and backstop what amounted to a branch office of the U.S. government, a large international organization, and a foreign government. Back in Washington, relations among State, Defense, and the NSC were strained, and the Department of Defense was not assiduous in keeping other agencies or even the White House informed of what was going on in Baghdad. While Bremer might have done more to correct this problem, any such attempt would likely have brought him the sort of rebuke he eventually received from Secretary Rumsfeld for communicating directly with the White House. Whatever the cause, the result was to leave the President's principal advisors, Rumsfeld aside, feeling ill-informed, unconsulted, and consequently uncommitted to the course Bremer was charting.

CHAPTER THREE

Creating the Governing Council

One striking feature of the planning for postwar Iraq was the prolonged uncertainty over basic goals. At the level of strategic objectives, administration decisionmakers displayed a remarkable lack of clarity on what the United States sought to achieve politically in Iraq. Administration officials, to be sure, expressed a general desire to see Iraq become a democracy. But there was no agreement on what this would mean in practice or how it could best be achieved. Should there be an extensive occupation under the direction of an American proconsul, should an Afghan-style big-tent meeting of Iraqi notables determine the new government, or should power simply be handed over to a group of Iraqi exiles? The interagency process under way in late 2002 and early 2003 skirted these questions and instead focused on producing a list of Iraqi émigré names for what was to become known as the Iraqi Interim Authority (IIA), an undefined entity that would serve as a bridge to a new Iraqi government.[1] The process did not define the responsibilities the IIA would assume in a liberated Iraq and eventually stalled in disagreements between the State Department and the Pentagon about which names should be included. As late as May of 2003, as Bremer was brought on board, there was still no agreement on the composition or function of this entity.

Prewar planning for the governance of Iraq took place in a variety of parallel and often disconnected settings. At the Pentagon, Paul Wolfowitz, one of the most forceful proponents within the administration

---

[1] Author interview with Roman Martinez, July 1, 2007.

for creating a democratic system of government in Iraq, had convened an informal, ad hoc group of advisors to discuss postwar governance issues in the months leading up to the war.[2] It is not clear whether that group generated any substantive policy proposals, but several of its members went on to play key roles in the CPA's democratization program. Wolfowitz chose Deputy Assistant Secretary of State Scott Carpenter to coordinate its sessions. Separately, the State Department had convened a large group of Iraqi exiles in a series of workshops known as the Future of Iraq Project. The entire project was coordinated by Thomas Warrick, and the prominent Iraqi exile Kanan Makiya played a leading role in the project's Democratic Principles working group. The establishment of ORHA in January 2003 appears not to have had a significant impact on planning for postwar governance, as that new entity focused instead on preparing for a humanitarian or refugee crisis in the aftermath of conflict.

Forming the Governance Team

For his initial meetings at the Pentagon on these issues, Bremer brought with him Ryan Crocker, a State Department Arabist who had been involved with the Future of Iraq Project and became his senior advisor on governance issues. Following his arrival in Baghdad, Bremer integrated members of Wolfowitz's working group into his staff. Thus, the CPA's governance team was born. Crocker directed the unit, Carpenter became his deputy, Roman Martinez was detailed from the Pentagon, and Meghan O'Sullivan, a State Department employee who had been working with ORHA, joined. In addition, the team included several more junior U.S. Foreign Service officers and Arabic-speaking British diplomats. Hume Horan, a senior State Department Arabist and former ambassador to Saudi Arabia, also worked with the team but reported directly to Bremer.

---

[2] The following description of the development of the Governance Team draws primarily on an interview on July 1, 2007, with Roman Martinez.

The entire State Department component of this team were on short-term assignments, and most had committed to staying only through the selection of the Governing Council, which took place on July 13, 2003.[3] Most crucially, Crocker, an Arabic-speaker who had spent his professional life in the Islamic world, left at this stage. Bremer sought to have him stay in Baghdad but did not succeed.[4] His departure posed a choice for the CPA. In a memo to Bremer, Crocker laid out two options for his own replacement: (1) keep the structure of having a "senior, Arabic speaking officer with extensive area experience as director" and keep Carpenter as deputy or (2) elevate Carpenter to director (whom he noted was "certainly capable" of the job) and bring in an area specialist as deputy.[5] Bremer chose to promote Carpenter. As a consequence of the departure of Crocker and other Foreign Service officers, a great deal of influence in the political decisionmaking process was concentrated in the hands of three of Bremer's aides who had been involved in the Iraqi political process from the earliest days of the occupation: Carpenter, O'Sullivan, and Martinez.[6] Carpenter had experience overseeing democracy-promotion efforts both at the State Department and as a staffer for the International Republican Institute in Eastern Europe. O'Sullivan had previously worked on Middle Eastern issues at the Brookings Institution, a Washington think tank, and with the State Department's Policy Planning Staff. Martinez had worked in the Pentagon's Office of Special Plans. All three were competent and energetic, but none spoke Arabic. Their efforts were supplemented by a number of more senior American and British Arabists who came and went over the succeeding 14 months. The functioning of the governance team changed again in the fall of 2003 as the NSC asserted a more direct role on their issues. Robert Blackwill, named a deputy to Condoleezza Rice in the late summer of 2003, came to play

---

[3] Author interview with Roman Martinez, July 1, 2007.

[4] Author interview with L. Paul Bremer, July 30, 2007.

[5] Ryan Crocker to Paul Bremer, Governance Staffing, July 6, 2003.

[6] The most concrete measure of their influence is that nearly all governance memos sent to Bremer bore some combination of their names. One might also note that Bremer mentions each a dozen or more times in his memoirs. See Bremer, *My Year in Iraq*.

a very influential part thereafter, providing guidance from Washington and joining Bremer in Baghdad at particularly critical moments.[7] Late 2003 also saw the arrival of Richard Jones and Ronald Schlicher, two senior U.S. Foreign Service officers with extensive experience in the Middle East.

Planning for an Iraqi Interim Authority

The core strategic question was how long, if at all, the United States should directly govern Iraq. There is a widespread view that the Pentagon strongly favored a short occupation followed by a handoff to its favored exile leaders, most notably Ahmed Chalabi of the Iraqi National Congress.[8] Douglas Feith acknowledged deep divisions between Defense, on the one hand, and State and CIA, on the other, about the pace and modalities of such a transfer of power. Earlier in 2002, for instance, Feith recalled an interagency debate in which State representatives opposed a proposal to convene a meeting of Iraqi émigré opposition figures, citing a list of questions such as: Which Iraqis would take the lead in organizing the conference? Where would it be held? "It became increasingly clear," stated Feith, "that Richard Armitage and others at the State Department were concerned about the possibility that the Iraqi National Congress would play the leading role, which could make individuals such as Ahmad Chalabi look good." Consequently, Feith argued, little in the way of serious interagency planning happened. "Key steps never occurred. State and CIA endorsed a strategy of liberating Iraq, but opposed every element of it at the tactical

---

7   Author interview with Andrew Rathmell, July 13, 2007.

8   Notably, this claim is echoed by senior CPA figures. See Bremer, *My Year in Iraq*, p. 44, and Larry Jay Diamond, *Squandered Victory: The American Occupation and the Bungled Effort to Bring Democracy to Iraq* (New York: Times Books, 2005), p. 29. However, Douglas Feith argues: "Of the thousands of pages of material that senior Defense Department officials wrote for interagency meetings on post-Saddam Iraqi governance, I know of *not one* supporting this charge. Even in informal meetings and conversations, I never heard anyone at the Defense Department make an argument or suggest a plan for Chalabi into power in Iraq" (Feith, *War and Decision,* p. 255).

and operational levels. It was the job of the National Security Council staff to discipline the process."[9]

One reason the White House was reluctant to begin formal planning for post-Saddam Iraq was its desire to maintain the focus on its diplomatic efforts to secure UN Security Council endorsement for an ultimatum to Saddam that would authorize military action if he failed to disclose all information regarding his WMD programs. As a result, planning throughout 2002 was heavily compartmentalized, with little cross-fertilization between Defense and State, or even between the civilian and military arms of the Pentagon. Agency differences were neither resolved nor even identified in a systematic fashion.

While Pentagon officials did, at times, promote an "early transfer" of power to Iraqis, there was considerable vacillation and internal division even within the Department of Defense on this point. In January 2003, Bill Luti, a deputy to Under Secretary of Defense Douglas Feith, accompanied by Jay Garner, briefed a group of outside national security analysts on U.S. plans for postwar Iraq. The briefing was organized by the Pentagon Public Affairs Office and held in Rumsfeld's personal conference room. Luti's presentation indicated that Garner's ORHA mission would be short and would focus on humanitarian affairs. Garner would then be superseded by another senior American official who would govern Iraq through a transition to democracy, during which a constitution would be drafted and elections held. Rumsfeld joined the meeting an hour or so later and offered to answer questions. Asked about how Iraq would be governed after the fall of Saddam, he showed no familiarity with the details of the Luti briefing, responding instead that he favored a rapid Afghan-style handover of authority. [10]

Paul Wolfowitz, a figure often associated with favoring an early transfer of power, also had shifting views. When asked in a briefing in early 2003 whether the proposed IIA would have the ability to pass

---

[9] Author interview with Douglas Feith, November 4, 2008.

[10] James F. Dobbins, *After the Taliban: Nation-Building in Afghanistan* (Dulles, Va.: Potomac Books, 2008), pp. 149–150. See also pp. 146–148 for a contemporaneous discussion between the author and NSC staffer Zalmay Khalilzad along the same lines.

laws, he said no.[11] This position clearly presumed that de jure and de facto authority would rest with American forces. Thus, in the months and even weeks leading up to the invasion, there were at least two contending models even within the Pentagon: a formal occupation in the mode of post–World War II Japan and Germany or a quick handover to an indigenous regime following the precedent of Afghanistan in 2001–2002.

Independent of the Pentagon's deliberations, the State Department had convened a large group of Iraqi exiles under the rubric of the Future of Iraq Project. The exiles divided into a variety of working groups focusing on the various issues that would arise in postinvasion Iraq, including 32 individuals who convened in the Democratic Principles working group. That group produced a report authored principally by Kanan Makiya, in collaboration with a few other individuals, entitled "The Transition to Democracy in Iraq."[12] The report was in large part aspirational, detailing the many problems of governance created by decades of Ba'athist dictatorship and offering a vision for the new structures Iraqis should build in terms of the judiciary, security institutions, civil society, and constitutional arrangements. At the time, the State Department apparently took the view that the principal value of the exercise was in its bringing exiles together to discuss Iraq's future and not in its plans per se.[13] This view was not entirely fair to the report by the Democratic Principles working group. Though not a comprehensive roadmap for political transition in Iraq, it did highlight many of the most salient challenges that would be thrown up by a postwar administration and offered some plausible ideas for how to structure a transitional government. In particular, it called for creating a transitional authority elected by a conference of Iraqi exile groups and

---

[11] Author interview with Roman Martinez, July 1, 2007.

[12] Conference of the Iraqi Opposition, *Final Report on the Transition to Democracy in Iraq*, as amended by the members of the Democratic Principles Work Group, November 2002. Kanan Makiya offers a description of its drafting in an interview entitled "Putting Cruelty First: An Interview with Kanan Makiya," with the online journal *Democratiya* on December 16, 2005.

[13] Makiya interview, December 16, 2005. See also Bremer, *My Year in Iraq,* p. 25.

a strictly defined timetable for creating a representative government.[14] The plan also emphasized the importance of expanding the initial transitional authority to bring in internal Iraqi political actors who had opposed Saddam Hussein from within Iraq. Significantly, it recommended holding local elections throughout Iraq within 12 months to create a legitimate, grassroots basis for a future government.[15] Moreover, the report explicitly stated that neither the United States nor the United Nations would be required to "police or manage into existence the new and budding democratic institutions."[16] Whatever its merits, the report's recommendations were not embraced by the administration or by the Iraqi exile community as a whole, some of whom derided the report as reflecting the views of only a small minority.[17]

A January 2003 report from the National Intelligence Council staff stressed the need to quickly shift political controls to Iraqi leaders, warning that "Attitudes toward a foreign military force would depend largely on the progress made in transferring power."[18]

Uncertain about how, and how quickly, to empower a new Iraqi government, U.S. officials sought to engage the often fractious Iraqi exile community in a series of conferences. The first opened in London on December 14, 2002. Zalmay Khalilzad, who had been recently named presidential envoy to "Free Iraqis," headed a large American observer delegation to the conference. Though there were considerable disagreements among the various Iraqi parties, the meeting's principal achievement was the selection of a 65-member committee that would serve as the nucleus of a new political order.[19] This committee in turn met on the eve of hostilities on February 25, 2003, in Salah ad Din in Iraqi Kurdistan. Khalilzad told the delegates, "The United States has

---

[14] *Final Report on the Transition to Democracy in Iraq*, pp. 21–22.

[15] *Final Report on the Transition to Democracy in Iraq*, p. 24.

[16] *Final Report on the Transition to Democracy in Iraq*, p. 5.

[17] Allawi, *The Occupation of Iraq*, p. 84.

[18] Quoted in Michael R. Gordon and General Bernard E. Trainor, *Cobra II: The Inside Story of the Invasion and Occupation of Iraq* (New York: Pantheon Books, 2006), p. 468.

[19] Gordon and Trainor, *Cobra II*, p. 86.

no desire to govern Iraq. The Iraqis should govern their own country as soon as possible."[20] Although the United States was adamant that no provisional government would be declared from the Salah ad Din meeting, this issue dominated discussions.

At the March 10 NSC meeting, President Bush approved the Defense Department's proposal to establish an Iraqi Interim Authority made up of exiles, Kurds, and other internal Iraqi leaders "as soon as possible after liberation." A political conference would be convened in liberated Iraq to ensure "internals" were represented in the IIA, which would work with the U.S. government to appoint ministry officials and control the foreign affairs, justice, and agriculture ministries almost immediately. The remaining ministries would be transferred to full Iraqi control as soon as possible, and the IIA "would serve only in the interim, until a more fully representative government can be established through elections."[21]

Shortly after the fall of Baghdad, Garner and Khalilzad organized a conference in Nasiriyah on April 15 and then a 300-person meeting in Baghdad on April 28. Garner hoped to hold elections for a transitional government within 90 days of his arrival in Iraq and publicly announced his attention to do so. Khalilzad, in his meetings with the Iraqi exile leadership, delivered messages that gave at least some of them the impression that the United States favored a quick handover to a transitional government.[22] Even as Iraqis anticipated an early transfer of authority, however, American officials perceived, in the divisions among Iraqis that became apparent at the Baghdad meeting, the difficulty in doing so.

The possibility of a quick transfer to Iraqi governance remained in play in the immediate aftermath of the regime's fall. Although Garner

---

[20] Quoted in Patrick Cockburn, *The Occupation: War and Resistance in Iraq* (London: Verso, 2006), p. 39.

[21] National Security Council, "Summary of Conclusions: NSC Meeting on Regional Issues," March 11, 2003, quoted in Feith, *War and Decision,* p. 408. Feith argues (pp. 402–409) that the State Department and CIA were opposed to this plan and suggests that this is why Khalilzad, an NSC staffer, was chosen to inform Iraqi opposition leaders of the plan rather than a State Department diplomat.

[22] Feith, *War and Decision,* p. 103; Bremer, *My Year in Iraq,* p. 44.

told Kurdish leaders Massoud Barzani and Jalal Talabani on April 21 that they would not be allowed to set up an interim government, he made a number of statements that appeared to downplay ORHA's central role in the governance of Iraq. After a trip to Basra in the first week of May, he declared "Next week, or by the second weekend in May, you'll see the beginning of a nucleus of a temporary Iraqi government, a government with an Iraqi face on it that is totally dealing with coalition."[23] The Garner-Khalilzad consultations with Iraqis—on April 15 in Nasiriyah and April 28 in Baghdad—appeared to be the first two steps of three to the formation of a temporary Iraqi government. Under the auspices of Garner and Khalilzad, the 300 representatives at the Baghdad Conference, drawn from Iraq's various ethnic, religious, and political groups, voted overwhelmingly to form an Iraqi-led government. They called for another, larger conference in a month's time to select the postwar transitional government, a resolution endorsed by presidential envoy Khalilzad.[24] Among the principles agreed to by the conference's participants was "the need to begin a process that will lead to a broad based national conference to be convened in a period of not more than four weeks from April 27th to form a transitional government."[25]

From Interim Authority to Governing Council

The growing chaos on the ground in Iraq seems to have caused the administration to retreat from this plan and choose what had earlier been the lead option, the creation of an American occupational authority led by a senior political figure. Rumsfeld's thinking, as outlined in a "pre-decisional" memo on May 8, 2003, suggests a shift in this

---

[23] Quoted in Patrick Tyler, "Aftereffects: Postwar Rule: Opposition Groups to Help Create Assembly in Iraq," *New York Times*, May 6, 2003. On Garner's meeting with Barzani and Talabani, see LTG (Ret.) Jay Garner, interview with PBS Frontline: "The Lost Year in Iraq," August 11, 2006.

[24] Allawi, *The Occupation of Iraq*, pp. 103–104.

[25] Quoted in "Historical Documents Requested by Ambassador Bremer," October 22, 2003.

direction. In the memo, he argued that the coalition ought to promote Iraqis who share its goals, noting that "regardless of what the Coalition does, it will be assumed that the Coalition set up the Interim Iraqi Authority. . . . Therefore, we should accept that fact, not worry about that, and get on with the task and make sure it succeeds."[26] He went on to say that the Coalition will engage in "hands-on political reconstruction . . . consistently steer[ing] the process in ways that achieve stated U.S. objectives. The Coalition will not 'let a thousand flowers bloom.'"[27] He noted that the process will be "inherently untidy" and require trial and error.[28] Finally, he also urged respect for Iraq's "singular character," saying that the transition to democracy will take years and that "rushing elections could lead to tyranny of the majority."[29] In sum, the memo laid out a rationale for an extended and deeply engaged American occupation.

What Rumsfeld's memo did not do was lay out a plan for a political transition in Iraq. When Bremer arrived at the Pentagon for his initial meetings, Martinez told him that no specific decisions had been made in terms of governance in Iraq.[30] As Bremer recalls the discussions over the next several days:

> The direction that all of us followed was from the President, and his direction was quite clear: that we were going to try to set the Iraqis on a path to democratic government and help them rebuild their country. Now, none of us at that time [knew]—certainly I didn't know—what that would entail. The general guidance I had from the President and others was, "Get over there and give us your recommendation."[31]

---

[26] Donald Rumsfeld, "Principles for Iraq—Policy Guidelines," Draft Working Paper, May 8, 2003, p. 2.

[27] Rumsfeld, "Principles for Iraq," p. 2.

[28] Rumsfeld, "Principles for Iraq," p. 4.

[29] Rumsfeld, "Principles for Iraq," p. 4.

[30] Author interview with Roman Martinez, July 1, 2007.

[31] Paul Bremer, interviews with PBS Frontline, "The Lost Year in Iraq," June 26 and August 18, 2006.

U.S. government indecision on precisely what sort of transition would take place in the wake of the invasion had two consequences. First, it left the CPA bereft of plans, the preparations done by ORHA having been premised on an entirely different and a much more abbreviated vision of America's responsibility for the country's postwar governance. Second, and arguably more important, it left Iraqis with the impression that the United States had initially intended to hand over sovereignty quickly and then had gone back on its word, sowing the seeds of distrust between Iraqis and Americans.

Within days of his arrival in Iraq on May 12, 2003, Bremer concluded that any interim authority in Iraq could not consist only of the leadership council of Iraqi exiles that had emerged from the conferences of opposition groups. That council consisted of Kurdish leaders Jalal Talabani and Massoud Barzani, secular Shi'ites Ahmad Chalabi and Ayad Allawi, religious Shi'ites Abdul Aziz Hakim and Ibrahim Jaafari, and secular Sunni Nasser Chadirchi. All of these men had spent decades outside of Iraq or in Iraqi Kurdistan, and a number of them had strong links with Iran. It was not clear what support, if any, they would command among the newly liberated population of Iraq. Moreover, as Bremer has acknowledged, the CPA "wanted more control over creating the interim government than the [leadership council] wanted us to have."[32] In a briefing paper in advance of Bremer's first meeting with the council, Carpenter noted "a disconnect between what [the council's] expectations are and what our current position is vis-à-vis the pace of the Interim Authority's creation."[33] The memo went on to urge Bremer to emphasize his authority to the group, outline his priorities, and listen to their concerns, but not to commit to any particular program of action. The administrator did just that in his meeting with the council, known within the CPA as the G7, on

---

[32] Bremer, *My Year in Iraq*, p. 44.

[33] Scott Carpenter to Paul Bremer, "Tonight's Meeting with the Iraqi Leadership Council," May 16, 2003.

May 16, 2003.[34] In the days and weeks after this meeting, Bremer and his team sought to develop a transition plan. In a communication to President Bush a week after his arrival, Bremer wrote, "My message [to Iraqi leaders] is that full sovereignty under an Iraqi government can come after democratic elections, which themselves must be based on a constitution agreed by all the people. This process will take time."[35] In that brief passage, the administrator signaled his determination to oversee an extensive process of institution-building and his willingness to delay direct elections to do it. This position developed and solidified over the first weeks of the CPA. It reflected procedural, practical, and political concerns.

The CPA took the view that elections had to be delayed to allow the basis for representative government to develop in terms of political parties, civil society, and democratic habits more generally. Rushing into elections, on this view, would only harden divisions in a fragile society and empower extremist parties, opening the door to further instability and undemocratic outcomes. As Rumsfeld noted, "We need to lay a foundation for self-government. The way to get a nontheocratic system is to go slowly. That suggests we should not rush to have elections. . . . Otherwise, the fundamentalists will very likely sweep."[36] Bremer made a similar point in emphasizing the importance of developing what he termed the "shock absorbers" of civil society that would mediate the power of the state and help protect individual rights.[37] These concerns about quick democratization reflected the prevailing wisdom drawn from international experiences with nation-building in the 1990s as well as political scientists' efforts to theorize the process of democratic transition.[38] Every postconflict specialist Bremer spoke

---

[34] Six of the seven members attended. Adel Mahdi and Hamid Bayati attended on behalf of Abdul Aziz Hakim, who was reportedly ill. Bremer speculates that his illness might have been "diplomatic" due to his mistrust of the coalition (Bremer, *My Year in Iraq*, p. 46).

[35] Paul Bremer to George Bush, May 20, 2003.

[36] Donald Rumsfeld to Doug Feith, "Oil and Democracy," May 21, 2003.

[37] Bremer, *My Year in Iraq*, pp. 12, 19.

[38] See, for example, James Dobbins, Seth G. Jones, Keith Crane, and Beth Cole DeGrasse, *The Beginner's Guide to Nation-Building* (Santa Monica, Calif.: RAND, 2007), pp. 189–211;

to urged him not to rush into holding elections.[39] Drawing on this consensus, the CPA in its directives to governorate coordinators cited the example of Bosnia, in particular, where early national elections had further empowered ultranationalist parties.[40]

On top of this, there were practical issues. In Bremer's words, "There was no electoral law. There were no political parties' laws. There were no electoral constituencies. There had been no geographies that had been defined. There were lots of mechanical problems."[41] Organizing an election that met international standards would require building the apparatus of electoral democracy, and that would take time. Moreover, the situation in those first months remained chaotic, and the CPA was simultaneously trying to stabilize the country.

Finally, specific political concerns drove CPA decisionmakers. In particular, they sought to forestall the rise to power of the Iranian-linked Islamist parties, Hakim's Supreme Council of the Islamic Revolution in Iraq (SCIRI) and Jaafari's Da'wa Party. Hume Horan, in recommending that the CPA intervene to cancel a rudimentary election that U.S. Marines had initiated in Najaf, noted that SCIRI was by far the best-organized and best-funded political organization in the city and therefore would have an undue advantage over any other groups.[42] Given the Islamist parties' presumed strength in the Shi'ite south, some in the CPA worried they would dominate any election held before other political parties could establish themselves.[43] A different political concern had to do with preserving the CPA's own control over the political process in Iraq until an elected government could be formed. When discussing plans for creating local governance structures, Carpenter argued that it was "critical that no elections take place in the

---

and Jack Snyder, *From Voting to Violence: Democratization and Nationalist Conflict* (New York: W. W. Norton, 2000).

[39] Author interview with L. Paul Bremer, July 30, 2007.

[40] Rory Stewart, *The Prince of the Marshes: And Other Occupational Hazards of a Year in Iraq* (Orlando, Fla.: Harcourt, 2006), p. 214.

[41] Bremer, Interviews with PBS Frontline, June 26 and August 18, 2006.

[42] Hume Horan, "Draft Najaf Trip Report," May 28, 2003.

[43] Author interview with Scott Carpenter, August 2, 2007.

interim period prior to the ratification of a constitutional framework by the Iraqi people. Elections could create a legitimate counter authority to the CPA, making its ability to govern more difficult."[44] Another memo warned that local elections would "largely sacrifice Coalition control over the outcome."[45] Undergirding these views on democratization was another assumption that a later CPA document made explicit: "The Iraqi people will accept the legitimacy of the Coalition and the Interim Administration."[46]

The first task for the CPA's democratization plan was to create a larger and more representative group to supersede the exile-dominated leadership council. It was hoped that the new council would help put an Iraqi face on the occupation. At the start of June, Bremer met with the G7 and told them of his vision of an expanded consultative group that would fully represent Iraq's diversity and speak on behalf of the Iraqi people to the CPA.[47] He made clear, however, that this council would not be a government. The two tasks that the council would take on would be naming ministers to work with the CPA in running the bureaucracy and offering advice to the CPA on the political process.[48] Bremer invited the G7 to provide him the names of individuals who could populate the enlarged grouping. The members of the leadership council did not respond to this invitation with much enthusiasm, and it soon became apparent that the council would not simply expand itself. In Bremer's view, "the G7 had flunked the test."[49]

---

[44] Scott Carpenter to Paul Bremer, "Interim Local Selection Processes (ILSP)," May 20, 2003; author interview with Scott Carpenter, August 2, 2007.

[45] Ryan Crocker and Scott Carpenter to Paul Bremer (Drafted by Roman Martinez), "Building Towards a New Iraqi Constitution," undated.

[46] CPA, "Achieving the Vision to Restore Full Sovereignty to the Iraqi People," Working Document, September 4, 2003, Part 1, p. 8.

[47] Scott Carpenter and Ryan Crocker to Paul Bremer (Drafted by David Pearce), "Your Meeting with the G7," May 31, 2003.

[48] CPA HQ to Secretary of State and Secretary of Defense, "Iraqi Leaders Discuss Political Process, Currency, and Demobilized Military with Amb. Bremer," June 19, 2003.

[49] Bremer, My Year in Iraq, p. 89.

The CPA itself began a complicated process of seeking representative figures of some stature from around the country to populate the new council. This entailed governance team staffers traveling to different provinces and seeking out suggestions from regional CPA officials, coalition commanders, and local Iraqi notables.[50] Bremer and the CPA then met with the political figures they were already working with, both to persuade them to join the council and to vet new names.[51] When some of the Iraqi political leaders voiced opposition to specific candidates for the council, the CPA was often in no position to determine whether these complaints reflected legitimate objections or merely the rivalries of a new political order. For instance, Carpenter related an interview with candidate Yonadam Kanna of the Assyrian Democratic Movement in which he confronted Kanna with the fact that his name was on a list of Mukhabarat informers.[52] Finding good information—especially historical information—was not always easy. In a memo to Bremer, Carpenter detailed a conversation he had had with Kanna, who was being considered for a position on the Governing Council. U.S. intelligence had acquired a document that listed Kanna as a paid informer of the Iraqi Intelligence Service under Saddam Hussein. "As soon as he understood what I was accusing him of he began to sweat profusely," Carpenter noted. After regaining his composure, however, Kanna adamantly rejected that he had ever received payments from the Iraqi Intelligence Service, stating that he supposed the Kurdistan Democratic Party was behind the creation of the list since it had a strong incentive to undermine him. Carpenter told Bremer that he was in a bind because it was "extraordinarily difficult to assess the veracity of Kanna's assertions." Defense Intelligence officials told Carpenter that while the file was an authentic Iraqi Intelligence Service document, it

---

[50] Author interview with Roman Martinez, July 1, 2007.

[51] See, for example: Roman Martinez to Paul Bremer, "List of Names to be Shared with Leading Iraqis," July 1, 2003; Political Team to Paul Bremer, "Read-ahead for Your July 5 Meeting with Adnan Pachachi," July 5, 2003; Hume Horan to Paul Bremer, "Talking Points for Meeting with Imam Husayn al Sadr," July 4, 2003; Roman Martinez to Paul Bremer, "Kanan Makiya Meeting—July 5," July 4, 2003.

[52] Scott Carpenter to Paul Bremer, "Meeting with Yonadam Kanna," July 7, 2003.

did not necessarily mean much in itself. After considering the information, Carpenter recommended Kanna for the Governing Council "absent corroborating evidence he was a paid informer."[53]

As they gathered more names and interviewed more candidates, the CPA sought to achieve a council that balanced Shi'ite and Sunni, Arab and Kurd, religious and secular, exiles and "internals," and also included minorities and women. Bremer and his political team considered a variety of hypothetical sizes and compositions for the council.[54] Finally, after six weeks of nearly nonstop effort, they announced a multiethnic, 25-person council on July 13, 2003. The council's first order of business was to select a president. In a sign of challenges to come, the members could not agree on a single candidate and chose instead to have a nine-man presidency that would rotate monthly.

A large minority of the Governing Council were expatriates. Women and Kurds were also represented—a gender and ethnic minority that had been underrepresented in the past. Shi'ites were in the majority. The Governing Council was largely an advisory group, although Bremer did accord it the authority to nominate ministers. Through February 2004, the council's de facto power to veto or shape policies grew. The CPA made a conscious decision to accord the council this growing authority. Once serious negotiations began, in the late winter of 2004, on the Transitional Administrative Law (TAL), the interim constitution under which the interim government would operate, Iraqi political forces outside the Governing Council also demonstrated an ability to block or require change in key provisions. As a result, political influence moved away from the Governing Council and toward leaders of the most important ethnic and religious communities, to include in particular the Shi'ite Grand Ayatollah Ali Husseini al-Sistani.[55]

---

[53] Memo from Scott Carpenter to Ambassador Bremer, "Subject: Meeting with Yonadam Kanna," July 7, 2003.

[54] CPA Political Team to Paul Bremer (Drafted by Roman Martinez), "Size of the Governing Council," July 4, 2003.

[55] Bensahel et al., *After Saddam*, pp. 106–107.

In August 2003, the Governing Council chose and Bremer installed ministers in each of the Iraqi ministries. Up to this point, the ministries had been run by international senior advisors working for the CPA. But the Iraqi bureaucracy had difficulty working effectively for several reasons. First, most ministries were damaged or destroyed in the looting during and following the capture of Baghdad. In a number of instances, staff had no way to work following the end of the conflict. Second, civil servants had been discouraged from taking initiative under Saddam Hussein's regime. CPA staff often commented that most Iraqi civil servants were inefficient. They put in truncated hours, frequently failed to fulfill assignments, and in a number of instances, did not have any clear responsibilities. For example, the tax agency, which employed a few thousand people, had nothing to do for several months after the CPA decided not to collect taxes in 2003. Nevertheless, employees continued to be paid.[56]

Increases in government salaries provided by the CPA made government employment more attractive. As a consequence, applicants queued for these jobs. Civil servants and ministers frequently rewarded friends and relatives in this manner. For example, the number of directors general in the Ministry of Electricity rose from 12 to 80 between August 2003 and February 2004. Many of these individuals had ties to members of the Governing Council or ministers.

Conclusion

Bremer regarded the decision to mount an extended occupation rather than immediately accord power to an Iraqi interim government as having been made, in principle, prior to his appointment. He believed it embodied in the general guidance he had received from the President and Rumsfeld. The record on this point is unclear. The continuing debate over when and by whom the decision to mount a lengthy occupation was made reflects the general lack of clarity characteristic of the administration's planning for, and early management of, its inter-

---

[56] Bensahel et al., *After Saddam*, pp. 107–108.

vention in Iraq. Deputy Secretary of State Armitage has since indicated that he and other participants in the NSC process were surprised by this turn of events, but he also acknowledges that others, including Secretary of Defense Rumsfeld, and perhaps the President, were not.[57] Given that neither the President, nor any of his principal advisors had so much as met Bremer prior to his selection, something more than simple confidence in Bremer's judgment seems to have been in play. It seems likely, therefore, that the decision to supersede Garner almost immediately on his arrival in Baghdad was occasioned by the mounting chaos there and was accompanied by an inclination to assert a firmer American grip, one result of which was selecting and dispatching Bremer with a mandate to that effect.

Among others taken by surprise by the decision to mount an extended occupation were the U.S. military. According to Ricardo Sanchez, the armed services were left "completely in the dark. Throughout the summer of 2003," he recalls, "the services were operating under the guidance and expectation that a rapid withdrawal was to be expected. Chaos ensued in early July, 2003, when Abizaid stopped the redeployment of forces and required a replacement for any unit departing the country."[58] In Sanchez's view, the decision to mount a lengthy occupation was made by default as a result of de-Ba'athification, disbandment of the Iraqi army, and Abizaid's decision to halt the withdrawal of U.S. troops. This may be an oversimplification. But the fact that the commander on the ground was unable to discover when such a decision had been made—and by whom—is indicative of the confusion that surrounded these events.

It is impossible to know what would have happened if the United States had empowered an unelected Iraqi government in the spring of 2003. Perhaps the Iraqi leaders would have risen to the challenge. It seems equally possible, however, that the sectarian fighting that erupted in 2005–2006 would still have come, perhaps all the sooner, and at a time when Iraqi institutions would have been even less able to cope than they proved to be a couple of years later. Certainly an Iraqi gov-

---

[57] Author interview with Richard Armitage; Ferguson, *No End in Sight,* pp. 268–269.

[58] Author interview with Lieutenant General Ricardo Sanchez, January 27, 2009.

ernment formed in the spring of 2003 would have enjoyed an even narrower political base than the one empowered a year later, would have been more dominated by émigré leaders long absent from the country, and would have faced all the challenges that the CPA encountered.

What is certain is that this shift in policy left Iraqi leaders feeling deceived and senior people in the administration feeling unconsulted and uncommitted to the path Bremer was following.

CHAPTER FOUR

Establishing Security

Before leaving Washington, Bremer learned that the U.S. military was still operating on the basis of an order from CENTCOM Commander General Tommy Franks that aimed to withdraw most American troops from Iraq over the next few months.[1] Concerned, Bremer raised the matter of what he considered to be inadequate troop levels with both Rumsfeld and the President. On the evening of his May 12 arrival in Baghdad, Bremer told a somber gathering of senior staff, "Establishing law and order will be our first priority."[2] He repeated this statement in a letter to President Bush a week later, noting that it was critical to "impose law and order on the streets of Baghdad. This, far more than the much-discussed evolution of political structures, is what dominates the life of the average urban resident. . . . People must no longer fear to send their children to school or their wives to work."[3]

Bob Gifford, a U.S. State Department official who had been advising the Iraq Ministry of Interior, told Bremer when he arrived that whatever law and order existed under Saddam had broken down. Three weeks of largely unchecked looting had destroyed many government buildings in Baghdad. Gifford continued that the police had largely disappeared: "In theory there are about four thousand poorly trained

---

[1] According to Sanchez, Franks' order would have reduced the American force level in Iraq to 30,000 by August 1, 2003. Sanchez, *Wiser in Battle*, p. 168.

[2] Bremer, *My Year in Iraq*, p. 17.

[3] Letter from L. Paul Bremer to President George W. Bush, May 22, 2003.

officers on duty in Baghdad. But they're armed only with pistols. Most of them have just disappeared, like the army."[4]

President Bush had authorized Bremer to oversee, direct, and coordinate all U.S. government activities in Iraq—including in the security sector—*except* personnel under the authority of the local American military commander.[5] This meant that Bremer had responsibility for the Iraqi army, police, and other Iraqi security services, which came to include the Facilities Protection Service, Department of Border Enforcement, and Iraqi Intelligence Service. Bremer placed Walter Slocombe in charge of these institutions, assisted by Bernard Kerik for the police.

Disbanding the Army

The discussions about whether to disband the Iraqi army evolved over the course of the winter and spring of 2003. "In January 2003," noted Under Secretary of Defense for Policy Douglas Feith, "we presented to Secretary Rumsfeld a briefing proposing what to do with the Iraqi army." By March, senior Pentagon officials, having concluded that it made sense to try to keep a portion of the army intact and then to downsize it and reform it, presented that concept to the President. Garner produced a plan for the Iraqi army, which Feith briefed to the President at a March 10, 2003, National Security Council meeting. "I laid out the pros and cons of using the army, as well as the pros and cons of disbanding it." On the plus side, the army was viewed as having discipline, infrastructure, vehicles, and skilled personnel. It also made little sense to throw thousands of soldiers onto the street. At the same time, however, the army was poorly organized, corrupt, brutal, and anti-democratic, and it would be difficult to reform. "I told the President that after weighing the pros and cons, Secretary Rumsfeld

---

[4]   Bremer, *My Year in Iraq*, p. 18.

[5]   On the chain of command, see Letter from President George W. Bush to Bremer, May 9, 2003. Also see Secretary of Defense, "Memorandum for Presidential Envoy to Iraq, Designation as Administrator of the Coalition Provisional Authority," May 13, 2003.

supported keeping the army, but it was a difficult call."[6] President Bush backed Rumsfeld's recommendation.

But the situation on the ground began to change. During an April 17 secure video conference, General John Abizaid, then deputy to General Franks, reported to Deputy Secretary of Defense Wolfowitz that "there are no organized Iraqi military units left." The Iraqi police had also deserted their posts in all major cities. Looting was widespread and was doing billions of dollars of damage. All major government ministries, police stations, and government buildings sustained major destruction. Because the Iraqi army had "self demobilized," as the Pentagon put it, prewar plans to use the Iraqi military for postwar stability operations were rendered impractical, at least in the short term.

The dissolution of the army had been encouraged by the U.S. military, which lacked the manpower necessary to capture and intern Iraqi army members and therefore urged them to disperse to their homes, threatening to treat anyone armed and in uniform as hostile. This disjuncture between combat- and postcombat-phase planning was symptomatic of the larger failure to align ends and means through the transition from conventional combat to postconflict reconstruction.

In the weeks after Abizaid's briefing, Slocombe, who was preparing to take up his assignment in Baghdad, discussed options with top officials in the Pentagon, including Wolfowitz. These officials believed that an early recall of the former army would be a practical and political mistake for at least three reasons.

First, most believed that the Iraqi army had already self-demobilized. As Slocombe pointedly remarked: "Demobilization had already happened."[7] Senior policymakers argued that when Saddam's regime was toppled, there was not a single intact Iraqi military unit anywhere in the country. In a memo to Bremer, for example, Slocombe noted that "the old regular army has ceased to exist; a fortiori, there never was a civil MOD bureaucracy to call back to work. . . . Moreover, any such reconstituted units would have to be retrained into a more flexible, modern force with different ethos (and different officer-

---

6   Author interview with Douglas J. Feith, November 4, 2008.

7   Author interview with Walter Slocombe, May 5, 2008.

enlisted relations) than those that prevailed in the past."[8] With no army left in Iraq, this argument continued, the U.S. government was not really disbanding the Iraqi army. It was merely recognizing what had already happened: The army had disbanded itself.[9]

Second, CPA officials believed that disbanding the army had an important symbolic purpose. As Bremer later noted:

> It's absolutely essential to convince Iraqis that we're not going to permit the return of Saddam's instruments of repression—the Ba'ath Party, the Mukhabarat security services, or Saddam's army. We didn't send our troops halfway round the world to overthrow Saddam only to find another dictator taking his place.[10]

This meant building a new army from scratch that would include more representation from Iraq's diverse ethnic groups. Saddam's officer corps had been disproportionately Sunni, with almost all senior positions assigned to Saddam's loyalists. The rest of the army had been made up of draftees, many of whom were Shi'ites. In an email to Scott Norwood, Walt Slocombe noted that it was "right and necessary to dissolve the old army formally to clear the way to create any army suitable for the new, free Iraq."[11]

Third, CPA officials argued that infrastructure problems precluded standing up the old Iraqi army. "There was not a single unit or barracks left intact," McManaway noted. "So it was not a question of standing up a few old battalions."[12] When Saddam's military melted

---

[8]   Walt Slocombe to Amb. Bremer, "Involving Iraqi Ex Officers and Others in NIC and Defense Planning," May 25, 2003.

[9]   Walter B. Slocombe, "To Build an Army," *Washington Post*, November 5, 2003. Also see L. Paul Bremer, "How I Didn't Dismantle Iraq's Army," *New York Times*, September 6, 2007; Dan Senor and Walter Slocombe, "Too Few Good Men," *New York Times*, November 17, 2005; L. Paul Bremer, "What We Got Right in Iraq," *Washington Post*, May 13, 2007.

[10]   Bremer, *My Year in Iraq*, p. 54.

[11]   Email from Walt Slocombe to Scott Norwood, "Subject: Paper for USDP," July 15, 2003.

[12]   Author interview with Clayton McManaway, December 5, 2007.

away, barracks and bases had been demolished, stripped of all usable arms and equipment down to the wiring, plumbing, and even bricks.

The Iraqi army was also top-heavy. Saddam's army had been about the size of the American army, but CPA officials found that it had 11,000 generals compared to roughly 300 in the U.S. Army. (This figure apparently included colonels, and many officers who were effectively retired, but it is still comparatively large.[13]) Slocombe and other CPA officials argued that they could have offered positions to only a small percentage of the old officer caste, leaving the vast majority disgruntled. And the CPA assumed that few draftees would return to the military voluntarily. "To get them back we'd have to go into their homes and drag them out," argued Slocombe.[14] Slocombe's consultations with Americans officials in Washington and Baghdad convinced him that most agreed that the only viable course was to build a new, all-volunteer, professional force open to members of the former army against whom there was no persuasive evidence of major abuse. He drafted an order to accomplish these objectives. On May 10, drafts of this order were forwarded to the Secretary of Defense; the Defense Department's general counsel; Deputy Secretary of Defense Wolfowitz; DoD's Under Secretary for Policy, Douglas Feith; the head of Central Command, General Tommy Franks; and the coalition's top civil administrator at the time, Jay Garner, asking for comments.

This approach was consistent with at least some prewar thinking. The State Department's "Future of Iraq" study had concluded in May 2002, "the Iraqi Army of the future cannot be an extension of the present army, which has been made into a tool of dictatorship," although this position was at variance with most of the administration's prewar planning. On May 9, 2003, Secretary of Defense Rumsfeld circulated to other members of the National Security Council a memo titled "Principles for Iraq-Policy Guidelines" specifying that the coalition "will actively oppose Saddam Hussein's old enforcers—the Ba'ath Party, Fedayeen Saddam, etc." and that "we will make clear that the

---

[13] Author interview with Omar Al-Shahery, October 22, 2008.

[14] Bremer, *My Year in Iraq*, p. 55.

coalition will eliminate the remnants of Saddam's regime."[15] That same day, on the eve of his departure to Iraq, Bremer sent a memo to Secretary Rumsfeld and his general counsel, William J. Haynes, summarizing his own conclusion that dissolving Iraq's army and other security-related institutions would "reinforce our overall policy messages and reassure Iraqis that we are determined to extirpate Saddamism."[16]

On May 12, Garner and Bremer (the latter having just arrived in Baghdad) jointly sent Stephen Hadley, Deputy National Security Advisor, a memo noting their intention to pay all former Iraqi government employees except uniformed military and members of the Iraqi Intelligence Service. On May 13, en route to Baghdad, Slocombe briefed senior British officials in London on the proposal. They told him they recognized that "the demobilization of the Iraqi military is a fait accompli." His report to Washington following that visit added that "if some U.K. officers or officials think that we should try to rebuild or reassemble the old R.A. (Republican Army), they did not give any hint of it in our meetings, and in fact agreed with the need for vigorous de-Ba'athification, especially in the security sector." On May 15, two days after Bremer's arrival, Garner learned of the planned order to disband the army and remonstrated with him. "We have always made plans to bring the army back," he insisted. Bremer remained adamant. Garner did persuade Bremer to take the Ministry of the Interior, which oversaw the police, off the list of institutions to be dissolved.[17]

Over the following week, Slocombe continued consultations about the planned order with top Pentagon officials, including Feith. During that same period, Lieutenant General David McKiernan, the field commander of the coalition forces in Iraq, was sent the draft order disbanding the army, and his staff seem to have reluctantly cleared it,

---

[15] Bremer, *My Year in Iraq*, p. 39.

[16] Memo from Paul Bremer to Jim Haynes, "Subject: Proclamation on Dissolved Institutions, CC: Paul Wolfowitz, Doug Feith, Gen Franks, Gen Garner, Jaymie Durnan, Walt Slocombe," May 10, 2003.

[17] Woodward, *State of Denial,* p. 195, It seems probable that Garner had not received the draft of the order dissolving the army, which had been sent to him on May 10 before Bremer arrived on May 12.

although Mckiernan later stated that he had neither seen nor approved the order.[18] On May 19, Rumsfeld received a final draft of the proposed order for his approval. As Feith explained, "the changing situation on the ground led us to a different analytical conclusion than what we had come to in March," when he had briefed the President. "The pros—the arguments for trying to keep the army intact—had largely disappeared. For example, there was no discipline left in the army and it had, in fact, disbanded. And all of the cons remained."[19]

Apart from minor edits to the order, no senior military or civilian officials other than Garner formally raised objections to the proposal to dissolve most of Saddam Hussein's security apparatus. On May 22, the full National Security Council, with President Bush in the chair, was briefed on the plan. No one raised objections. However, this apparent unanimity masked serious reservations and misunderstandings.

On Friday, May 23, 2003, Bremer signed CPA Order Number 2, "Dissolution of Entities." The order formally dissolved a wide range of Iraqi institutions, including the Ministry of Defense and the Iraqi Intelligence Service. It terminated the service of all members of the former military and announced that the coalition planned to create a New Iraqi Army (NIA) "as the first step in forming a national self-defense capability for a free Iraq."[20] Bremer's press spokesman, Dan Senor, stayed up the entire night coordinating the text of the announcement and press plans with Rumsfeld's special assistant, Larry Di Rita, who was in Baghdad at the time.[21]

As noted, Bremer had informed the President and the other members of the National Security Council of his intended action on May 22, the day before the order was signed. "No one at the meeting said 'don't do it,'" noted Frank Miller, the senior NSC staffer responsible for coordinating policy toward Iraq. "To be clear, though, most of us had

[18] Michael R. Gordon, "Fateful Choice on Iraq Army Bypassed Debate," *New York Times*, March 17, 2008.

[19] Author interview with Douglas Feith, November 4, 2008.

[20] Coalition Provisional Authority, Order Number 2: "Dissolution of Entities," May 23, 2003.

[21] Author interview with Dan Senor, October 31, 2008.

no advanced warning that it was coming. No one from the Pentagon had brought this to our attention. It was blown through the system."[22] Bremer had also sent a letter to President Bush the same day noting that he would parallel dissolving the Ba'ath party "with an even more robust measure dissolving Saddam's military and intelligence structures to emphasize that we mean business."[23]

Colonel Paul Hughes had been acting as ORHA's principal liaison with remnants of the Iraqi army. He was in touch with officers who had in turn registered 137,000 former soldiers who were applying for the $20 payment that Garner had promised all Iraqi government employees, most of whom had not been paid for the past several months.[24] General John Abizaid, then Franks' deputy at CENTCOM, had also met with several Iraqi army generals.[25] Hughes and other U.S. military officers believed, based on these contacts, that the bulk of the Iraqi army would respond positively to a recall. Garner's decision on a one-time payment to government employees had specifically excluded former military personnel. Bremer's May 23 order dissolving the various security-related entities promised a one-time termination payment of unspecified amount to dismissed employees but did not indicate whether this applied for former soldiers. The CPA then spent the following month debating whether, how, and how much to pay dismissed soldiers.

Although London had not objected to the dissolution of the army, British officials were concerned over the manner in which this was done. During his stop in London, Slocombe was told that the "reintegration of former military into society will be an issue. Large numbers of unemployed former soldiers have created crime problems in other places." Slocombe's initial response was curt: "[T]he military was not an appropriate tool to solve the unemployment problem." He continued that the "new military should not be expected to sop up

---

[22] Author interview with Frank Miller, June 6, 2008.

[23] Letter from Jerry Bremer to President George W. Bush, May 22, 2003.

[24] Author interview with Paul Hughes, November 21, 2008.

[25] Sanchez, *Wiser in Battle*, p. 176.

unemployment—if only because at any plausible size, it would not sop up very much."[26] Bremer and his British colleague, John Sawers, met with a group of Iraqis that included political leaders, independents, government bureaucrats, and professionals. According to Bremer, they had a uniform request: "All argued that we should continue to pay 'the salaries' of former military members. . . . Participants felt that such payments made sense on security, humanitarian, legal, and other grounds."[27]

CPA officials recognized some potentially negative consequences of disbandment. In a May 19 message to Secretary of Defense Rumsfeld, Bremer explained that "the order will affect large numbers of people."[28] A month later in another note to Rumsfeld, Bremer stated that when the CPA dissolved the Ministry of Defense and the old armed forces, it left roughly 230,000 officers and noncommissioned officers unemployed, "some of whom have been demonstrating in cities around Iraq protesting their not having been paid. This discontent among a respected group with training in weapons and with networks of contacts and loyalties presents a significant threat."[29] The CPA and U.S. Central Command began to focus on a two-phased approach to reintegrate demobilized Iraqi soldiers. Slocombe announced that the CPA intended to have a full division of 12,000 New Iraqi Army soldiers trained and operational in one year, and three divisions a year later. In addition, the CPA announced on June 23—a month after the dissolution order—a program of transition payments to former career military personnel. The CPA refused to pay anything to those in the top four Ba'ath party ranks, who numbered about 6,000. Other career personnel would begin to receive a monthly stipend, although it took

[26] Walt Slocombe to Amb Bremer, "Results of Slocombe Meetings at NATO on 12 May and in London 13 May 2003," May 14, 2003.

[27] Memo from L. Paul Bremer to Secretary Rumsfeld, "Subject: June 19 Political Consultation with Iraqis," June 20, 2003.

[28] Memo from Paul Bremer to Secretary Rumsfeld, "Subject: Dissolution of the Ministry of Defense and Related Entities," May 19, 2003.

[29] Memo from Ambassador Bremer to SecDef, "Subject: Should We Pay the Ex-Military?" June 15, 2003.

yet another month to begin making the promised payments because of the lack of a functioning banking system and difficulty in identifying those eligible.

Slocombe and other officials initially resisted paying soldiers who only weeks earlier had been in arms against the United States, although most of them had, in fact, simply gone home as they had been urged to do by the U.S. military. Also, as a practical matter, it was not until the middle of June that the coalition was able to obtain the personnel roster of the prewar Iraqi army, which was needed to make payments. Had the CPA so chosen, it could have begun with the list of 137,000 soldiers assembled by Hughes, although doing so could have caused problems with those not so paid. Once the full roster was obtained, the CPA established stipends roughly equivalent to the base pay of the former soldiers. A smaller one-time payment was eventually made to former conscripts. Payments were begun in early July 2003. They continued for the entire 14 months of CPA's existence and have been sustained since under successive sovereign Iraqi governments.[30]

As soon as the coalition announced its intention to pay Saddam's soldiers, organized demonstrations by former military personnel stopped, leading coalition military intelligence to drop "former officers" as a threat category. Nevertheless, it was during this period that insurgent groups began to pay young men $100 to kill a U.S. soldier and $500 to disable a Bradley or Abrams armored vehicle. How many of these volunteers were former soldiers is unknown. Financial uncertainty among former career soldiers, most of whom had retained their weapons, may well have contributed to support for the emerging insurgency.

The CPA developed, but proved unable to implement, a plan to help former soldiers gain skills and find jobs in the civilian economy. ORHA plans in this regard had assumed most of the army would remain under arms. In fact, even as Bremer was announcing his new policy, Hughes was back in Washington negotiating with the contractors who were to implement ORHA's original, rather limited reintegration scheme. The CPA subsequently contracted with the International

---

[30] Author interview with L. Paul Bremer, August 12, 2008.

Organization for Migration (IOM), a nongovernmental organization, to undertake a comprehensive program of reintegration of former army soldiers. But after the August 2003 bombing at the UN headquarters in Baghdad, the IOM, like all UN-related organizations, was ordered to withdraw from Iraq in accordance with UN security procedures and was therefore unable to implement this contract.

Building the New Army

While senior U.S. military officers did not formally object to Bremer's order dissolving the Iraqi army, they did expect army members to be subsequently recalled in large numbers fairly quickly.[31] CPA officials, on the other hand, foresaw a much more deliberate process for raising the new army, one which would rely largely on former Iraqi army veterans but would remobilize them much more slowly in much smaller numbers. Colonel John Agoglia, then a CENTCOM liaison officer with the CPA, recounted a mid-June conversation between Bremer and Major General Paul Eaton, who had been brought in to assist Slocombe in building a new army. Bremer explained, "Listen, we're going to recall the army, but we are not going to do it in three months: we're going to do it in two years, and we're only going to recall three divisions over those two years."[32]

Bremer was not averse to having the U.S. military take over responsibility for training this new army. Shortly after Sanchez's arrival to take command of coalition forces in Iraq, Slocombe briefed him on plans for training the Iraqi army. Sanchez was puzzled as to why Eaton, a major general on active duty, should be working for Slocombe rather than himself and suggested that it would make more sense to put this program under the military chain of command, with CPA responsible for setting the overall policy. Slocombe agreed, and so did Bremer. But the agreement was not implemented, apparently due to lower-level resistance and a lack of higher-level follow-through, until Rumsfeld

---

[31] Sanchez, *Wiser in Battle,* p. 190; Ferguson, *No End in Sight,* p. 210.

[32] Ferguson, *No End in Sight,* p. 225

dictated the transfer in March 2004. Instead, the U.S. military concentrated on creating the Iraqi Civil Defense Corps, elements of which were raised locally, given limited training and light arms, and attached directly to U.S. military units, with whom they operated and on whom they depended for support and oversight.[33]

CPA officials formally established the NIA through Order Number 22, which set out the mission, conduct, discipline, terms and conditions of service, rank structures, and administrative arrangements for the army.[34] In a June 3 briefing to senior Pentagon officials, Slocombe stated that the CPA would seek to build "an Iraqi army that would fit professionally and affordably into a new, democratic Iraqi government."[35] This army would focus on external, not internal, security.

Bremer intended that Slocombe should be in charge of all security-related lines of operations. Kerik chafed at the subordination, however, and Bremer gave him independent authority, leaving Slocombe to concentrate on rebuilding the Iraqi military. Slocombe's chief helper in this task was Eaton, who headed the Coalition Military Assistance Training Team (CMATT). Eaton had virtually no time to prepare for the mission, having received a call on May 9, 2003, from General Kevin Byrnes, commander of the U.S. Army Training and Doctrine Command. "We just kind of looked at each other," Eaton noted, "and I said 'it's a little late, getting this kind of notification. I would have figured the guy to do that would have been on station already.'" Eaton further noted that "In the beginning, there was no, zero, urgency on the part of the Secretary of Defense to provide the requisite resources to truly develop the Iraqi security force."[36]

---

[33] Author interview with Walter Slocombe, May 5, 2008; author interview with Lieutenant General Ricardo Sanchez, January 27, 2009.

[34] Coalition Provisional Authority, Order Number 22: "Creation of a New Iraqi Army," August 18, 2003.

[35] Feith, *War and Decision*, p. 434.

[36] "Iraq: Three Years, No Exit: Rebuilding Iraq Has Been Tougher Than Expected," CBS News Online, March 13, 2006.

Despite these difficulties, Slocombe informed Bremer on July 16 that training of the army's 1st battalion was under way and he was "heading for 18 infantry [battalions] trained and operating, with support elements, and higher echelon HQs and staffs named and at work" by the end of 2004.[37] By September and October, the CPA hoped to have the first battalion commissioned and operating; the second, third, and fourth battalions engaged in training; and equipment needs defined for the entire army. In a departure from the Saddam era, the CPA decided the new Iraqi army would be an all-volunteer force. CMATT began accepting volunteers with service in the old army up through the rank of lieutenant colonel and planned to promote from within to create new general officers. The CPA also decreed that the army—and other security forces more broadly—would reflect the ethnic, religious, and regional diversity of the country. In addition to training efforts, the CPA also created the Office of Security Affairs to build the Ministry of Defense. It served as the CPA's defense policy office and as the actual Ministry of Defense until the new one was established in May 2004.[38]

In July, a team of contractors from the Vinnell Corporation arrived in Iraq to help with the training effort. Composed of retired Army and Marine Corps personnel, the team was supposed to begin planning and preparations to train the new army. The Vinnell contract provided planners, operations officers, unit trainers, and translators, but the U.S. government had not asked the company to provide drill instructors—the trainers who work directly with military recruits to instill fundamental skills and knowledge. Instead, CMATT assumed that U.S. and coalition forces would provide the soldiers to serve as drill sergeants for the New Iraqi Army's basic training. U.S. Central Command, however, never tasked that mission to CJTF-7, and the drill sergeants did not materialize until much later. This situation put

---

[37] Memo from Walt Slocombe to Ambassador Bremer, CC: LTG Feliu, MG Eaton, Jerry Thompson, Dean Popps, "Re: Timelines for National Security and Defense," July 16, 2003.

[38] Andrew Rathmell, Olga Oliker, Terrence K. Kelly, David Brannan, and Keith Crane, *Developing Iraq's Security Sector: The Coalition Provisional Authority's Experience* (Santa Monica, Calif.: RAND, 2005), p. 26.

CMATT in a bind. While its staff had grown to 18, it was still far too small to provide drill instructors from within its own organization. Assistance came from the British and Australian armies, which provided officers to support the basic training mission. CMATT eventually received seven U.S. officers from the 3rd Infantry Division for a two-week period that summer.[39]

The Vinnell Corporation subcontracted its training component to MPRI and recruitment to SAIC. Other private security firms, such as Blackwater and DynCorp International, also became involved in a range of such security tasks as training, convoy security, and protective security.

In late summer 2003, American commanders began forming and training their own Iraqi paramilitary units, called the Iraqi Civil Defense Corps (ICDC). The ICDC included a panoply of Iraqi units that individual U.S. divisions and brigades recruited to assist in such tasks as trash cleanup, construction, base security, and even patrolling. CJTF-7 eventually added other units. This was intended only as a temporary force. As a CPA analysis of the ICDC noted, "it is not a permanent institution: Enlistment is for a year; program to be evaluated and reviewed in 6 months."[40]

The ICDC end state was not defined. The CPA felt that this was a decision for an Iraqi government to make. Responsibility for authorizing, funding, and equipping the forces remained with CPA, while CJTF-7 assumed responsibility for their training and operational employment. As Sanchez later remarked, "Some of the initial training programs were absolutely abysmal and some were superb." This led CJTF-7 to standardize the program in the spring of 2004, something Sanchez acknowledged, "we should have done from the very beginning." The ICDC remained a major bone of contention between CJTF-7 and the CPA, and, at least in Sanchez's view, the CPA per-

---

[39] Wright and Reese, *On Point II*, p. 435.

[40] Email from Meghan L. O'Sullivan to Scott Norwood, "Subject: ICDC," August 6, 2003.

ceived the ICDC "as a direct competitor for funding with the New Iraqi Army and as a major detractor."[41]

In September, the Department of Defense added the creation of an Iraqi Air Force and an Iraqi Coastal Defense Force to the CPA's responsibilities.[42] In November 2003, the CPA tasked the Combined Joint Special Operations Task Force–Arabian Peninsula under CJTF-7 to begin training the 36th Iraqi Commando Battalion. The commander of the Combined Joint Special Operations Task Force–Arabian Peninsula assigned three U.S. Army Special Forces Operational Detachments–Alpha to the mission. More changes were ordered in spring 2004, following the Sunni and Shi'ite uprisings across Iraq in anticipation of the turnover of sovereignty to the Iraqis.[43]

Once CMATT had trained and equipped army recruits, their battalions joined coalition forces in security missions. The 4th Infantry Division, operating in central Iraq, employed the first battalion; the 1st Armored Division in Taji employed the second battalion; while the third battalion deployed to Mosul with the 101st Airborne Division.

As the insurgency began to gather momentum in the late summer and early fall of 2003, there was growing alarm at the slow pace of Iraqi army training.[44] At the time Bremer dissolved the old Iraqi army in late May, the U.S. military had still been operating under guidance that assumed a reduction in its force levels in Iraq to as few as 30,000 by year's end. This objective was soon abandoned, further withdrawals were canceled, and some reinforcements were sent; but the U.S. military and its civilian masters still hoped that Iraqi forces could quickly assume greater responsibility for public security. Pentagon officials consequently pushed for using larger numbers of Iraqi army and police units to combat the growing insurgency. In a brief trip to Iraq

---

[41] Author interview with Lieutenant General Ricardo Sanchez, January 27, 2009.

[42] On the Coastal Defense Force see, for example, Information Memo from Paul D. Eaton to the Administrator, "Subject: Iraq Coastal Defense Force Naval Base Site," November 9, 2003.

[43] Wright and Reese, *On Point II*, pp. 430–431.

[44] Info Memo from Walt Slocombe to Amb. Bremer, "Subject: Welcome Home," August 31, 2003.

in August, Secretary Rumsfeld advocated an extended training program that would stand up 27 battalions of the Iraqi army in one year
instead of the two to four initially outlined by Slocombe.[45] Senior CPA
officials objected, noting that Iraq's security forces were not yet able
to take on this burden. In a memo to Secretary Rumsfeld, for example, Bremer argued that "none of Iraq's security institutions—the New
Iraqi Army, the Iraqi Police Service, the Iraqi Civil Defense Corps or
the Iraqi Border Police—are ready to assume full responsibility" for
establishing security. Consequently, he concluded, "Iraq's security will
rely on foreign forces in the immediate future."[46] But the pressure from
Washington continued, and Rumsfeld wrote a memo to Bremer and
Abizaid a week letter noting that "our goal should be to ramp up the
Iraqi numbers, try to get some additional forces and find ways to put
less stress on our forces, enabling us to reduce the U.S. role."[47]

Some CPA officials quipped in frustration that the Pentagon was
trying to make Iraqi forces into bionic men, after a U.S. television
series of the 1970s whose motto was to make its hero, Steve Austin,
"better, stronger, faster." Rumsfeld sent a memo to Bremer about training of the Iraqi army that stated, "I am concerned about the pace of
the recruiting and training of the Iraqi army. It feels slow. I wonder if
we could consider requiring each U.S. division to recruit and train a
brigade of former soldiers every quarter."[48] Members of Iraq's Governing Council pushed for a quicker handover to Iraqi forces. One of the
most vocal was Ahmad Chalabi, who argued that "only Iraqis would
be capable of improving the security situation."[49]

---

[45] Author interview with L. Paul Bremer, July 30, 2007.

[46] Memo from Paul Bremer to Secretary Rumsfeld, "Subject: Your Meeting with the Governing Council," September 6, 2003.

[47] Memo from Donald Rumsfeld to Gen John Abizaid and Paul Bremer, CC: Gen Dick
Myers, Paul Wolfowitz, Doug Feith, and Reuben Jeffery, "Subject: Reporting on Security
Issues," September 12, 2003.

[48] Memo from Donald Rumsfeld to Jerry Bremer, CC: Gen. Dick Myers, Paul Wolfowitz,
Gen. John Abizaid, and Doug Feith, "Subject: Iraqi Army," October 15, 2003.

[49] Memo from Roman Martinez through Meghan O'Sullivan to the Administrator, "Subject: CODEL Meeting with Governing Council," October 27, 2003.

On September 5, 2003, Secretary Rumsfeld ordered the first of two major expansions of the original plan for Iraq's armed forces. Driving this expansion was the realization among senior U.S. policymakers by early fall 2003 that a coherent insurgency was emerging in Iraq and that Iraqi security forces had to play a critical and immediate role in engaging that threat. The new program called for the original 27 battalions and three divisions of the New Iraqi Army to be operational by September 1, 2004, two years earlier than the original June 2003 plan. Based on this new timeline, Eaton requested a significant "augmentation of forces in order to accomplish the accelerated mission," which included money and personnel to help recruit the force, staff a training academy, and establish a military advisor and training assistance group.[50]

Rumsfeld's September push also included the creation of an Iraqi Coastal Defense Force for river and coastal patrolling, and an Iraqi Air Force, to be initially equipped with eight C-130 transport aircraft and 12 UH-1 "Huey" helicopters. Iraq's expanded military forces would be stationed at brigade-size garrisons, at one air base, and at one naval base with supporting recruiting offices, training centers, and support facilities of all types. The plan required a major increase in U.S. and coalition support, including mobile training teams, embedded unit advisors, equipment fielding teams, and significantly greater military and civilian construction capability. The cost of the program ballooned from the $173 million in the first phase to just over $2.2 billion in the second.[51]

In December 2003, Eaton sent a memo to CJTF-7 and U.S. Central Command outlining progress in building the army. He noted that the first Iraqi brigade headquarters, which included the first, third, and fourth battalions, would be operational by January 2004, and the entire brigade would be operational by April 2004. He then explained that two more brigades would be operational by July, another three

---

[50] Memo from Paul D. Eaton to the Chairman, Joint Chiefs of Staff, through Administrator Coalition Provisional Authority and Commander USCENTCOM, "Subject: Request for Forces to Assist with Training and Equipping the New Iraqi Army," November 6, 2003.

[51] Wright and Reese, *On Point II*, pp. 441–442.

brigades by August, and three more by October 2004. The army's mission, he pointedly remarked, was both internal and external, including "defense of the national territory and the military protection of the security of critical installations, facilities, infrastructure, lines of the communication and supply, and population."[52] This reflected a lengthy debate between CJTF-7 and the CPA over the role of the new Iraqi military, with the former adamant that it should include internal security, and the latter resistant.[53]

By January 2004, Eaton's CMATT staff had grown to 200. Spain, the United Kingdom, Australia, and Poland contributed officers to CMATT, and a British brigadier general, Jonathan Riley, served as CMATT's deputy chief.[54] Several countries, including Jordan, also provided assistance in rebuilding the Iraqi army. As Eaton explained in a memo to Bremer, "I went to Jordan 25 August to look into buying used military equipment and selective senior officer training opportunities (Major, Lieutenant Colonel and Colonel). The more I observed Jordanian officer behavior the more I was convinced that I had an opportunity to accelerate Iraqi Army development in a substantial way and expose Iraqi officers to a good Army with good leadership in a rational actor state."[55] Eaton argued that the Jordanian army was one of the best in the Arab world; there was no language barrier; and there was little opposition from Sunni, Kurdish, and Shi'ite officers in the army.

But there was "intense" pressure from some in the Governing Council against Jordanian involvement. As David Gompert, who succeeded Slocombe as senior advisor for defense and security affairs, explained to Bremer, "Points of criticism include that exiled Ba'athists and others with connections to insurgents have gained a foothold in Jordan and would infiltrate and influence [Iraqi Air Force] officers

---

[52] Memo from Paul D. Eaton to COMUSCENTCOM, CG CJTF-7, "Subject: The New Iraqi Army—Capability Statement," November 22, 2003.

[53] Author interview with Lieutenant General Ricardo Sanchez, January 27, 2009.

[54] Wright and Reese, *On Point II*, p. 446.

[55] Information Memo from MG Paul Eaton to the Administrator, "Subject: Officer Training Initiatives—Jordan," January 5, 2004.

training there. Regardless of the validity of these complaints, members of the [Governing Council] are mounting concerted criticism of training in Jordan."[56]

Despite CMATT's efforts to improve its structure, senior Pentagon officials found that the overall coalition effort needed to be revamped. In January 2004, Rumsfeld sent Army Major General Karl W. Eikenberry to assess the Iraqi security force training programs. Eikenberry had just completed more than a year in Afghanistan as head of the Office of Military Cooperation–Afghanistan, which was in charge of building the Afghan National Army. Eikenberry found CPA training efforts underresourced and disorganized and recommended that both police and military be consolidated for efficiency and effectiveness and placed under the command of CJTF-7.[57]

Bremer had no difficulty with transferring army training to the U.S. military, a move that he had favored all along, but he was opposed to doing the same with the police. He also objected to Eikenberry's suggestion that the planned size of the Iraqi army be reduced in favor of a more rapid buildup of the Iraqi Civil Defense Corps, lightly armed and quickly trained auxiliary troops attached directly to U.S. units across the country. As Bremer wrote in a memo to Rumsfeld, "I do not agree with the plan to reduce the Iraqi Armed Forces to a single division. Such a decision essentially overturns everything we have said to the Iraqi people about our intention to produce an army for Iraq's legitimate defensive needs. And it undercuts our consistent message that we want Iraqi security forces to assume responsibility for Iraq's security."[58] In addition, Bremer and John Abizaid wrote a joint memo to Rumsfeld noting that, while they agreed that "coalition forces should play an expanded role in the execution of ISF [Iraqi security force] training,"

[56] Action Memo from David C. Gompert to the Administrator, "Subject: Egyptian Military Training Assistance to the IAF," January 6, 2004.

[57] Sanchez, *Wiser in Battle*, p. 315.

[58] Memo from Paul Bremer to Secretary Rumsfeld, "Subject: Security Assessment," February 3, 2004.

they believed "policy and resource authority should remain with CPA (and then the US Mission to Iraq)."[59]

Despite significant increases in the number of recruits churned out by CMATT, several major challenges ensued in early 2004. In the spring of 2004, on the first day of the coalition offensive to retake Fallujah from Sunni insurgents (see Chapter Ten), CJTF-7 ordered the second battalion of the New Iraqi Army, then operating with the 1st Airborne Division north of Baghdad in Taji, to support the offensive led by the U.S. Marine Corps. The Iraqi unit was to man checkpoints and form a cordon around the city. Of the five companies in the battalion, two were on leave. The three companies on duty boarded trucks for the move accompanied by a new ten-person CMATT advisor team from the U.S. Marines. As they drove through a Shi'ite neighborhood in Baghdad, a large crowd accosted them about the immorality of attacking fellow Iraqis. Shots rang out and seven Iraqi soldiers were wounded, whereupon the convoy returned to camp. CJTF-7 then provided helicopters to move the Iraqi companies to Fallujah later that night. By that time, however, the unit had begun to dissolve as groups of soldiers refused to take part in an operation that would pit them against other Iraqis. Major General Eaton recalled the situation:

> At the Pickup Zone, in the dark, blades turning on several CH-47s, about 70 Iraqi Soldiers became demonstrably upset . . . the situation was chaotic and the senior Marine, Major Chris Davis, called me to inform me of what was going on. . . . Major Davis indicated he was about to stand the unit down, and ultimately did so. I met him at dawn the next morning after a dangerous trip from Baghdad to review the situation. We dismissed the 70 Iraqi soldiers who were the greatest problem, and changed out three company commanders and the battalion commander, replacing them from within the battalion.[60]

---

[59] Memo from Paul Bremer and Gen. Abizaid to Secretary Rumsfeld, "Subject: Eikenberry Report," February 12, 2004.

[60] Wright and Reese, *On Point II*, p. 449.

In March 2004, Rumsfeld transferred CMATT from CPA to CJTF-7 supervision. Under U.S. military oversight, the numbers trained grew quickly, but quality did not keep pace. Desertion was a problem because of low salaries and collapsing security in the country. And the army's mission remained somewhat uncertain between protecting the country from external threats and countering a spiraling insurgency within its border. One assessment concluded that the Iraqi army "suffered most, perhaps, from the unclear nature of its mission. . . . While CENTCOM planners had from the start expected to use Iraqi military personnel for internal security, the force being built was designed with an external defense role in mind."[61]

Reforming the Police

In a May 19 memo to Rumsfeld, Bremer argued that disbanding the army was critical "to destroy the underpinnings of the Saddam regime and to demonstrate to the Iraqi people that we have done so, and that neither Saddam nor his gang is coming back."[62] Unlike the Iraqi army, however, the U.S. military and the CPA believed the police needed to be retained. In a subsequent email, Bernard Kerik contended that the United States needed to get "police officers back to work" and to recruit additional officers "as quickly as possible."[63]

One of the initial challenges, however, was the poor state of the police. "Police were at the bottom of the barrel in Saddam Hussein's Iraq," remarked Douglas Brand, who served as the chief police advisor for Iraq under Kerik. "There had been little development of the Iraqi police since the first Gulf War. The police had little equipment and

---

[61] On desertion rates, see, for example, Rathmell et al., *Developing Iraq's Security Sector*, p. 40.

[62] Memo from Paul Bremer to Secretary Rumsfeld, "Subject: Dissolution of the Ministry of Defense and Related Entities," May 19, 2003.

[63] Email from Bernard B. Kerik to Paul Bremer, CC: Patrick Kennedy, "Subject: Recruitment of Police Personnel," June 5, 2003.

training, and didn't know how to do basic forensics."[64] A May 2003 CPA document concluded that "the Iraqi Police, as currently constituted and trained, are unable to independently maintain law and order and need the assistance and guidance of Coalition Force assets (or some appropriate follow on force) to accomplish this task." It continued that they "have suffered years of neglect, coupled with a repressive command structure that prohibited training, proactivity, initiative and stifled attempts toward modernization of the police."[65] Kerik's assessment was equally grim, concluding that the "Iraqi police are not competently trained or constituted to accept the task of providing executive law enforcement without the assistance of . . . substantial peace keeping/training forces. Most Iraq citizens view the police as unprofessional and corrupt." He continued that the police infrastructure was in dismal condition, there was no culture of human rights, the police lacked sufficient equipment, and there was no external oversight of the police.[66]

Despite these concerns, Kerik made a push to get police onto the street. Bremer also asked General Abizaid for help via secure telephone: "I need as many American Military Police as you can get me as fast you as you can get them here." Abizaid thought for a moment and then responded: "I might be able to scrape up about 4,000 more," noting that he could start flying them into Iraq from the United States and Europe within 48 hours.[67]

UN peacekeeping missions routinely deploy one international police officer for every ten soldiers. In Kosovo only three years earlier, some 5,000 UN international police (of whom around 500 were American) had been deployed, alongside nearly 50,000 NATO troops. Despite this recent experience, there was considerable skepticism

---

[64]   Author interview with Douglas Brand, January 16, 2009.

[65]   Coalition Provisional Authority and Iraqi Interior Ministry, *Iraq Police: An Assessment of the Present and Recommendations for the Future* (Baghdad: Coalition Provisional Authority and Iraqi Interior Ministry, May 30, 2003), p. 4.

[66]   Memo from Bernie Kerik to Ambassador L. Paul Bremer III, "Subject: Iraqi Police Service Talking Points," July 13, 2003.

[67]   Bremer, *My Year in Iraq*, p. 32.

within the Bush administration about deploying American or other international civilian police to Iraq. In the months leading up to the invasion, the State Department had proposed sending 5,000 armed civilian police once the fighting had stopped. The White House cut that number to 1,500 and decreed that they should all be unarmed.[68] Garner went to Rice in early March to appeal this decision, but succeeded only in getting her agreement to reconsider it at a later date once he got on the ground in Iraq.[69] In an August 8 memo to Bremer, NSC officials Elliot Abrams and Frank Miller argued against trying to deploy significant numbers of civilian police. "While we understand the urgent need to create a functional police force in Iraq and we fully support your efforts to do this," they noted, "we have some misgivings that a full-scale CivPol [civilian police] plan would contribute significantly to accomplishment of that goal." Among their most salient concerns was that it would be impossible to get the number of officers called for in the plan. The solution to this problem was not the UN, Abrams and Miller argued, since many UN police were of poor quality and had poor human rights records.[70] Instead, CPA and the U.S. military scrambled to build a police force. Kerik began a shoestring program to retrain existing police in the basics of what he called "modern policing."[71] By mid-July more than 15,000 officers had returned to duty, and by August there were 32,000 police. But Kerik was still concerned that the numbers were too low, and in a briefing to Bremer he argued that "for adequate security, a country needs a policeman for every 300 to 350 inhabitants. So Iraq needs something like 65,000 to 75,000 of them. And today we've got, at best, about 32,000."[72] The

---

[68] The State Department had great difficulty deploying even a small fraction of this reduced number through the end of 2003.

[69] LTG (Ret.) Jay Garner, interview with PBS Frontline: "Truth, War & Consequences," July 17, 2003; Ferguson, *No End in Sight*, p. 237.

[70] Memo from Elliot Abrams and Frank Miller to Ambassador L. Paul Bremer, "Subject: Civilian Police Mission in Iraq," August 8, 2003.

[71] Memo from Bernard B. Kerik to Ambassador L. Paul Bremer, "Subject: Recommendations and Strategies for 'Standing Up an Interim Police Service,'" May 30, 2003.

[72] Bremer, *My Year in Iraq*, p. 128.

U.S. military also began to recruit, train, and equip police forces, typically with little coordination with CPA. "The U.S. military was always polite when we talked to them about policing," noted Brand. "But they did their own thing."[73]

CPA officials had initially identified Taszár Air Base in Hungary as a potential site for Iraq's police training facility. A CPA on-site assessment indicated that it had adequate barracks, mess halls, classrooms, and firing ranges.[74] Hungarian government officials had initially been supportive of the plan, but the Hungarian parliament stalled on the CPA's request. In August, Bremer informed Kerik that the "police training deal in Hungary is dead," noting that "their parliament has to 'study' and debate the issue. Meaning no decision before the end of the year at the earliest."[75] This required looking elsewhere for a suitable facility. Bremer dispatched McManaway to Amman to meet with King Abdullah. The king agreed that Jordan could provide a locale for police training, and in September the CPA formally signed an agreement with Jordan to build a police academy.[76]

The CPA wanted a secure site that would enable the United States to train up to 35,000 new recruits in basic policing skills and democratic policing.[77] The program's goal was to establish a 70,000-member Iraqi police force over a period of 18 to 24 months and also to reform the Ministry of Interior.[78] But the bulk of Iraq's police training pro-

---

[73] Author interview with Douglas Brand, January 16, 2009.

[74] Coalition Provisional Authority and Iraqi Ministry of Interior, "Taszár Air Base, Hungary—Facility Inspection Report & Plan," July 10, 2003.

[75] Bremer, *My Year in Iraq*, p. 152.

[76] Memo from Walt Slocombe to Ambassador Bremer, CC: LTG Sanchez, Ambassador McManaway, Commissioner Kerik, "Re: Meeting in Jordan," August 5, 2003; author interview with L. Paul Bremer, November 15, 2007.

[77] "Agreement between the Government of the United States of America and the Government of Jordan Concerning the Use of a Jordanian Site for Iraqi Police Training," September 2003.

[78] Info Memo from Ambassador Bremer to Secretary of Defense, "Subject: Iraqi Police Training and Development," September 29, 2003.

gram did not begin until late November 2003, an inordinately long delay that had significant and enduring effects.

There were occasional bright spots. Police apprehended a number of key insurgents and criminals, including Muhammed Isa Jodeh Audeh al-Sa'adi, the Ba'ath chairman of the Karbala region. Iraqi police, assisted by CPA advisors and elements of the 82nd Airborne Division, arrested Ryadh Al-Ani, the former director general of the Iraqi secret police in Kirkuk.[79] "There were a number of heroic police," noted Brand. "And their performance got better over time."[80] Yet gauging success was problematic, since the CPA—and the U.S. government more broadly—lacked good measures of effectiveness. They frequently evaluated the police and other security services based on the number of soldiers or police trained. These data were virtually useless in analyzing their performance, since it provided no insight into their capacity to enforce the law or maintain order. Consistently lacking in CPA memos were linkages between the actions of Iraqi security forces and statistics on the level of criminal activity, insurgent violence, or public perceptions of security.[81]

Problems remained acute. As late as September, Iraqi Police Service personnel levels were 50 percent or less than what the CPA had estimated it needed.[82] In a memo to Bremer, Ramadi Governance Coordinator Keith Mines noted that "local sheikhs and politicians describe

---

[79] Memo from Bernard B. Kerik to L. Paul Bremer, "Subject: Recent Ministry of Interior Enforcement Ops.," July 5, 2003.

[80] Author interview with Douglas Brand, January 16, 2009.

[81] Author interview with Matt Sherman and Joshua Paul, May 8, 2008. Also see memo from Walt Slocombe through Amb Kennedy to Amb Bremer, "Subject: Response to Your Questions on Police, Security and Border Guards," July 6, 2003; memorandum from Walt Slocombe to L. Paul Bremer, "Subject: Security Forces Summary," August 22, 2003; Coalition Provisional Authority, "Reconstruction Progress Report: Progress Made in August 2003," September 9, 2003; Coalition Provisional Authority, "Weekly Report Status: ICDC/NIA/ Border Guards Hired/Total Police in 4 Major Cities/Detainees by Camp and Third Country Detainees/Tribal Levies," September 12, 2003.

[82] Info Memo from Ambassador Bremer to Secretary Rumsfeld, CC: Secretary Powell and Dr. Rice, "Subject: Iraqi Police Training and Development," September 19, 2003; and Info Memo from Douglas Brand for Administrator, "Subject: Iraqi Police Training and Development," September 19, 2003.

a police force that is undermanned, poorly-led, mis-armed or under-armed, and without vehicles and equipment." He continued that the challenge with the police "is the most serious issue we face here, and is directly related to the ongoing deaths of American soldiers in this sector." Al Anbar leaders asked Mines and other CPA officials to provide sufficient money and equipment so that they could build an adequate police force themselves. After reading the memo, Bremer penned a note in the margin to McManaway, noting that "we cannot simply have every body in the country training his own police force."[83]

The CPA initially relied on the State Department's Bureau of International Narcotics and Law Enforcement Affairs, which turned to private contractors to actually implement police training. This had worked adequately in prior postconflict environments, but in those smaller societies, the United States had other allies helping with the training and the American or NATO militaries were deployed in adequate numbers to provide a secure environment throughout the time needed to train new security forces. In Iraq, none of those conditions applied. In addition, police training efforts had to be improvised long after the intervention, rather than having been planned and organized before it, as had occurred in earlier cases.

Kerik has been roundly criticized by senior and junior CPA officials for being a terrible manager and planner, and costing the CPA significant time in building police capacity. "We lost several critical months under Kerik in rebuilding the police program," said Fred Smith, a CPA official.[84] "We didn't realize until late in the game that Kerik didn't have any interest in administrative details," recalled Clayton McManaway. "He was running around on operations, and he didn't like to do planning and budgeting. We lost months with the police under Kerik. Equipment hadn't been ordered and contracts hadn't been put

---

[83] Memo from Ramadi Governance Coordinator Keith Mines to Ambassador Bremer, "Subject: 'Give Us the Tools'—Al Anbar Leaders Willing to Work on Security, But Need A Serious Police Force to Do So," August 27, 2003.

[84] Author interview with Fred Smith, July 23, 2008.

in place."[85] Brand further noted that "there was no strategic plan to develop the police and implement it."[86]

It took far too long for the police training program to begin, and basic police equipment was not being ordered. There were too few competent international police trainers available to work with the Iraqi police, and it took CPA far too long to establish a police training academy. As Bremer later acknowledged: "We had a particular problem with the police. We had tried to secure a police training site in Hungary, but that didn't work out. So we negotiated an arrangement with Jordan. But the whole process took too long. We didn't really get the police program started until after Ramadan at the end of November 2003."[87] Kerik left Iraq for good in September. "He paraded around the palace with a full entourage of photographers as he bid Jerry Bremer and others farewell," recalled Fred Smith. "He was headed back to the States and a Rose Garden event with the President."[88]

Bremer began to come under pressure to turn police training over to the U.S. military. Rumsfeld sent Bremer and General Abizaid a memo in late September 2003 noting that while he understood their interest in making sure the police are sufficiently trained, "it is urgent that we get a rapid expansion of the police capability in Iraq." Rumsfeld suggested that U.S. Central Command become the "executive agent" for police training, and that the Coalition Provisional Authority retain control over the curriculum, type of training, and trainers. "Since Bernie Kerik left," Rumsfeld complained, "I understand things have slowed down on police training."[89] In another memo to Bremer, Rumsfeld said that Abizaid's concern "is that the CPA lacks the resources

---

[85] Author interview with Clayton McManaway, July 22, 2008.

[86] Author interview with Douglas Brand, January 16, 2009.

[87] Author interview with L. Paul Bremer III, November 15, 2007.

[88] Author interview with Fred Smith, December 14, 2008.

[89] Memo from Donald Rumsfeld to Jerry Bremer and Gen. John Abizaid, CC: Gen. Dick Myers, Paul Wolfowitz, and Doug Feith, "Subject: Training Iraqi Police," September 29, 2003.

and administrative capacity at the present time to adequately support the activity—the police, the training and the like."[90]

But there were also reservations within both the CPA and the U.S. military about training forces too quickly. According to one memo from the Joint Staff, "if we accelerate too fast, the quality of training and subsequent force capabilities could be negatively impacted."[91] In September, General John Abizaid and Lieutenant General Ricardo Sanchez briefed Bremer on their strategic plan. "Jerry," Abizaid said, "I recommended that Rick's people take over police training from the CPA." Bremer had been expecting something like this proposal. He was opposed. "Although our soldiers were the best combat troops in the world, they had been trained and equipped for fast-moving operations where they killed the enemy, not for community policing and criminal investigations." Bremer continued: "We've been around this track before, John. I am fully on board with moving as fast as we can to stand up Iraqi security forces. . . . But I'm really not convinced that the Army knows how to train professional police, and now that we finally have the Jordanian options worked out, I don't want to switch tracks again."[92] Bremer told Rumsfeld that "we welcome assistance from the military in the police program. The single most useful service they can immediately provide is helping us with the recruitment of police."[93] But he succeeded in keeping CPA control over the police program, at least temporarily.

Throughout the fall, there continued to be serious problems. The CPA reported that the Iraqi Police Service was short of basic equipment, and only 29 percent of required uniforms were on hand. There were also only 2,500 police vehicles, and few had headlights. As one

---

[90] Memo from Donald Rumsfeld to Jerry Bremer, CC: Paul Wolfowitz, October 2, 2003. In fact, as indicated above, Kerik made little progress on police training during his four-month stint in Iraq.

[91] Joint Staff, Memo for the Military Assistant to the Deputy Secretary of Defense, August 2003.

[92] Bremer, *My Year in Iraq*, p. 168.

[93] Memo from Paul Bremer to Secretary Rumsfeld, "Subject: Training Iraqi Police," October 1, 2003.

CPA document concluded, this "limits effectiveness and contributes to potential blue-on-blue engagements. An immediate stop-gap measure is needed."[94] In early January, Bremer raised salaries for the Iraqi Police Service. Their pay scale was inadequate and most had little incentive to risk their lives in the face of a mounting insurgency. Bremer also authorized hazard pay. But salaries were still inadequate; and three weeks later, when the police were threatening to strike, he increased their total take-home pay by another 65 percent.[95]

CPA assessments of police performance continued to be negative. Iraqi police were often cowed by local militia forces and insurgents and were involved in criminal activity.[96] As one memo cogently noted, the police "scarcely deserves the name."[97] In a memo to Rumsfeld, Bremer acknowledged that "fielding properly-trained Iraqi policemen accompanied by capable civilian mentors is going slower than we would like, partly because of the slowness in getting CivPol in place and partly because even basic training takes time (8 weeks is the bare minimum)."[98] Police increasingly became a target of insurgent activity. A CPA assessment concluded that that "anti-coalition elements appear intent on demoralizing police in the region." The police chief in Kharma was assassinated, police in Hit were ambushed by insurgents, and police headquarters in Fallujah were targeted.[99]

---

[94] Coalition Provisional Authority, *Iraqi Police Service Stop Gap Proposal* (Baghdad: Coalition Provisional Authority, December 20, 2003), briefing slide two.

[95] Bremer, *My Year in Iraq*, p. 273.

[96] Coalition Provisional Authority, "South Central Region/Governorate Al Qadisiyah, Weekly Situation Report," February 11, 2004; Coalition Provisional Authority, "North Region/Sulaimanya Governorate, Weekly Situation Report," February 19, 2004.

[97] Info Memo from Mark Etherington to the Administrator, "Subject: Weekly GC Update—Wasit," February 29, 2004.

[98] Memo from Paul Bremer to Secretary Rumsfeld, "Subject: Improving Iraqi Police Capabilities," February 4, 2004.

[99] Info Memo from Stuart Jones to the Administrator, "Subject: Weekly GC Update—Al Anbar," February 20, 2004. Also see Memo from Dr. Liane Saunders to the Administrator, "Subject: Weekly RC Update—North," January 31, 2004; Info Memo from Ronald E. Neumann to the Administrator, "Subject: Highlights of the June 24th MCNS Meeting," June 24, 2004.

Continuing problems with the police eventually led Rumsfeld to overrule Bremer in February 2004 and hand over responsibility for police (and army) training to the U.S. military. One of the final straws was the assessment done at Rumsfeld's request by Lieutenant General Karl Eikenberry, which was deeply critical of the CPA's efforts to build the Iraqi police. Secretary Rumsfeld issued an order on March 8, 2004, giving responsibility for training both the army and police to the U.S. military. This led to the creation of the Multi-National Security Transition Command–Iraq, which took control over these functions.

Just prior to Rumsfeld's February decision, Bremer sent him a memo stating that "I do not agree with placing the Iraqi police program under the military command" arguing that it would "convey to the Iraqis the opposite of the principle of civilian standards, rules and accountability for the police."[100] Civilians should be in charge of the police, not the military. But there was support among some CPA police advisors, including Steven Casteel, who had replaced Kerik at the head of the CPA police effort in September 2003. Casteel noted to Lieutenant General Sanchez: "Boy, am I glad we're working with you all now. We can't get a damn thing done over at CPA. Maybe now, we'll get this stuff moving."[101]

This expectation was not soon fulfilled. Under U.S. military management, numbers eventually increased, but quality did so much more slowly, and the former could not make up for the latter. In a strongly worded memo, Casteel noted that the deteriorating security situation in April demonstrated the poor performance of the police: "the previous training provided by the Major Subordinate Commands was deeply flawed, with the major emphasis on quantity, not quality. We also recognized that mid-level and senior leadership was non-existent."[102]

---

[100]Bremer's note was in response to Eikenberry's report. Memo from Paul Bremer to Secretary Rumsfeld, "Subject: Security Assessment," February 3, 2004. Also see, for example, Memo from Paul Bremer and Gen Abizaid to Secretary Rumsfeld, "Subject: Eikenberry Report," February 12, 2004.

[101] Sanchez, *Wiser in Battle*, p. 317.

[102]Info Memo from Steven W. Casteel to Chief Operating Officer, "Subject: Status of the CPA Ministry of Interior," April 23, 2004.

The February handover still left the CPA with three police-related responsibilities: setting police training standards, providing civilian trainers to the military, and reforming the Ministry of Interior.[103]

Dealing with Neighbors: Iran, Syria, and Turkey

Regional states are the most affected by conflict in neighboring societies, suffering most directly the consequent flow of refugees, endemic disease, criminality, illegal drugs, terrorism, and commercial disruption. Regional states often have the greatest potential influence on the society in conflict, by reason both of their proximity and their long-standing commercial, cultural, tribal, and political and economic ties. When regional states act together, they can have a powerful calming effect by exercising convergent pressures on the warring factions. When they do not act together, they tend to exacerbate the conflict by backing different contenders for power, thereby often extending a conflict that it may be in their best interest to help terminate.

Thus, neighboring states have an irresistible incentive to involve themselves in the affairs of failing states, and their involvement can be a powerful force for good or bad. Given these considerations, it is important for an intervening power to engage with the neighbors and secure their support. This was difficult for the United States in Iraq, however, given the controversy surrounding the invasion and the stated American objective, which was to make Iraq a model democracy, with the avowed intention of undermining the legitimacy of nearby, non-democratic regimes, ultimately leading to a change in their own form of government. This was not a project likely to appeal to neighboring governments, and most of them were consequently opposed to the American intervention.

---

[103] Memo from David Gompert to L. Paul Bremer, "Subject: Secretary Rumsfeld's Instruction," February 21, 2004.

Iran

The revolutionary regime in Tehran had no love for Saddam. Indeed, he had once been their deadliest enemy. In the aftermath of the U.S. invasions of Afghanistan in 2001 and again in the spring of 2003, the Iranian government had made overtures of cooperation to Washington, which the Bush administration chose to ignore. Fearing that they might be next on Washington's target list and not wanting to see the U.S. military ensconced on both its eastern and western flanks, the Iranian regime then moved to enhance its own influence in post-Saddam Iraq and reduce that of the United States.

In a May 29 briefing to President Bush, Bremer noted that he faced emerging threats that included "Iranian-sponsored Islamic extremism."[104] The United States tracked the movement of Iranian intelligence and Al Quds Force officials into and out of Iran and occasionally picked some up.[105] These concerns were frequently conveyed to Iraq's Shi'ite groups. For example, Ryan Crocker bluntly told Abdul Aziz Hakim that the United States "had reliable information about close involvement in the Badr Corps of Iran's Revolutionary Guards and its Al Quds Force. We were aware of direct contact between Iranian officials and Badr Corps members."[106] British Ambassador John Sawers told Hakim in a separate meeting that he was "concerned by continuing reports that the Badr Corps were receiving weapons from Iran; and crossing the Iranian border on a regular basis."[107] Reports were widespread of Iranian infiltration into other Iraqi Shi'ite parties, such as the Islamic Da'wa Party.[108] Da'wa and SCIRI vehemently disputed these accusations. As Hakim told CPA officials, he wanted

---

[104]Coalition Provisional Authority, "Presidential Update," May 29, 2003.

[105]Author interview with L. Paul Bremer, November 15, 2007.

[106] Coalition Provisional Authority, Memo on meeting between Ambassador Ryan Crocker and Abdul Aziz Hakim, June 9, 2003.

[107]Memo from Julie Chappell to Ambassador Sawers, "Subject: Political Process; Call on Abdul Aziz al Hakim," June 20, 2003.

[108]Memo from Darrell Trent to Ambassador Kennedy, "Subject: Islamic Dawa Party and Dr. Haidar al-Abadi, Minister of Communications," September 14, 2003.

"nothing to do with Iran."[109] In a meeting with Meghan O'Sullivan, for example, SCIRI's Hamid Bayati "insisted there were no Iranians in the Badr Forces" and "claimed that to SCIRI's knowledge, there were no Iraqis in the Badr Corps working for Iranian intelligence." O'Sullivan responded by arguing that the CPA viewed "Iranian influence in the Badr Corps as threatening" the partnership with the United States.[110] In a subsequent meeting with Bremer, Hakim noted that anti-Americanism was high within SCIRI because of constant U.S. accusations about Badr Corps links with Iran, even though SCIRI had repeatedly assured the coalition that ties did not exist. Bremer, however, pushed back on Hakim's assertion and reiterated coalition opposition to Iranian interference in Iraq.[111]

Despite Shi'ite denials, there were numerous reports of Iranian infiltration into Iraqi provinces.[112] According to a memo from Bremer to Secretary Rumsfeld in July 2003, the "Iranians have moved some 3 kms into southern Iraq. I have asked JTF to investigate. Recommend Iranians be ordered out immediately."[113] The CPA put together a memo on "Moves to Counter Pro-Iranian and Ba'ath Elements," which was supported by Secretary Rumsfeld and others in the Pentagon.[114] According to Liane Saunders, the Regional Coordinator for CPA North, "every time I have traveled to the Iranian or Syrian borders, both at official and unofficial crossing points, I have found a complete

---

[109] Email from Joanne Dickow to Scott Norwood and Patrick Kennedy, "Subject: Background for Meeting with Abdul Aziz Al-Hakim," July 1, 2003.

[110] Memo from Meghan O'Sullivan to Ambassador Ryan Crocker, CC: Scott Carpenter, Ambassador Hume Horan, Political Team, "RE: 28 May Meeting with Hamid Bayati, SCIRI," May 29, 2003.

[111] Memo from Meghan O'Sullivan to Political Team, "Re: Meeting between Ambassador Bremer and Abdul Aziz Hakim, SCIRI," June 24, 2005.

[112] See, for example, email from Ehad Dhia to Amb. Patrick Kennedy, August 17, 2003.

[113] Memo from L. Paul Bremer, III, to Secretary Rumsfeld, "Re: CPA Priority Issues," July 9, 2003.

[114] Memo from Donald Rumsfeld to Jerry Bremer, "Subject: Memo on Counter Moves," June 2, 2003.

absence of U.S., IBP [Iraqi Border Police], or ICDC forces."[115] One of the biggest threats in eastern Iraq, noted Wasit governorate coordinator Matthew Goshko, was "the continuing flow of illegal border-crossing by the Iranians. Money and weapons support is provided to many of the insurgent groups in Iraq."[116] Consequently, the CPA closed several major border posts along the Iranian border to decrease the flow of foreign fighters and material into Iraq. CPA officials also helped the Ministry of Interior develop an Iraqi Department of Border Enforcement. Key steps included reinforcing and rebuilding major border posts along Iraq's borders, training and recruiting Department of Border Enforcement employees, and developing a visa and passport system in accordance with international standards.[117]

In one instance, UK and Danish patrols identified seven positions directly to the east of Al Qurnah, approximately 3–4 kilometers apart, consisting of rudimentary huts displaying Iranian flags. Coalition patrols could see defensive positions and infantry weapons. Initial assessments were that the Iranian activity was localized and appeared to involve no more than a company-sized group (around 100 men). Consequently, the United States sent a demarche to Iran through the Swiss Embassy protesting the Iranian border posts, since Bremer was prohibited by the White House from talking to Iran. It noted that "the United States strongly protests the establishment of Iranian border posts that encroach into Iraqi territory and wishes to convey the seriousness of the issue. . . . These Iranian border positions [that] occupy former Iraqi border posts include those in the Basrah and Maysan Provinces.

---

[115] Email from Liane Saunders to Scott Carpenter, "Subject: Re: Re: Arbil attack," February 2, 2004. On the porous Iranian border, also see Memo from Edward Messmer to Ambassador L. Paul Bremer, "Subject: 'Diyala Comes to Baghdad,' Scene Setter—14 May 2004," May 10, 2004.

[116] Memo from Matthew Goshko to the Administrator, "Subject: Al-Hilla Visit," June 17, 2004.

[117] Info Memo from Steve Casteel to the Administrator, "Subject: Border Policy Update," June 7, 2004.

This forward movement of Iranian border posts is unacceptable."[118] The U.S. State Department also asked Saudi Arabia, Kuwait, UAE, and Britain to demarche the Iranians.

While U.S. officials declined to talk directly to Iran, British officials were willing to do so. British diplomat and CPA official John Sawers traveled to Tehran and met with a number of Iranian officials, including the Deputy Intelligence Minister, who ran the Ministry of Intelligence and Security's (MOIS's) external operations, and officials from the Iranian Revolutionary Guard Corps (IRGC). His message was blunt: "I gave him clear messages on the IRGC's presence, the hostile MOIS activity, Iranian support for Ansar al-Islam, and links to Muqtada al-Sadr and other Shia extremists." All of these developments were unacceptable, Sawers noted. And Iran could not expect cooperation on issues of concern to it—such as pilgrims, border contacts, and new consulates—while elements of the regime were acting in a way that undermined the coalition in Iraq.[119] The trip appeared to have a positive impact. By September, the number of cross-border Iranian incursions appeared to decrease, at least temporarily. In a meeting with Bremer in September, for example, Lieutenant General Graeme Lamb of the British Army argued that "there were still cross border incursions by Iranians, particularly in the southern portion of the border," but that he did not see "the same large numbers crossing as before."[120] Unfortunately, progress was only temporary.

In February 2004, John Berry wrote an alarming analysis of Iranian involvement, arguing that Iran was eager to split Iraq into three semi-autonomous zones, "which in addition to emasculating its old enemy, would guarantee its dominion over the two holy cities, the income they derive from pilgrims, and the prestige of having them

---

[118] Message from the United States of America to the Islamic Republic of Iran, July 15, 2003; memo from William J. Luti to Under Secretary of Defense for Policy, "Subject: Iranian Border Encroachment into Iraq—Status of Demarche," July 17, 2003.

[119] Memo from John Sawers to Ambassador Bremer, "Subject: Visit to Tehran," July 31, 2003.

[120] Memo from Maj. Patrick J. Carroll, USMC, to Ambassador Paul Bremer, "Subject: Notes of Meeting between Amb. Bremer, Sir Hillary Synnott, and General Lamb on 18 Sep 03," September 18, 2003.

back in the same orbit as Qom."[121] In March, Mike Gfoeller reported from CPA South Central Region that "the Iranians—in particular, the IRGC and the MOIS—are working solely with three groups opposed to the democratic future we are seeking to help the Iraqis achieve: SCIRI, the Da'wa Party, and Muqtada al-Sadr's organization."[122]

There was also deepening concern among Iraqis in areas such as Karbala that Iranian influence was growing. As Berry argued, "After events in Najaf, the Arab faction is convinced that SCIRI-Sistani and Moqtada Sadr are making common cause to advance Iranian interests by creating a semi-autonomous Shi'a entity dominated by pro-Iranian mullahs." He continued that Iranian families in Najaf were creating an economic free trade zone that was increasing the price of food and housing. "Especially telling," Berry continued, was that "Iranian agents have lured one of our translators to take a free trip to Iran, and other members of our locally hired staff (translator and receptionist) have been approached by the same people to provide information on CPA staffers' doings. Clearly, then, we are under surveillance by Iranian agents." He recommended putting controls on the border with Iran and, at the very least, requiring foreign visitors to have passports with their date of entry stamped into them. He also recommended empowering local police to expel those who "overstayed their welcome."[123]

In discussion with Bremer and David Gompert, Ayad Allawi argued that there was documented evidence that the Iranians were working with al Qaeda. In response, Bremer commented that "our analysis is not dissimilar" and it "pointed to Al Zarkawy and Al Qaeda being behind the suicide attacks." But Bremer also noted that Iranian strategic objectives were complicated, since he believed Iran did

---

[121] Email from John Berry to Dick Jones, "Subject: Najaf Situation and Ramifications for Karbala and Shi'ite Heartland," February 5, 2004.

[122] Email from Mike Gfoeller to Paul Bremer, "Subject: South Central Region: Progress, Opportunities, and Risks," March 2, 2004.

[123] Info Memo from John F. Berry to the Administrator, "Subject: Weekly GC Update—Karbala," January 31, 2004. Better border control arrangements came up repeatedly in CPA documents, including Info Memo from Steven W. Casteel to Deputy Administrator Jones, "Subject: Talking Points—Border Update; IPS Training (Egyptian involvement); Jordan Academy," March 17, 2004.

not want Iraq to fail because the Kurds would then declare independence.[124] The CPA ultimately realized it had to live with some Iranian presence in Iraq. "Some level of Iranian influence, direct and indirect, is an unavoidable reality," wrote David Richmond, the United Kingdom's special representative for Iraq. "But it is balanced in part in a number of ways. In Maysan, for example, the Badr police chief is held in check by the virulently anti-Iranian GC [Governing Council] representative Abu Hatim and his brother the Governor."[125]

The CPA had no means of engaging with the Iranian government, either to simply complain or to seek to co-opt it in stabilizing Iraq, and Washington refused to do so. Whether such contacts would have yielded better behavior is hard to say, but the failure to try was certainly a missed opportunity. Iran played only a marginally damaging role during the CPA period, but it became a much more serious source of trouble in succeeding years.

Syria

"I was actually more concerned about the Syrians than the Iranians," Bremer recalled, "because of the transit of foreign fighters coming into Iraq from Yemen, Saudi Arabia, Sudan, and other countries, who came through Damascus."[126] An assessment produced for the Ministerial Committee for National Security, which was chaired by Bremer, concluded that "there is a substantial body of evidence that Syrian nationals, and possibly agents of the Syrian government, are providing aid to insurgents in Iraq. This assistance," it continued, "is coming in a number of forms including weapons, cash, aid and refuge, and training." Despite repeated U.S., Iraqi, and other coalition requests to the Syrian government that it cease government-sponsored support to insurgents,

---

[124]Info Memo from Peter Khalil through David Gompert to the Administrator, "Subject: Meeting with Dr Iyad Alawi [sic], Ambassador Bremer and Mr. David Gompert," February 9, 2003.

[125]Email from David Richmond to BAGHX-e Telegrams, "Subject: Iraq/Iran: Assessment of Current Iranian Activity: Southern Iraq," March 31, 2004.

[126]Author interview with L. Paul Bremer, November 15, 2007. Clayton McManaway also argued that problems with Syria throughout 2003 were much greater than with Iran. Author interview with Clayton McManaway, December 12, 2007.

the assessment concluded that the flow of assistance to Iraqi insurgents had not slowed.[127] U.S. and Iraqi intelligence estimates reported that Syria was providing assistance to groups in Fallujah: "There is sufficient evidence that they are receiving help from the Syrian intelligence services."[128]

Syrian intelligence agents also operated front companies in Iraq, and Iraqi officials allowed fighters to come into Iraq from Syria.[129] In a memo to Bremer, Ambassador Ron Neumann noted that Iraqi intelligence estimates "reported that enemy cells, associated with the Zarqawi Group, were moving out of Syria and possibly had the Green Zone on their target list."[130] In briefings for the new Iraqi leadership, CPA officials concluded that Syria "has been the most politically antagonistic of Iraq's neighbors since liberation" by providing refuge to former regime members and smuggling weapons, money, and fighters into Iraq.[131] The CPA's response included a range of steps: monitoring the activities of companies associated with the Syrian government, especially those that had "connections with the Syrian intelligence services"; intensifying border patrols along the Iraqi-Syrian border; periodically shutting down Syrian border crossings; and filing lawsuits in international courts against Syria's seizures of Iraqi assets.[132] None of this proved very effective, and Syria remained the principal pipeline for foreign terror-

---

[127] Ministerial Committee for National Security, "Curtailing Syrian Assistance to Iraqi Extremists: Draft Analysis of Options," May 2004.

[128] Ministerial Committee for National Security, "Security Situation—Updates and Analysis," May 6, 2004.

[129] Info Memo from Fred Smith to the Administrator, "Subject: Update on May 20th Meeting of Ministerial Committee for National Security," May 19, 2004; Info Memo from Fred Smith to the Administrator, "Subject: Materials for May 29th Meeting of Ministerial Committee for National Security," May 27, 2004; author interview with Frank Miller, June 6, 2008.

[130] Info Memo from Ronald E. Neumann to the Administrator, "Subject: Highlights of the June 24th MCNS Meeting," June 24, 2004.

[131] Coalition Provisional Authority, "Briefing Materials for the New Iraqi Leadership: Key Issues," May 2004.

[132] Coalition Provisional Authority, Working Group on Threat from Saddamists, "The Evolving Security Threat/Check List, Conclusions and Actions," May 10, 2004.

ists (although Saudi Arabia was the largest single source) for years to come.

Turkey

Turkey presented a different sort of challenge. Its incursions into Iraq were even more blatant than those of Iran and Syria, since it alone, among Iraq's neighbors, occasionally sent its military forces across the border. Unlike Iran and Syria (and Saudi Arabia), Turkey was not ideologically opposed to a democratic Iraq, but Ankara was opposed to the likely consequence, a highly autonomous, or even independent Kurdistan. Turkey had refused to allow American forces to transit its territory to invade Iraq, and remained concerned about Kurdish separatist militants operating out of Iraq against targets in Turkey.

Throughout the CPA's lifespan, the United States was eager to encourage other governments to contribute military contingents in Iraq. In June 2003, Deputy Secretary of Defense Wolfowitz met with Turkey's undersecretary for foreign affairs, Ugur Ziyal, who proposed to contribute a military contingent to Iraq. Wolfowitz responded that he would talk to Bremer and consider Turkey's offer.[133] Bremer was initially interested; in a memo to Secretary Rumsfeld he noted, "there may be some value in getting Turkish assistance to train police outside the north and in securing Turkish involvement in an international police force."[134] In another memo to Rumsfeld a few weeks later, Bremer wrote that "in the short run it is desirable to have Turkey and Pakistan contribute troops to assist in stabilizing Iraq."[135] The Pentagon then began talking to Turkey about deploying troops to Iraq. In late July, Vice President Dick Cheney told senior Turkish officials in Washington that Turkish troops would be welcome in northern Iraq.[136]

---

[133] Memo from Lisa Heald to OPCA Staff, "Subject: Turkey Paper on Iraq Reconstruction," June 19, 2003.

[134] Memo from L. Paul Bremer to Secretary of Defense, "Re: Turkish Proposals," July 5, 2003.

[135] Memo from Paul Bremer to Secretary Rumsfeld, "Subject: Governing Council and Foreign Peacekeeping Troops," August 5, 2003.

[136] Author interview with Frank Miller, June 6, 2008.

A number of officials in the Governing Council, including the Kurds, were deeply opposed to a Turkish military presence in Iraq. So were a growing number of U.S. officials. Ryan Crocker warned Bremer that while there was "continuing interest in Washington in having Turkey contribute troops to a stabilization force in Iraq," there were significant downsides. "For obvious historical reasons, Turkish forces on Iraqi soil would be poorly received by virtually all elements of the Iraqi population."[137] Crocker subsequently recommended that Bremer tell Secretary Rumsfeld bluntly that "we should not use Turkish troops in the Stabilization Force." Instead, Crocker recommended that the CPA limit neighboring-state contributions to Jordan.[138] The deployment of Turkish troops to Iraq was particularly problematic since one of the areas under discussion was Anbar Province. The Turkish military said it would support deploying forces to Anbar, but it would have to run the supply lines through Kurdish territory. "I saw the maps," noted Bremer, which included supply lines running from the Turkish border through Mosul and down to Anbar. "It was a non-starter."[139]

During the fall of 2003, U.S. and Turkish concerns with the Kurdistan Workers Party (PKK) and the Kurdistan Freedom and Democracy Congress (KADEK) operating on Iraqi soil became more acute. The PKK was founded in 1974. Advocating the establishment of an independent Kurdish state, it conducted a violent insurgency against the Turkish government. The PKK changed its name to KADEK in 2002 and proclaimed, unconvincingly, a commitment to nonviolent activities in support of Kurdish rights. As the Iraq insurgency continued to worsen, the U.S. military focused on dealing with Sunni and Shi'ite groups and paid little attention to the PKK or the KADEK. This infuriated Turkey. As a Pentagon memo explained, Turkey "expects concrete action by the US in eliminating PKK/KADEK from Iraq. There is already the impression in Turkey that the US is not moving

---

[137] Memo from Ryan Crocker to Presidential Envoy L. Paul Bremer, "Subject: Memo to SecDef on Police and Stabilization Force Issues," June 8, 2003.

[138] This was formalized in a memo from Presidential Envoy L. Paul Bremer to Secretary Rumsfeld, "Subject: Police and Stabilization Force Issues," June 8, 2003.

[139] Author interview with L. Paul Bremer, November 15, 2007.

fast enough in this regard." It continued that further delays "will create tremendous pressures on the Turkish Government" to send troops into Iraq.[140]

In October, the Turkish parliament authorized the government to deploy troops to Iraq, noting that they preferred to go to Salah ad Din Province. But in discussions with the Governing Council, Bremer encountered substantial opposition, especially from the Kurds. Massoud Barzani threatened to resign, and a number of CPA officials believed that deploying Turkish troops to Iraq would be counterproductive, warning that approval of such a deployment would emasculate the Governing Council, since it would have to be done over their objection.[141] In a memo to Bremer, Scott Carpenter argued that "in our view Turkish troops will likely exacerbate the security challenges facing the coalition." Key concerns were that Syria and Iran might view the development as a provocation and encourage further interference inside Iraq. It might also push some individuals such as Barzani into closer cooperation with Iran.[142]

CPA assessments also concluded that there was general Kurdish ambivalence toward dealing effectively with the PKK and KADEK, forcing Turkey to take matters into its own hands. U.S. military forces came across "Turkish flying roadblocks inside Iraq aimed at interdicting PKK movements," and received reports that "Turkish [Special Forces] have worn U.S. Army uniforms when ambushing PKK units, apparently to try to provoke PKK attacks on Coalition Forces." There were also tense standoffs between Turkish and coalition military forces. As one CPA report noted, "Turkish [Special Forces] have clandestinely surveilled CF [coalition forces] in Dohuk" and "Turkish regular forces

---

[140] The memo on "Talking Points" (October 23, 2003) was included as an attachment in an email from Scott Norwood to Executive Secretariat CPA, "Subject: Turkish Demarche from Grossman HOT," October 30, 2003.

[141] Author interview with Meghan O'Sullivan, January 29, 2008.

[142] Memo from Scott Carpenter to the Administrator, "Subject: Turkish Troops," October 10, 2003.

have pointed guns at CF, including following CF vehicles with their tank tubes."[143]

While CPA officials believed that the PKK was a terrorist organization, they argued that U.S. and Iraqi military forces should not be used to target PKK forces. For the CPA, "adding burdens to our efforts in Iraq in order to 'repair' relations with Turkey would be a mistake of historic proportions."[144] The U.S. military shared this view: "CJTF-7 has opposed this option on the grounds that it would require at least a division of troops to carry out."[145] Instead, the U.S. advocated increasing pressure on Kurdish leaders such as Massoud Barzani and Jalal Talabani to eliminate PKK safe havens in Iraq.[146]

Countering the Insurgency

At a meeting between senior CPA officials and Iraqi leaders shortly after he arrived in May, Bremer had acknowledged "the urgent need to establish greater law and order in Baghdad and beyond."[147] CPA threat warnings noted alarming trends in violence. As one assessment concluded, insurgent attacks were

> centered on US/coalition targets, ranging from checkpoints, and increasingly, to higher profile targets such as helicopters and tactical level military headquarters. These are still locally orga-

---

[143]Info Memo from Philip Remler to the Administrator, "Subject: Weekly GC Update—Dohuk," May 15, 2004; Info Memo from Philip Remler to the Administrator, "Subject: Weekly GC Update—Dohuk," April 17, 2004.

[144]Info Memo from Philip Remler to the Administrator, "Subject: Weekly GC Update—Dohuk," May 15, 2004.

[145]Info Memo from Roman Martinez through Scott Carpenter to the Administrator, "Subject: Elements of a Kurdish Strategy," January 14, 2004.

[146]Info Memo from Meghan O'Sullivan and Roman Martinez through Scott Carpenter to the Administrator, "Subject: Friday Meeting with Barzani and Talabani," January 22, 2004.

[147]Meeting with members of the Iraqi Opposition Leadership Council, "Subject: Countering Wahhabi Infiltration in Iraq," Friday, May 16, 2003.

nized, indigenous attacks. However, all current indicators point to ingress of experienced external forces in the V Corps/OCPA area (i.e., Iraq) that may be planning for more intensified attacks or for organization of disparate groups.[148]

The coalition force's significant acts database, which was compiled from military reporting, included all attacks (such as small arms fire, antiaircraft fire, indirect fire, and improvised explosive device attacks) against coalition forces, as well as against civilian "neutrals" and Iraqi security forces.[149] The data shown in Figure 4.1 indicate that the daily

Figure 4.1
Significant Attacks, June 2003 to June 2004

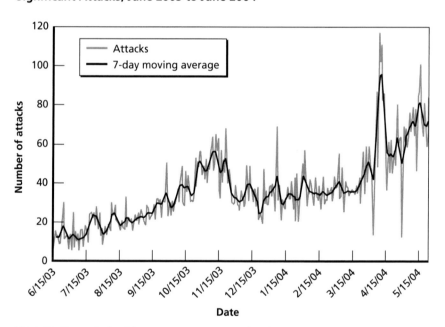

SOURCE: Charts derived from U.S. Department of Defense, CPA Daily Threat Updates (unclassified).
NOTE: Includes attacks against all types of targets (civil and military).
RAND MG847-4.1

---

[148] Coalition Provisional Authority, "CPA Fusion Cell Threat Warning," June 4, 2003.

[149] The data may underestimate the number of attacks. For instance, in periods of major unrest, reporting units tend to combine multiple incidents into one report.

average number of attacks carried out against all targets in Iraq roughly quadrupled from June 2003 to June 2004. Figure 4.2 shows a similar increase in attacks against coalition forces.

While CPA and senior U.S. government officials publicly assured Americans that the security situation was not as bad as press reports indicated, internal CPA documents showed growing alarm. In a memo to Secretary Rumsfeld, Bremer said the threat to U.S. forces came from several sources. The first included elements of the former regime, such as Ba'athists, Fedayeen Saddam, and intelligence agencies. They focused their attacks on three targets: coalition forces, infrastructure, and Iraqi employees of the coalition. "To date," Bremer wrote, "these elements do not appear to be subject to central command and control. But there are signs of coordi-

Figure 4.2
Attacks on Coalition Forces, June 2003 to June 2004

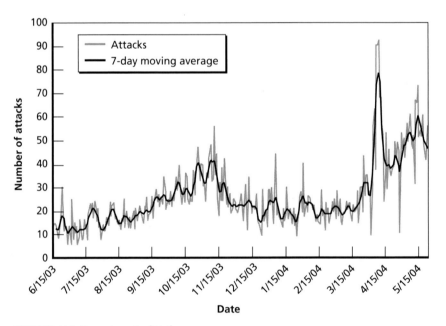

SOURCE: U.S. Department of Defense.

RAND *MG847-4.2*

nation among them."[150] Former Mukhabarat officers were active in a number of ways, including making radio-detonated Kaneen bombs. They used money channeled through radical Islamic clerics, who had been funded by wealthy donors in the United Arab Emirates and other Gulf countries.[151] The second threat was Iranian subversion: "Elements of the Tehran government are actively arming, training and directing militia in Iraq. To date, these armed forces have not been directly involved in attacks on the Coalition. But they pose a longer term threat to law and order in Iraq." The third threat included terrorist groups, especially jihadists from Saudi Arabia, Syria, and Yemen. Bremer noted that there were clear indications that Ansar al-Islam, a terrorist group associated with al Qaeda, was operating inside Iraq and actively surveilling coalition targets.[152]

CPA memos suggested possible connections between Saddam Hussein and the insurgency, although the former dictator was not seen to be coordinating insurgent efforts. According to one memo, for example, Saddam's point of contact in Diyala Governorate was his deputy in the former Revolutionary Command Council, Izzat Ibrahim Al-Douri. "Izzat Ibrahim visited twice Diyala," it noted, "and met the Ba'ath local leaders in Tchiaila village, conveying to them former dictator's orders."[153] During these visits, Izzat Ibrahim called on the Ba'athists to resume their work, meetings, and military activity within the party's cells and made decisions concerning their reorganization and combat. In a briefing to President Bush, Bremer argued that the security situation was "not acceptable."[154] U.S. security assessments also suggested an increasing tempo of attacks against coalition forces

---

[150] Email from Paul Bremer to Jaymie Durnan, "Subject: Message for SecDef," June 30, 2003.

[151] Coalition Provisional Authority, "Summary: Bomb-Making Tips, Mukhabarat Habits, Views from the Street," July 15, 2003.

[152] Email from Paul Bremer to Jaymie Durnan, "Subject: Message for SecDef," June 30, 2003.

[153] "Trips Notes, Hilla, Coalition Provisional Authority—South Central (CPA-SC)," June 26–27, 2003.

[154] "Iraq Security and Military Issues," briefing, NSC Meeting, July 1, 2003.

using small arms, mortars, rocket grenades, and improvised explosive devices. The "attack patterns show emerging regional coordination in: Baghdad, Karbala, Fallujah, Mosul, and Tikrit."[155] In response, the coalition conducted a series of offensive operations to pressure insurgent groups and disrupt their activities, for example, operations Desert Scorpion, Sidewinder, and Soda Mountain.

Reports of the deteriorating security environment were reinforced by public opinion polls that were circulated throughout the CPA. According to one poll commissioned by the State Department's Bureau of Intelligence and Research, "Results of the first American public opinion poll to be completed in Iraq confirm the view that Iraqis are unhappy with the conditions in their country after the end of Saddam's regime."[156] In a Gallup poll, 94 percent of Iraqis in Baghdad believed the city was a more dangerous place to live after the U.S.-led invasion. Majorities also said they were afraid to go outside their home during the day (70 percent) and night (80 percent) because of safety concerns. And anti-American sentiments in much of Iraq were extremely high.[157]

Yet polls also indicated that as many as 83 percent of Iraqis wanted coalition forces to stay in the country because they were concerned that a premature withdrawal would trigger an anarchic situation.[158] This was one irony of the U.S. military presence in Iraq: Most of the population argued that coalition forces were unable to protect them and were a critical part of the problem but noted that the security situation would get even worse if the United States left.[159] As Jalal Talabani

---

[155] Coalition Provisional Authority, "Security Update for Ambassador Bremer," July 18, 2003.

[156] Department of State, Office of Research, Bureau of Intelligence and Research, "Iraqis Offer Dim Evaluation of Reconstruction Effort Thus Far," August 22, 2003.

[157] "Iraqi Impressions of Coalition Forces and the Security Situation in Iraq: Office of Research Survey Results from 7 Cities in Iraq & Preliminary Results from Gallup Baghdad Survey," September 30, 2003.

[158] Memo from Don Hamilton for L. Paul Bremer, "Subject: Release of Gallup Polls," September 23, 2003.

[159] Memo from Don Hamilton to the Administrator, "Subject: Opinion Surveys," August 6, 2003; Coalition Provisional Authority, "Iraqi Opinion in Selected Cities, 1–7 January

stressed during a private meeting with Bremer, "a U.S. presence in Iraq would be critical for at least the next five years" to provide security, prevent the resurgence of the Ba'ath Party, and keep regional neighbors from interfering in Iraq.[160]

Senior Iraqi officials were increasingly concerned about the security situation. Samir Shakir Mahmood, who became the Minister of Interior and later ambassador to the United Nations and United States, argued in a memo to the CPA that the situation in Anbar Province "is very unsatisfactory. People are very resentful and fearful. Anger is mounting giving real opportunities to the remnants of the old regime to regroup and cause serious problems." He continued that "the Coalition Authority should change its approach."[161] Wolfowitz sent a strongly worded memo to Bremer in December complaining about Governing Council member Mohsen Abdul Hameed, who gave an interview to Baghdad newspaper *Al Ahali* noting that he accepted the legitimacy of resistance against coalition forces. Wolfowitz asked Bremer to have Hameed clarify his remarks. "If his views were correctly stated and he refuses to amend them," he said, "I suggest that at the very least he should be excluded from any security information which is made available to the Governing Council."[162] Bremer confronted Hameed, who said he had been gravely misquoted. When Bremer asked him to issue a retraction, however, Hameed responded that he could not for

---

2004"; and Info Memo from Don Hamilton to the Administrator, "Subject: 14 February Focus Group Report: Samarrahans Really Dislike the Coalition," February 22, 2004.

[160] Coalition Provisional Authority, "Amb. Bremer's June 16 Meeting with Jalal Talabani," June 19, 2003.

[161] Memo from Samir Shakir Mahmood to Roman Martinez, "Subject: Al-Anbar Province," June 11, 2003.

[162] Memo from Paul Wolfowitz to Administrator of the Coalition Provisional Authority, "Subject: Comments of Governing Council Member Abd al Hamid," December 26, 2003. Similar concerns were expressed in a memo from Scott Carpenter to Bremer. Memo from Scott Carpenter to the Administrator, "Subject: Readahead for your meeting with Mohsen Abdul Hameed," January 11, 2004.

political and security reasons. But he promised "not to make this kind of statement again."[163]

Despite mounting evidence to the contrary, U.S. military commanders were frequently optimistic about the security situation, especially in briefings to Pentagon leadership. During a secure video teleconference with Secretary Rumsfeld, CPA officials became incensed when General Abizaid reported that "all brigade commanders are confident they are winning."[164] Tensions occasionally surfaced between the CPA and the U.S. military. Abizaid told Bremer that he was fed up that Walt Slocombe continued to oppose rehiring field-grade army officers. In a memo to Secretary Rumsfeld and Bremer, former Deputy Secretary of Defense John Hamre, who visited Iraq in June 2003, argued that while the relationship between Bremer and Sanchez appeared to be good, there was major friction at lower levels. In response to the memo, Bremer agreed that "we need better coordination," and noted that he had already requested CPA personnel to act as political advisors to each division commander and to represent the CPA in each of the 18 governorates.[165]

There was a serious lack of intelligence on the nature of the insurgents. Senior CPA officials met regularly with Sanchez's J-2 staff, the CIA chief of station, and the FBI, but intelligence assessments of the insurgency and its causes were often inadequate.[166] "I was deeply frustrated with the lack of intelligence on the insurgency," recalled Bremer. "I never had a good handle on the insurgent command and control network. We had lots of good information at the tactical level, such as where a specific insurgent lived. But I didn't have a good picture at the strategic level." McManaway added that "we didn't know who we

---

[163] Memo from Jerry Bremer to Deputy Secretary Wolfowitz, "Subject: GC Member Mohsen Abdul Hameed's Media Comments," January 14, 2004.

[164] Memo from Ambassador Patrick Kennedy to Ambassador Paul Bremer, "Subject: Back-Brief on SecDef SVTCs," September 15, 2003.

[165] Memo from John Hamre to the Secretary of Defense and the Administrator, Coalition Provisional Authority, "Subject: Preliminary Observations Based on My Recent Visit to Baghdad," July 2, 2003; and Letter from Bremer to Hamre, July 2, 2003.

[166] Author interview with Clayton McManaway, December 12, 2007.

were fighting. Was the insurgency organized? We didn't know. Jerry Bremer and I would often talk at night, and I'd tell him that we don't even know who was killing us."[167] Even with all the data on detainees, Bremer saw precious little analysis on what was motivating insurgents or their support network to fight. "I received no useful products on what was motivating detainees. The interrogator would interrogate the person, write an intelligence report, and send it up the chain of command. They were primarily interested in immediate, actionable intelligence. But it wasn't clear to me that anyone was looking at broad trends across the intelligence reports. And, if they were, I didn't see any of the products."[168]

Part of the problem with intelligence was that the CIA was focused on finding nonexistent weapons of mass destruction. The CIA station chief oversaw the Iraq Survey Group, an intelligence organization under CIA official David Kay and Army Major General Keith Dayton, which had roughly 1,400 coalition civilians and military personnel searching for WMD. In a briefing to President Bush, Bremer argued that the United States needed to "increase and sharpen Intel collection—esp. local HUMINT [human intelligence]," which was lacking.[169] In August 2003 Bremer called CIA Director George Tenet, who was on vacation in New Jersey, and asked for help. The FBI had briefed Bremer that the United States was facing a more serious insurgent threat in light of such recent attacks as the car bomb outside the Jordanian embassy in Baghdad. Tenet responded that the CIA would do what it could to help, but was overworked because of the search for WMD. This was their primary focus, not the insurgency. "The imbalance was staggering," noted McManaway, "between the intelligence analysts working on weapons of mass destruction and those working on the insurgency."[170]

---

[167] Author interview with Clayton McManaway, July 23, 2008.

[168] Author interview with L. Paul Bremer, November 15, 2007.

[169] "Iraq Security and Military Issues," briefing, NSC Meeting, July 1, 2003.

[170] Author interview with Clayton McManaway, July 22, 2008.

The military faced the same intelligence challenges. There was a joint working group run by the CJTF-7 intelligence staff and the Defense Intelligence Agency in Baghdad, but both organizations struggled with a dearth of human resources. The military was dramatically short of people who spoke Arabic and Kurdish, which created a problem in interacting with locals and forced military intelligence officials to rely on translators. The paucity of good human intelligence created an overreliance on signals and other technical collection methods.[171]

On August 19, 2003, the CPA suffered a serious blow when the UN headquarters was targeted, killing Sergio de Mello, the special representative of the UN Secretary-General. The next day, Bremer called in his senior security and intelligence advisors, and requested the establishment of a fusion cell that would bring together all the U.S. government intelligence resources. "I want it done now, with no more delay," Bremer said. "Forget about agency boundaries and reporting channels. Combine all your assets. We're in trouble here. The terrorists have arrived in a deadly serious way, and we've got to be just as serious."[172] The attack caused the UN to eventually pull out of Iraq, which CPA officials viewed as a setback to stabilization and reconstruction efforts.[173] The UN bombing also led to a large-scale effort to protect Governing Council members, including the provision of body armor, vehicles, and security details.[174]

Throughout the fall, CPA reporting indicated that the security situation continued to deteriorate. An assessment from the CPA's Office of Policy Planning, for example, argued that "the security environ-

---

[171] Author interview with Frank Miller, June 6, 2008.

[172] Bremer, *My Year in Iraq*, p. 142. Also see, for example, Coalition Provisional Authority, Baghdad, Memorandum of Understanding (MOU), "Subject: Responsibilities and Relationships between CPA Intelligence Section and CJTF-7 J2," August 29, 2003.

[173] Letter from Philip Cooper, Director of Administration, United Nations Assistance Mission to Iraq, to Patrick Kennedy, Chief of Staff, Coalition Provisional Authority, October 31, 2003; and letter from Patrick Kennedy to Philip Cooper, November 2, 2003.

[174] Action Memo from Meghan O'Sullivan through Tom Krajeski for Ambassador Pat Kennedy, "Subject: Security Package for Governing Council Members," August 26, 2003; memo from the Administrator, Ambassador Paul Bremer III, for Governing Council, "Subject: Security for Governing Council Members," August 26, 2003.

ment is changing but remains very difficult." It noted that street crime had increased from prewar levels, and insurgents were getting smarter about conducting attacks.[175] Meghan O'Sullivan wrote to Bremer that "recent intelligence suggests that terrorists are seeking to use 'official' vehicles to target their victims."[176] Insurgents also targeted key infrastructure, such as pipelines, which disrupted oil exports, electricity production, and fuel distribution.[177] Railroads were targeted as well, and, as one CPA report concluded, "the extent to which normal operations are disrupted remains unacceptable if the railway is to function as a vital transport element in the movement of cargoes."[178] According to a U.S Central Command assessment for Secretary Rumsfeld, "the emerging threats and attacks against the Iraqi infrastructure are reaching a level that requires immediate and unprecedented action. We are losing the consent of the Iraqi people by failing to meet their expectations in some of the most basic areas of life support. As such we risk losing the peace."[179]

A CPA memo to Bremer from CPA advisor Lydia Khalil noted that "the opposition has expanded to include mostly those with ideological and practical grievances, not just spoilers and fundamentalists. Because they did not expect this to happen to them, tribes and notables who were initially supportive and cooperative have begun to turn against the Coalition." One of the reasons was that the U.S. military found itself dealing "with politically sensitive issues and administrative issues that it is not trained to deal with."[180] By the end of 2003,

---

[175] Info Memo from Office of Policy Planning for the Authority, "Subject: 30 Day Review of the CPA '60-Day Plan,'" September 17, 2003.

[176] Info Memo from Meghan O'Sullivan through Scott Carpenter to the Administrator, "Re: Readahead for August 23 Meeting with the GC," August 21, 2003.

[177] Coalition Provisional Authority, "Pipeline Sabotage," October 28, 2003.

[178] Info Memo from Darrell Trent to the Administrator, "Subject: Ministry of Transportation Issues Update," December 15, 2003.

[179] Memo from United States Central Command to Secretary of Defense, "Subject: Energy Systems Stability and Security in Iraq," August 28, 2003.

[180] Info Memo from Lydia Khalil through Ambassador Richard Jones to the Administrator, "Subject: Meeting with Sharif Ali and Ambassador Jones," December 21, 2003.

CPA assessments concluded that there was some coordinated organization, leadership, and planning behind insurgent attacks: "The enemy has shown signs of regrouping and becoming more sophisticated since September. He has been adapting his tactics, using more standoff weapons."[181]

CPA officials began pushing for more reliance on economic and political efforts, rather than simply military force, to deal with the insurgency. Keith Mines, the CPA's Al Anbar governance coordinator, argued that success required a multifaceted approach that included creating jobs, reconsidering CPA's de-Ba'athification policy, reviewing the composition of the Governing Council, increasing projects in Al Anbar, and equipping and training the Iraqi police. "We are dealing with an urban insurgency," Mines noted, "of the kind that successfully fought the British for decades in Northern Ireland and caused the British to quit Palestine, and continues to defy the Israelis throughout the West Bank and Gaza." It required not just military action, he argued, but aggressive new political and economic activities. "The widespread opposition to the occupation in Fallujah it seems to me will similarly always make military options problematic."[182]

Despite the growing concerns, there remained some optimism. Following a trip to Iraq in September, Secretary Rumsfeld sent a laudatory note to Bremer: "The progress I saw since my last visit in April is solid. The Iraqi people are seizing their opportunity to open a new positive chapter in their history through their participation in the local, regional, and national governing council organized by the Coalition."[183]

Conclusion

Bremer's order to dissolve the army had been intensively discussed within the Pentagon. It had apparently been cleared with the respon-

---

[181] Coalition Provisional Authority, "Draft Points for Message," December 30, 2003.

[182] Info Memo from Keith Mines.

[183] Letter from Secretary Rumsfeld to Ambassador L. Paul Bremer, September 15, 2003.

sible U.S. military command in Iraq and briefed to President Bush, his principal advisors, and the British before its promulgation. No formal objections seem to have been raised, other than by Bremer's predecessor, Jay Garner, when he finally learned of the plan. Yet the order had not been reviewed on an interagency basis until the President and his chief advisors were informed on the day before its announcement. Most were taken by surprise, and a meeting at that level is not the place to begin debating such a measure. This failure of consultation was part of a wider problem. "From April to June 2003," recalled Frank Miller, "the White House was getting virtually no information from CPA in Iraq. It was not getting to us."[184] Bremer and Slocombe might have been wise to involve Garner in the decision process earlier on, but they were not in a position to bring in other agencies. That would have required either a much more proactive Pentagon approach to interagency sharing, or a much more demanding National Security Council staff than then existed.

The lack of military objections to the order also masked a fundamental misunderstanding. The military officers who knew about the order and went along with it expected that large elements of the Iraqi military would be recalled in short order once the original dissolution had been formalized. This proved not to be the case, and Bremer and Slocombe can be criticized for not having made their full intentions plain from the beginning. As with the failure to bring Garner into the loop earlier, the result was long-term resentment and endless recriminations.[185]

---

[184] Author interview with Frank Miller, June 6, 2008.

[185] Although the record is murky, it does appear that the CENTCOM commander, General Franks, cleared the dissolution order, and that members of the Joint Staff also went over it and suggested minor revisions. General McKiernan has since denied clearing or even knowing in advance of the decree, but his staff seems to have reluctantly indicated their assent. (See Gordon, "Fateful Choice on Iraq Army Bypassed Debate," 2008.) Colonel Agoglia, then the CENTCOM liason officer with the CPA, has said that he advised General Abizaid, then the CENTCOM Deputy Commander and subsequently its head, to go along with the proposed CPA order on the grounds that the army would be recalled shortly after its formal dissolution. According to Agoglia, Abizaid agreed, seeing the CPA order as a necessary evil. (See Ferguson, *No End in Sight,* p. 211.)

The order was certainly remiss in at least one respect: It made no provision for payments to the dismissed soldiers. A month later, the CPA announced that stipends would be paid to former career personnel, and a month later still such payments actually began. Later still, one-time payments were made to dismissed conscripts. The CPA also tried to initiate, but proved unable to implement, a program to help former soldiers find civilian employment.

Given that disarmament, demobilization, and reintegration schemes had by 2003 become a standard part of most conflict reconstruction missions, there seems little good reason not to have incorporated all aspects of such a program into the original order, even if it had been necessary to delay promulgation to do so. Approaching the issue in this more comprehensive fashion could have attenuated the negative reaction among former soldiers and their families, provided those separated from the service a benign outlet for their continued activity, and facilitated recruiting some of them back into the new army in due course. Had the order to disband the army been subjected to more thorough interagency scrutiny and debate, it seems possible that such a program might have been added to the measure.

Five years after the issuance of CPA General Order Number 2, the Iraqi army is among that country's most effective institutions and the police are still among the worst. This does not mean that Bremer's decision to disband the army was necessarily right or that his decision to keep the police was wrong, but it does illustrate the advantages of making a clean start. Disbanding the army may have fueled the insurgency, but it is worth noting that most dismissed soldiers were conscripts who were probably happy to be released. In addition, the majority were also Shi'ites or Kurds, who were not susceptible to insurgent recruitment. On the other hand, the significant number of disgruntled former Sunni officers might have proved less troublesome had they been retained inside the army.

In retrospect, it probably would have been better to put all army personnel on inactive status, continue to pay them, and recall individuals incrementally and selectively. This is not far from what eventually occurred. Most former soldiers were eventually paid and some were recalled to duty, but doing so in this manner would have avoided the traumatic effect of abolishing a force and a national symbol which, unlike the Ba'ath party, was respected in some parts of the Shi'ite as well as Sunni communities in Iraq. It would also have allowed an accelerated recall of individuals and a selective recall of entire units as the need emerged.

If dissolving the old army was probably an unnecessary and counterproductive gesture, the failure to quickly constitute a new force—by recalling significant elements of the old, among other things—was a more serious error. It is clear that the CPA's early plans regarding the size, composition, and external orientation of the new Iraqi army failed to anticipate the mounting threat of insurgency and eventual civil war.

CPA efforts to rebuild the Iraqi police were slow getting off the ground. Kerik's focus was almost entirely short-term and tactical, and his absence of international and federal government experience was a serious drawback. The State Department's performance was also disappointing.

It seems to have made little difference whether the existing Iraqi forces were retained or dissolved, or whether American civilians or the military took the lead in developing a new force, because all of these alternatives produced unsatisfactory results. What is evident is that in 2003 the U.S. government needed, but did not have and proved unable to quickly assemble the capacity to train, equip, oversee, and support large numbers of Iraqi soldiers and police. The fact that similar programs in Afghanistan had floundered even more over the previous year suggests that this incapacity was by no means unique to the effort in Iraq.

CHAPTER FIVE

Governing Iraq

The administration's plans for post-Saddam Iraq had assumed that the Iraqi government would remain largely intact and willing to take orders from the American authorities. ORHA's preparations had focused on humanitarian relief operations. Garner later explained,

> The three things that worried us the most was the setting of the oil fields on fire, because [Saddam Hussein] had done that in Kuwait during the first Gulf War; large number of displaced people, refugees as a result of the war itself; or him using chemical weapons against the Shi'a or the Kurds, which he had done before several times. [Another] thing was a breakout of an epidemic, because there's a pretty high incidence of cholera in that part of the world, especially in Iraq, and we knew that after the bombing started, we'd have sewage problems.[1]

Garner nevertheless recognized that ORHA would have to ensure that the Iraqi ministries continued to function between the fall of Saddam's regime and the establishment of a new Iraqi government, which he anticipated would occur fairly quickly.[2] Saddam had run a highly centralized administration in which all important decisions were made in Baghdad. Garner's team assumed that the most senior levels of ministry leadership—the minister and a few senior Ba'athists—could be

---

[1]   Garner, interview with PBS Frontline: "The Lost Year in Iraq." Garner also listed paying civil servants and pensioners as a priority.

[2]   Garner, interview with PBS Frontline: "Truth, War & Consequences."

replaced without substantially undermining the work of the ministries. The large civil service staffs in the ministries would keep them running under new leadership. As Condoleezza Rice expressed the concept, "we would defeat the army, but the institutions would hold, everything from ministries to police forces."[3] American advisors and expatriate Iraqi technocrats would work with the most senior Iraqi officials remaining in the ministry after the top-level Ba'athists were removed. One such team would be established for each ministry. Beyond these provisions, ORHA paid little attention to the problems associated with restoring Iraqi public services. When ORHA deployed to Kuwait, less than a week before Baghdad fell on April 9, Section Eight of ORHA's "Unified Mission Plan for Post Hostilities Iraq"—the section dealing with civil administration—lacked a mission statement, a concept of operations, and key objectives. These functions also remained, like the rest of ORHA, only partially staffed.[4]

This inattention to public administration stemmed in large measure from the U.S. administration's initial uncertainty about how long, if at all, it would seek to directly govern Iraq before turning power back over to an indigenous regime. Similar uncertainty was exhibited about how to install or deal with local governments. In a March 5 presentation to the President and the National Security Council, CENTCOM commander General Tommy Franks grandly promised that he would have military "lord mayors" ready to take over management of every major Iraqi city and town.[5] In fact, no such assignments were ever made. As the 3rd Infantry Division's official after-action review makes clear, units were never told *how* local governments were to be established. The 3rd Infantry Division reported that it "transitioned into Phase IV SASO [Stability and Support Operations] with no plan from higher headquarters." This resulted in "a power/authority vacuum created by our failure to immediately replace key govern-

---

[3] Quoted in Michael R. Gordon, "The Strategy to Secure Iraq Did Not Foresee a 2nd War," *New York Times*, October 19, 2004, p. 1.

[4] Rajiv Chandrasekaran, *Imperial Life in the Emerald City: Inside Iraq's Green Zone* (New York: Alfred A. Knopf, 2006), pp. 31, 34–35.

[5] Gordon and Trainor, *Cobra II*, p. 160.

ment institutions."[6] Military units were left to establish local councils on an ad hoc basis. In some cases, meetings to select a local governing council were announced, and candidates were selected from whoever showed up. In other cases, local leaders were selected by the military authorities. Although the lack of a political program made the military's public administration efforts "haphazard and ineffectual," these problems of local administration were dwarfed by the broader collapse of the Iraqi state that awaited the CPA.[7]

A month after the fall of Baghdad, it was clear that conditions on the ground in Iraq were vastly different from what American planners had expected. At ORHA's February 20 rehearsal for postwar Iraq, the cost of reconstruction was put at $1 billion for three years.[8] A few weeks later, Andrew Natsios, USAID administrator, stated that the United States was only budgeting for $1.7 billion in economic assistance to Iraq.[9] Based on these optimistic assessments, President Bush requested only $2.4 billion for Iraq's reconstruction in a supplemental appropriation sent to Congress on March 2003, in addition to the $1.1 billion earlier approved by that body. The administration had earmarked just $230 million of this amount for the electric-power sector.[10] As Major General Carl Strock, who was with the Army Corps of Engineers unit attached to ORHA, explained, "Our whole focus on our reconstruction effort was really not to go in and fix this country, but to fix what we broke, and we sort of made the assumption that the country was functioning beforehand. I had a dramatic underestimation of the condition of the Iraqi infrastructure, which turned out to be one of our biggest problems, and not the war damage."[11]

---

[6] Quoted in Thomas E. Ricks, *Fiasco: The American Military Adventure in Iraq* (New York: Penguin Press, 2006), pp. 150–151.

[7] Allawi, *The Occupation of Iraq*, pp. 98, 118–119.

[8] Gordon and Trainor, *Cobra II*, p. 154.

[9] Andrew Natsios, interview with ABC News Nightline: "Project Iraq," April 23, 2003.

[10] Chandrasekaran, *Imperial Life in the Emerald City*, p. 151. By contrast, in October 2003 the World Bank assessed Iraq's total reconstruction needs at $35 billion, including $12.1 for electricity.

[11] Quoted in Gordon and Trainor, *Cobra II*, p. 150.

Planners for postwar Iraq had "very little knowledge . . . of what was going on in the Iraqi government," and failed to realize that the state had effectively withdrawn from the detailed management of the country, except in a few vital areas necessary for the immediate survival and continuation of the regime.[12] Almost immediately after the war started, it became apparent that initial cost estimates were wildly off the mark. Marines capturing the Rumaylah oil complex found the plant's equipment in such bad condition that they assumed it had been sabotaged before they arrived. Yet when they talked to workers at the complex, they were assured that everything was in order and that the badly damaged machinery simply reflected the state of disrepair of the Iraqi oil infrastructure.[13] In April, officers from the Army Corps of Engineers visited the Baghdad South power plant to determine the source of power outages in the capital. The plant was filled with broken pipes, frayed wires, and a control room without computers. Lieutenant Colonel John Comparetto, the army's chief electrical engineer in Iraq, recalls: "When I first looked around, I said, 'Holy moly.' This is not good. I hoped it was an isolated incident. But it wasn't true. It was typical. We were underestimating how bad it was, no doubt about it."[14] The electrical infrastructure was further damaged during the war by the Iraqi military's use of surges and rolling blackouts as a signaling tactic. The power surges led the chief engineer of the Baghdad South plant to shut operations down as a precaution, thereby crashing the entire Iraqi electrical grid. Consequently, in early May, only 300 megawatts (MW) of electricity were being generated in Baghdad and 500MW nationally, less than one-eighth the prewar level.[15]

Moreover, as one ORHA official noted, "The state disappeared. Either the people melted away or the institutions were melted down by

---

[12] Sherri G. Kraham, interview with the *United States Institute of Peace Iraq Experience Project*, November 5, 2004; Allawi, *The Occupation of Iraq*, p. 115.

[13] Gordon and Trainor, *Cobra II*, p. 193.

[14] Quoted in Chandrasekaran, *Imperial Life in the Emerald City*, p. 152.

[15] Bremer, *My Year in Iraq*, pp. 18, 26, 65; Gordon and Trainor, *Cobra II*, pp. 273–274, 467. Bremer cites the figure of 300MW as the national number, but the State Department's *Iraq Status Report* listed this as the number for Baghdad alone.

them."[16] The buildings of 17 of the 20 ministries ORHA intended to reestablish were destroyed by looting and fires after the fall of Baghdad.[17] CPA officials would later calculate that the economic cost of the looting in the initial weeks after Iraq's liberation was near $12 billion.[18] This estimate, which would have represented a loss of a third of Iraq's annual GDP, was probably somewhat high, but the damage was certainly very extensive. Most senior Ba'ath Party members who ran Iraq's ministries disappeared after the fall of Baghdad; and looters had methodically destroyed the files, records, documents, and databases of most ministries.[19] Additionally, because the 12 telephone switching centers in Baghdad were disabled by bombing during the war, Garner and his team were unable to communicate with officials across Baghdad.[20] Consequently, in mid-May no Iraqi ministry was working at more than 40 percent capacity.[21]

It was against this near-anarchic backdrop that the Coalition Provisional Authority was formed. On May 8, the United States and Great Britain notified the United Nations of the creation of the CPA "to exercise powers of government temporarily" and "provisional administration of Iraq."[22] In the letter officially designating Bremer as administrator of the CPA, Secretary of Defense Rumsfeld wrote, "You shall be responsible for the temporary governance of Iraq" overseeing

---

[16] Andrew Erdmann, quoted in George Packer, *The Assassins' Gate: America in Iraq* (New York: Farrar, Straus and Giroux, 2005), p. 141. Erdmann would serve as the senior advisor to the Iraqi Ministry of Higher Education.

[17] Garner, interview with PBS Frontline: "The Lost Year in Iraq."

[18] Packer, *The Assassins' Gate*, p. 139.

[19] Bremer, *My Year in Iraq*, p. 45; Allawi, *The Occupation of Iraq*, p. 115. See also Author unattributed, *Kirkuk: A Challenge for the Coalition*, Paper to Ambassador Bremer, August 24, 2003, p. 4; and Stewart, *The Prince of Marshes*, pp. 45–46.

[20] Gordon and Trainor, *Cobra II*, p. 465.

[21] Memo from Colonel Warner Anderson to ORHA Health Advisor, "Citizen Payment for Health Care," May 17, 2003.

[22] Letter from John D. Negroponte and Jeremy Greenstock to Mr. Mumi Akram, May 8, 2003.

all executive, legislative, and judicial functions of the state.[23] And on May 22, UN Security Council Resolution 1483 recognized the United States and Britain as "occupying powers" in Iraq, and called on them "to promote the welfare of the Iraqi people through the effective administration of the territory."[24]

De-Ba'athification

Bremer's first official act as CPA administrator was to determine which Iraqis would assist the United States in administering Iraq or, more specifically, which ones would not. CPA Order Number 1, entitled "De-Ba'athification of Iraqi Society," was issued on May 16. This decree excluded the top four levels of the party membership from public employment and also provided that the top three layers of management in every national government ministry, affiliated corporation, and other government institutions would be reviewed for possible connections to the Ba'ath party. Any managers found to be "full members" of the party would be removed from their government positions. The order provided that Bremer or any of his designees could grant exceptions on a case-by-case basis.[25]

Although de-Ba'athification would eventually become one of the most controversial aspects of the CPA's governance of Iraq, it initially had broad interagency support and was widely popular in Iraq, at least among the Shi'ite and Kurdish majority. The idea of purging senior Ba'athists from government employment had initially been put forward by the émigrés within the Democratic Principles Working Group of the State Department's Future of Iraq project. In a paper entitled "The Case for the De-Ba'athification of Iraq and the formation of a

---

[23] Letter from Donald Rumsfeld to L. Paul Bremer, May 13, 2003.

[24] United Nations Security Council Resolution 1483, May 22, 2003.

[25] Coalition Provisional Authority, Order Number 1, "De-Ba'athification of Iraqi Society," May 16, 2003. The CPA estimated the number of Iraqi officials affected by the order to be 20,000. Jay Garner and the former CIA Baghdad station chief put the number of banned Ba'athists closer to 50,000 and thought the order too severe. See Garner, interview with PBS Frontline: "The Lost Year in Iraq,"

National De-Ba'athification Council," Mowaffak al-Rubaie and Ali Allawi proposed holding culpable all Iraqis who

- held office or had been active at any level in the Ba'ath party or in organizations related to it
- authorized or participated in crimes
- supported the ideology or proselytized for the Ba'ath party
- gave substantial moral, political, or material support to the Ba'ath party or its officials and leaders.

Al-Rubaie and Allawi (Iraq's future national security advisor and Minister of Finance and then Defense, respectively) consciously modeled their policy on the de-Nazification of post–World War II Germany, which affected 2.5 percent of the population in the U.S. zone. Al-Rubaie and Allawi proposed that de-Ba'athification, like de-Nazification, should extend beyond government and party members to affect members of the professional and commercial classes "who had benefited from the Ba'ath Party and its programmes [sic] directly or indirectly."[26] At the time this idea was being crafted in the State Department's workshops, the CIA assessed that Iraqi ministries should be "purged of Saddam loyalists and restructured to eliminate the Ba'ath party oversight mechanism."[27]

The concept of de-Ba'athification was scaled back and adopted by the Department of Defense in its planning for postwar Iraq. At the March 10 NSC meeting, Frank Miller presented a draft approach for President Bush that was directed at the top-ranking 1 percent of the approximately two million members of the Ba'ath Party, or roughly 0.1 percent of the Iraqi population. As Miller recollects, this policy was "intended to be a lenient process only to get recidivists who would

---

[26] Mowaffak al-Rubaie and Ali Allawi, "The Case for the De-Ba'athification of Iraq and the formation of a National De-Ba'athification Council," undated paper in CPA Archives. See also Allawi, *The Occupation of Iraq*, p. 150.

[27] U.S. Central Intelligence Agency, *Iraq: The Day After*, October 18, 2002, p. 1, cited in U.S. Senate Select Committee on Intelligence, *Prewar Intelligence Assessments About Postwar Iraq Together with Additional Views* [undated, publicly released on May 25, 2007], pp. 100–101.

oppose us out of the government."[28] After Miller's March 10 briefing, the de-Ba'athification policy was refined in interagency meetings, and by early May it had interagency clearance.[29]

Reporting from the April 15 Nasiriyah meeting, Kanan Makiya wrote: "The overwhelming sentiment of the meeting was for a very strict and thorough De-Ba'athification program."[30]

On May 9, the day that he was officially designated the President's envoy to Iraq, Bremer met with Defense Department officials for final preparations before leaving for Baghdad. Under Secretary of Defense for Policy Douglas Feith showed Bremer a draft of the "De-Ba'athification of Iraqi Society" order, which had been cleared among the relevant agencies for Garner to issue. Bremer asked Feith to wait and let him issue it *after* he got to Iraq so that he could communicate both symbolically and substantively that the Ba'athists were gone for good.[31]

Garner was apprised of the draft order shortly after Bremer's arrival in Baghdad. Accompanied by the CIA station chief, he tried to talk Bremer into appealing this instruction, arguing that he had always envisaged a much more limited purge, leaving all but a couple of senior officials in each ministry. Bremer was not dissuaded, and on the following day, May 16, issued CPA Order Number 1.[32]

---

[28] Author interview with Frank Miller, April 18, 2008.

[29] Feith, *War and Decision*, p. 430. In his memoirs, George Tenet claims that the CIA knew nothing about de-Ba'athification until Bremer announced it in Baghdad in May. However, Miller says "Tenet is flat wrong in his memoirs," and he is supported by both Feith's account of the March 10 NSC and that of Gordon and Trainor. See George Tenet, *At the Center of the Storm: My Years at the CIA* (New York: HarperCollins, 2007), pp. 426–427; Miller interview, April 18, 2008; Gordon and Trainor, *Cobra II*, pp. 160–161. In Woodward's *State of Denial*, Stephen Hadley is cited as having learned of the de-Ba'atification order only after it was issued. If so, he must have meant the text of the order itself, since the intent to issue such an order was briefed to the President by Miller, Hadley's subordinate on the NSC staff, at a meeting that Hadley presumably attended.

[30] Kanan Makiya, "Memorandum on My Trip to Iraq and Kuwait," April 18, 2003, quoted in Feith, p. 417.

[31] Bremer, *My Year in Iraq,* p. 39; Feith, *War and Decision*, p. 429.

[32] Woodward, *State of Denial,* p. 193.

On May 22, Bremer wrote to President Bush that "The dissolution of [Saddam's] chosen instrument of political domination, the Ba'ath Party, has been very well received. Several Iraqis have told me, literally with tears in their eyes that they have waited 30 years for this moment."[33] Polling subsequently supported this assessment. Initial Iraqi support for the policy appears to have even transcended the country's sectarian divisions, making this the most widely popular action taken by the CPA during its 14-month lifespan. According to a poll released on August 6, 94.6 percent of Iraqis surveyed said either all (27.4 percent) or some (67.2 percent) Ba'athists should be removed from office.[34] Similarly, Zogby polling in August showed that 92 percent of Iraqis opposed the participation of former Ba'ath party members in Iraq's political institutions.[35] It is possible, of course, that Sunnis were under-counted in these polls, but given that they represent some 20 percent of the population, these figures suggest that many of the Sunnis who were interviewed supported some measure of de-Ba'athification or at least told the surveyors that they did.

Throughout the summer and fall of 2003 the CPA was consistently told by Iraqi leaders that it should be doing *more* to implement de-Ba'athification. At a June 24 meeting with Bremer, SCIRI leader Abdul Aziz Hakim complained, "Unfortunately, the brave decision you have made regarding de-Ba'athification has not been fully implemented."[36] In August, six Iraqi provincial governors wrote Bremer requesting that the CPA reinforce the implementation of Order Number 1 to ensure that senior Ba'ath party members were eliminated from decisionmaking in all ministries and other government departments.[37] And when Secretary of State Colin Powell visited the Baghdad City Council on

---

[33] Paul Bremer, Letter to The President of the United States, May 22, 2003.

[34] Iraq Center for Research and Strategic Studies, "Results of the First Public Opinion Poll in Iraq," August 6, 2003.

[35] Department of State, "Iraqis Offer Dim Evaluation of Reconstruction Effort Thus Far."

[36] Meghan O'Sullivan to CPA Political Team, Memo, "Meeting between Ambassador Bremer and Abdul Aziz Hakim, SCIRI," June 24, 2003.

[37] Iskander Jawad Witwit, Abdul Kareem Birgis, Ne'ama Sultan, Haider Mahdi, Hazim Al-Sha'lan, and Akram Al-Yasri, "Working Paper of the Governors," August 21, 2003

September 14, several speakers praised the CPA's de-Ba'athification policy but made emotional appeals for the coalition to pursue the policy more vigorously.[38]

The CPA's approach to de-Ba'athification was intended to avoid the excesses of the early phases of de-Nazification in Germany, which had comprehensively barred Nazi party members not just from government jobs but from a much wider range of employment—from all but manual labor in the case of the most senior members. In contrast, the CPA order affected only the top 1 percent of party members, or about .01 percent of the overall population, and barred them only from working for the Iraqi government. (Thus, the numbers purged in postwar Germany were some 25 times greater than in post-Saddam Iraq— 2.5 percent of the population versus .01 percent.) The vast majority of former Ba'athists (99 percent) could keep their government jobs; those dismissed would retain their property and freedom unless accused and convicted of specific crimes. Moreover, the CPA initially authorized scores of exceptions to this policy, permitting selected former ranking party members to stay in government jobs. The numbers so exempted receded under pressure from Iraqi political leaders but climbed significantly when Bremer moved to reinstate thousands of former Ba'athist schoolteachers over Governing Council objections.

In June 2003, Bremer rescinded all exemptions granted by civilian and military officials without his explicit approval. The next month, his senior governance advisor, Ryan Crocker, warned: "The principle behind the policy . . . has won widespread support within Iraq and abroad. Yet the implementation of the policy has generated considerable confusion among both Iraqis and military and civilian members of the coalition."[39] Crocker noted that de-Ba'athification was being imple-

---

[38] Coalition Provisional Authority Headquarters, Cable to SecState and SecDef, "Draft Report on the Secretary's Visit to the Baghdad City Council," September 17, 2003. Later, Ali Allawi would lament: "The unwillingness to treat the Ba'ath legacy for what it was—a totalitarian state with a privileged elite—and therefore in need of a radical overhaul, made the CPA reforms essentially tentative and nominal." Allawi, *The Occupation of Iraq*, p. 162.

[39] Ambassador Ryan Crocker to Ambassador Bremer, Memo, "Assessing De-Ba'athification," July 9, 2003. Ambassador J. Paul Bremer to Secretary of Defense, Memo, "Iraqi De-Ba'athification Council," May 22, 2003.

mented differently in different locations, the policy was affecting more than just hard-core Saddamists, and the procedures for exceptions were operating too slowly. The first step he recommended to address these confusions was to facilitate the creation of Iraqi institutions to deal with de-Ba'athification, an idea Bremer had first proposed in a May 22 memo to Secretary Rumsfeld. Yet whereas Bremer originally only sought to "put an Iraqi face on the de-Ba'athification process" to "increase the legitimacy of what might otherwise be perceived as an exclusively U.S.-led enterprise," Crocker stressed that an Iraqi body would be more sensitive to the nuances of the policy.[40] Bremer agreed, and on August 10 proposed the creation of the Iraqi De-Ba'athification Council to the Governing Council, noting that "Iraqis, not the CPA, are best positioned to continue de-Ba'athification and make any necessary changes to its implementation."[41] This judgment turned out to be incorrect, as Bremer later acknowledged.

On September 3, the Governing Council notified Bremer that it had formed the High National De-Ba'athification Commission. The Governing Council selected Ahmad Chalabi to head the Commission, and a then–little known official from the Da'wa Party, Jawad (Nuri) al-Maliki, was made his deputy. These Iraqi leaders promptly took a harder line regarding de-Ba'athification. On September 17, Chalabi informed Bremer that the commission's first two decisions were to declare null and void all exceptions issued to members of the Ba'ath Party belonging to the fourth level with the rank of division member, or *firqah*, prior to September 14. In addition, the ban on public employment imposed by Order Number 1 would be expanded to include public activities and positions in institutions of civil society, the press, and the media.[42] Bremer requested clarification on such questions as guarantees of due process and specifics about the proposed extension of de-Ba'athification

---

[40] Ambassador Ryan Crocker to Ambassador Bremer, Memo, "Assessing De-Ba'athification," July 9, 2003. Ambassador J. Paul Bremer to Secretary of Defense, Memo, "Iraqi De-Ba'athification Council," May 22, 2003.

[41] Ambassador Paul Bremer III to The Governing Council, Memo, "Proposal for Implementing the Iraqi De-Ba'athification Council," August 10, 2003.

[42] Ahmad Chalabi, Letter to Ambassador Paul Bremer, September 17, 2003.

and asked the commission to consider the revocation of exemptions on a case-by-case basis rather than a blanket one.[43] By November 4, Bremer was sufficiently reassured on these points that he issued CPA Memo Number 7 ratifying the commission's September 14 decisions and officially delegating to the Governing Council the authority to carry out the de-Ba'athification of Iraqi society consistent with CPA Order Number 1. Subsequently, on December 9, Bremer informed all CPA civilians and coalition military personnel, "De-Ba'athification is now an Iraqi process . . . immediately cease any involvement in de-Ba'athification."[44] Bremer issued this guidance even as reports from Anbar, Salah ad Din, and Ninewa were reporting increasing Sunni hostility toward de-Ba'athification.[45]

By late 2003, Bremer had begun to have doubts regarding the course de-Ba'athification was taking. He was also inclined to loosen the policy in the aftermath of Saddam's capture, feeling it no longer so necessary to reassure the populace that the dictator and his regime would not be coming back. On December 16, he asked the head of his governance team to consider whether or not the policy should be suspended. Scott Carpenter argued against suspension, saying that the Governing Council and the de-Ba'athification Commission had not done anything "egregious or in violation of your order delegating authority on de-Ba'athification."[46] Chalabi's talk on de-Ba'athification at the December 16 Coordinators and Commanders Conference was well received. And on January 11, Chalabi released new guidelines on de-Ba'athification that CPA governance described as "consistent with

---

[43] See Memo from Meghan O'Sullivan to Paul Bremer, "Update for CPA Senior Advisors on De-Ba'athification," September 25, 2003.

[44] Memo from Paul Bremer to CPA Senior Advisors, CPA Governance Teams, CJTF-7, "De-Ba'athification: An Iraqi Process," December 8, 2003. Although the memo is dated December 8, Bremer did not sign it until the next day.

[45] For example, see Memo from Keith Mines to Paul Bremer, "Meeting with Sunni Leaders—Toward a Post-Saddam Sunni Policy," November 8, 2003; William R. Stewart to Paul Bremer, "Weekly GC Update—Salah ad Din," November 21, 2003; and Herro Mustafa to Paul Bremer, "Weekly GC Update—Ninewa," December 6, 2003.

[46] Memo from Scott Carpenter to Paul Bremer, "Suspension of De-Ba'athification," December 17, 2003.

the delegation of authority and . . . an improvement on the earlier ones in so far as they provide for a more transparent process and open avenues for reconciliation."[47]

Despite these reassurances, CPA disenchantment with Iraqi implementation of the de-Ba'athification process continued to grow. While enthusiastically embracing its mission to root out residual Ba'athists from government service, the commission was slower to move on efforts that promoted national reconciliation. It was accused of trying to centralize all functions and of assuming "virtually universal powers" over the appeals process.[48] The CPA specifically attributed these shortcomings to Chalabi, accusing him of exploiting the process to further his own political ambitions.[49] This issue eventually came to a head over Bremer's decision, discussed later, to rehabilitate several thousand school teachers who had been dismissed under his original order.

Electricity

In his first meeting with the Iraqi leadership council on May 16, Bremer stressed the CPA's resolve to provide security and other public services.[50] Among the latter, the CPA's greatest challenge came in the area of electric power. Interrupted electric service threatened not just to undermine public confidence in the competence of the CPA but also to shut down Iraq's few working water treatment plants and to disrupt petroleum production, which was necessary for regenerating the revenue base necessary to rebuild Iraq. Consequently, next to the $990 million set aside for salary payments, the largest item in the Iraqi

---

[47] See Scott Carpenter to Paul Bremer, Memo, "Your Participation in the Coordinators and Commanders Conference Thursday, January 8, 2004, Iraqi Forum, 2nd Floor Conference Room," January 8, 2004; and Scott Carpenter to Ambassador Richard Jones, Memo, "Welcome Home," January 11, 2004.

[48] See Stuart Jones to Paul Bremer, "Weekly GC Update—Al Anbar," March 6, 2004.

[49] Michael Ratney to Paul Bremer, Memo, "Sending a Message on De-Ba'athification," March 19, 2004.

[50] Scott Carpenter to Paul Bremer, "Memo on Tonight's Meeting with the Iraqi Leadership Council," May 16, 2003.

budget that the CPA drafted for the remainder of 2003 was the $294 million committed for "Electrical Improvements."[51] Bremer was told by his electricity advisor Peter Gibson that they could achieve a tenfold increase in electricity generation by July 4 and could reach prewar levels—approximately 4,000MW—by September 30.[52] This goal was later revised upward to 4,400MW in the strategic plan that the CPA drafted in July.

Initially, there was dramatic improvement, as the Ministry of Electricity was repopulated and power plants were brought back on line. Iraq nevertheless continued generating electricity below prewar levels until September, unable to exceed an average of 3,300MW per day. Bremer explained to a visiting congressional delegation that this failure to improve generation was due to three factors: antiquated and poorly maintained power plants that break down frequently; a "fragile" transmission network highly prone to full system collapse, thereby causing damage to power plants; and a slower-than-expected tempo of starting up the oil fields and producing natural gas to operate gas-fired generating units.[53] Damage to the power distribution system was also greater than had been previously thought, with the most significant destruction not from looting but from "intelligent and targeted sabotage" aimed at isolating Baghdad. In late June, John Sawers sent a confidential report to the British government saying: "The new threat is well-targeted sabotage of the infrastructure. An attack on the power grid last weekend had a series of knock-on effects which halved the power generation in Baghdad and many other parts of the country."[54] Within ten weeks of the fall of Baghdad, 120 large masts carrying elec-

---

[51]   David Oliver, "Briefing on 2003 Iraqi Budget," briefing slides, July 16, 2003.

[52]   Bremer, *My Year in Iraq,* p. 69.

[53]   Tom C. Korologos to Ambassador Bremer, "Memo on CODEL McCain," August 16, 2003. A team from General Electric visiting power plants in the Baghdad area in late June estimated that most equipment was "in poor condition due to sanctions" and "looting of spare parts" and key equipment was "preventing quick recovery." See General Electric, "Iraq Exploration Team Trip Report," briefing slides, June 2003.

[54]   Quoted in Gordon, "The Strategy to Secure Iraq."

tric cable had been stolen.[55] As of August 23, 623 high-voltage transmission towers were down, and wires had been cut on 145 towers.[56] A CJTF-7 briefing titled "Security of the Iraqi Energy Infrastructure" estimated that 1,000 tons of copper had been looted since the war.[57] In sum, the CPA estimated that more than $1 billion worth of damage had been done to Iraq's electrical sector by looters and saboteurs.[58]

The effects of these attacks against the power infrastructure, and the subsequent power shortages, were felt by the Iraqi people. According to a survey conducted by Zogby International from August 3 to August 19, 81 percent of Iraqis said the coalition was doing "a very/fairly poor job" at providing electricity.[59] In a survey of Baghdad residents conducted later that month, 99 percent of the 1,178 respondents said they had gone without electricity for long periods of time, compared with 42 percent who acknowledged that this had also happened before the war.[60] Colonel David Teeples, commander of the 3rd Armored Cavalry Regiment, which was responsible for Al Anbar Province, took the unusual step of writing directly to Bremer: "Our largest concern continues to be the amount of electricity that has been allocated to the Al Anbar Province. This is causing the rolling black outs with the exception of essential locations and is creating turbulence within the community." Teeples noted that the current power grid in Anbar was not sufficient to stand up the local factories and that the decreased electrical power correlated to an increase in unemployment.[61] A month

---

[55] Cockburn, *The Occupation*, p. 75.

[56] Stephen Browning to Ambassador Bremer, "Memo on Power Distribution System—Iraq," August 23, 2003. In an aerial survey from Baghdad to Basra conducted on April 16, Bechtel Corporation counted 15 electrical transmission towers down. A similar survey done in mid-June showed 65 towers down. See Andrew Natsios, "Trip Report," July 18, 2003.

[57] CJTF-7 Briefing, "Security of the Iraqi Energy Infrastructure," September 16, 2003.

[58] Author unattributed, "Proposed Governing Council Press Release: Penalties for Crimes," September 10, 2003.

[59] Department of State, "Iraqis Offer Dim Evaluation of Reconstruction Effort Thus Far."

[60] "Baghdad Living Conditions 6 Months after the War: Preliminary Results from Gallup Survey in Baghdad," Fieldwork dates 8/28–9/04.

[61] Letter from Colonel David A. Teeples to Ambassador Bremer, August 4, 2003.

later, a situation report from Najaf dated September 9 made a similar point: Inadequate electrical power was forcing the closure of local factories, exacerbating the unemployment problem, which presents "a real opportunity for easy exploitation by groups hostile to the Coalition."[62] And in October, the governor of Karbala, Akram al-Yasiri, warned the CPA coordinator of the South-Central Region: "Who do you think the young men in Sadr's Mahdi Army are? They are the unemployed, the men who lost their jobs when the factories here shut down because of the electricity shortage."[63]

To deal with these shortages and improve generation, Bremer took several steps. Acknowledging that the electrical grid was insufficient to meet the increased postwar demand for electronic goods, on July 7 the CPA initiated a policy of prioritizing distribution of electricity to health care facilities and water and waste disposal systems and instituting "rolling blackouts" to distribute power in an equitable manner.[64] On August 14, Bremer established a special task force under Steve Browning to achieve the 4,400MW goal and made updates on electricity his first meeting of the day. The task force received two engineers from each of the 12 army brigades in Iraq, who were subsequently deployed to power stations across the country on August 24 to work with the plant managers to conduct a detailed assessment of what repairs could be made in two months to increase output. By mid-September, the Task Force had produced a 52-page briefing outlining the challenges and plans for achieving a 4,400MW output by September 30.[65] Based on the task force's recommendations, Bremer authorized an additional $25 million to be spent on repair parts and supplies

---

[62] Author unattributed, "Situation in Najaf as of 1200 September 9, 2003."

[63] Email from Mike Gfoeller to Paul Bremer, "Achieving Victory in South-Central Region: Course Correction," October 18, 2003.

[64] Coalition Provisional Authority, "Statement of Policy: Allocating Electrical Power," July 7, 2003. This policy was not always observed by the governorates or the Iraqi officials comprising the Electric Energy Commission, however. See "CPA Strategic 60-Day Plan 30 Day Review," September 1, 2003; William Bassford to Amb. L. Paul Bremer, "Memo on Program Highlights, Implementation concerns, Baghdad Central," September 7, 2003.

[65] Chandrasekaran, *Imperial Life in the Emerald City*, pp. 155–156; "Electricity in Iraq," Unattributed Briefing, September 16, 2003.

to meet the goal, and on September 10 the CPA and Governing Council jointly announced a penalty of ten years imprisonment for the theft or destruction of power sources.[66]

The crash program worked, at least briefly. On October 1, electrical power generation reached a peak of 4,217MW, and on October 6, peak generation reached 4,518MW.[67] Basra enjoyed electricity 24 hours a day, seven days a week, from late September into October.[68] However, Iraq's electrical plants began shutting down shortly thereafter to perform their regularly scheduled fall maintenance. The average daily generation in October was only 3,948MW, which fell to 3,582MW in November.[69]

Electricity occupied an important place in the administration's request to Congress for supplemental funding in November 2003. Whereas the administration asked for $4.2 billion total to fund and equip Iraq's police and army, the request for "Rehabilitating Electric Power Infrastructure" totaled $5.7 billion.[70] Broken down further, $2.8 billion (or 51.1 percent) of this sum was devoted to new generation projects and the rehabilitation of the existing thermal units that were the backbone of the Iraqi electrical system; $1.55 billion (27.3 percent) was to be spent on the high-voltage 132- and 400-kilovolt (kV) systems, initially involving repairing or replacing the existing lines that had suffered decades of neglect; and $1 billion (17.6 percent) to cover projects on low voltage (33kV and below) distribution networks that

---

[66] LTC Randy Richardson to Ambassador Bremer, "Memo on Your Meeting Today with Dr. Aiham Alsamarrae, Interim Minister of Electricity," September 5, 2003; "Proposed Governing Council Press Release: Penalties for Crimes," September 10, 2003.

[67] Bremer, *My Year in Iraq,* p. 179; memo from Paul Bremer to Governing Council, "Update on Iraq Infrastructure Reconstruction," October 7, 2003.

[68] Andy Bearpark, "Directorate of Operations and Infrastructure Update," September 29, 2003; Ambassador Henry Hogger, "Basrah Governorate Weekly Situation Report," October 15, 2003.

[69] Brookings Institution, Iraq Index/State Department, *Iraq Weekly Status Report.*

[70] Memo from David Oliver to Paul Bremer, "Supplemental Priorities," September 12, 2003; Chandrasekaran, *Imperial Life in the Emerald City,* p. 160.

led directly into Iraqi homes.[71] Even before these supplemental funds became available, Bremer allocated $756 million to Task Force Restore Iraqi Electricity to fund 26 projects to maintain the momentum the electricity infrastructure program had gained by October.[72]

Most of these projects were geared toward raising Iraq's generation output to 6,000MW by June 1, 2004. The CPA planned to achieve this goal by funding maintenance programs—installing 830MW of new generation, increasing imports of electricity to 300MW, and meeting the fuel requirements established jointly by CPA-Oil and the Ministry of Oil.[73] As ambitious as these plans were, the NSC staff urged the CPA to aim significantly higher. An NSC-commissioned study noted that "the 6,000MW standard was developed by CPA some time ago; it is somewhat meaningless." It continued that the goal should be "meeting demand" and calculated the demand to be 7,200MW. Army Lieutenant General (Ret.) Joseph Kellogg argued that goal was not feasible given the state of the electrical system, and that it would cost another billion dollars just to increase oil production goals and import enough 40-MW generators necessary to reach 7,200MW.[74]

In October 2003, Brigadier General Steve Hawkins of the Army Corps of Engineers explained to the visiting Deputy Secretary of Defense that there was no direct correlation between money and megawatts.[75] This assessment proved to be tragically accurate. In December, the CPA briefed Congress that "The electrical power infrastructure of Iraq remains in a perilous condition." The planned autumn maintenance was "having only limited success due to the long lead-times required to obtain materials and supplies."[76] Additionally, the CPA's

---

[71] See "Analysis of the FY04 Supplemental," October 3, 2003; and Johnson to Michael Adler, "Electricity Supplemental," February 19, 2004.

[72] Memo from David Oliver to Paul Bremer, "PRB Allocation Request #38," October 29, 2003.

[73] "Electricity Generation Goals and Milestones," briefing slides, February 19, 2004.

[74] Waddell to Jackson "Meeting Electricity Demand," February 19, 2004.

[75] Memo from Jessica LeCroy to Clayton McManaway, "Themes from Wolfowitz Trip," October 27, 2003.

[76] CPA Baghdad, "Section 2207 Report to Congress," Draft, December 29, 2003.

program for outage support was not meeting targets to ensure that critical parts were available for Bechtel Corporation, the U.S.-based engineering and construction company, and Ministry of Electricity staff. Of the 366 total purchase orders necessary to perform the spring 2004 scheduled maintenance program, only 166 had been placed. As a result, USAID warned, "parts will not be available in the spring to improve capacity, reliability, and availability or existing generation units for the summer of 2004."[77]

The CPA identified three other problems that jeopardized the goal of generating 6,000MW by June.[78] First, security concerns were increasing. During the April 2004 uprisings, thousands of workers for private contractors were confined to their quarters in the Green Zone and unable to repair power plants and other infrastructure. General Electric and Siemens suspended their operations in Iraq. In May, insurgents killed several Russian technicians working on the Doura power plant, leading to pressure for all Russian employees in Iraq to withdraw.[79] Second, Iraq's fuel infrastructure was slow to recover from international sanctions and Ba'athist mismanagement. Consequently, when the Minister of Electricity proposed importing generators capable of producing 1,000MW to get power onto the grid quickly and relieve pressure on the operating power stations, Bremer vetoed the plan, stressing the high capital costs and insufficient diesel stocks necessary for these generators.[80] Finally, even when generation was not a problem, the security of the transmission lines was. As Bremer recollected, "The distribution system for electricity is badly designed. It is

---

[77] Bechtel's strategy in the face of these problems was to focus on the repairs necessary to return key turbines, such as those at Doura 5 and 6, to reliable service for at least one year, and even then at only 95 percent of nameplate capacity. See Memo from Chris Milligan to Paul Bremer, "Weekly Infrastructure Update," February 7, 2004.

[78] Memo from Robyn McGuckin to Paul Bremer, "Accelerating Work to Meet Summer Peak," December 6, 2003.

[79] Diamond, *Squandered Victory*, p. 237; Bremer, *My Year in Iraq*, p. 370.

[80] Author interview with L. Paul Bremer, November 15, 2007; CPA Headquarters Cable, "Iraq: Bremer Meets New Interim Iraqi Minister of Electricity," September 7, 2003; and Memo from Robyn McGuckin to Paul Bremer, "Read ahead for December 3 meeting between Ambassador Bremer and Dr. Aiham Al Sammarae," December 3, 2003.

brittle and susceptible to attacks." From March 20 to March 26, the CPA reported all eight 400kV transmission lines and all ten 132-kV lines as being out of service.[81]

The improbability of achieving 6,000MW by June 2004 became increasingly apparent to CPA officials. One internal report noted, "In certain critical areas," such as electricity, "it is becoming evident that CPA targets may be unachievable."[82] Therefore, instead of focusing on symbolic targets, "the most important measure is *Iraqi satisfaction through a program of managed expectations.*"[83] However, in May and June 2004, large majorities of Iraqis polled said that the electricity situation had gotten worse, even though the average generation of 4,293MW was Iraq's highest during CPA's existence and exceeded prewar levels of generation.[84]

Health Care

Iraq's health care infrastructure was decimated in the final years of Saddam Hussein's rule by war, sanctions, and misgovernment. During the 1990s, annual per-capita spending on health care had fallen from the equivalent of $17 to about 50 cents. Half of Iraq's primary health care facilities had closed in that decade, and the country's infant mortality rate was five times that of neighboring Saudi Arabia. The CPA's initial overview of the Ministry of Health noted: "Extensive looting and a decade of no maintenance have caused a breakdown in capital equipment, facilities and system failures. Public laboratories are in poor

---

[81] Author interview with L. Paul Bremer, November 15, 2007; CPA, "Administrator's Weekly Report, March 20–26, 2004," March 31, 2004.

[82] Memo from Andrew Rathmell to Paul Bremer, "Where Are We? A Review of the CPA Mission," April 14, 2004.

[83] Memo from Andrew Rathmell to Paul Bremer, "CPA Priorities Towards Transition," April 5, 2004.

[84] Memo from Gene Bigler to Paul Bremer, "First Results from June 10–15 Poll in Baghdad, Basrah, Mosul, Babylon, Diyala, Ramadi and Sulaymaniyah," June 27, 2004.

condition."[85] Although Iraq had 39,000 nurses, only 300 of them were university trained.[86]

In his memoir, Bremer notes that he only realized the enormity of Iraq's health care problem during an early visit to Baghdad's Children's Hospital. On seeing the decrepit condition of the facilities, Bremer told the staff that "The Ministry of Health is one of our top priorities." He approved a 3,200 percent increase in health care spending, with the immediate focus on a rapid return to at least prewar levels throughout Iraq. The CPA raised doctors' monthly salaries by 800 percent and ordered the purchase and distribution of over 22 million doses of vaccines for children. Later, the CPA's Strategic Plan, released in July, set the goal for reopening all of Iraq's 240 hospitals by October 1.[87]

The CPA's efforts to improve Iraqi health care made slow but steady progress. Whereas on April 9 only 30 percent of hospitals in Iraq had been functioning, by early July, almost all of Iraq's hospitals and 1,115 clinics were open to patients. However, the Ministry of Health reported that in Baghdad Iraqis were receiving only 70–75 percent of prewar basic service. The figure was 80 percent in the South, and even Kurdistan was receiving only 90 percent of prewar basic health care. The decade of no maintenance of equipment and facilities had caused systemic failure that needed to be redressed. Similarly, looting of facilities after the regime's fall caused a loss of capital equipment and technology and exasperated the infrastructure's existing challenges.[88] In August, the Ministry of Health estimated that 50 percent of Iraq's medical equipment was broken or in need of maintenance. The ministry also reported that although 4,270 tons of pharmaceuticals

---

[85] James Haveman, "Overview of the Ministry of Health," July 9, 2003.

[86] CPA Headquarters Cable, "Iraq's New Minister of Health: Rebuilding a Shattered Ministry," September 11, 2003. See also CPA Headquarters Cable, "Najaf Trip Report," September 30, 2003, for a Spanish assessment of the poor quality of nursing in the south.

[87] Author interview with L. Paul Bremer, November 15, 2007; Bremer, *My Year in Iraq*, pp. 32–35, 64, 69, 115. On the degradation of Iraq's health care system in the 1990s, see also Allawi, *The Occupation of Iraq*, p. 129.

[88] Memo from Paul Bremer to Reuben Jeffrey, "Material for Speeches," September 5, 2003; James Haveman to Ambassador Bremer, "Strategic Objectives for the Ministry of Health, Iraq: Status and Accomplishments," July 2, 2003.

and medical supplies had been delivered since it took over the Kima-dia pharmaceutical distribution system, the bureau lacked indigenous leadership and its warehouses were in "disarray."[89]

Despite these challenges, by September the CPA could point to some significant indicators of progress in the health care sector. Since the Ministry of Health was reformed, 9,000 tons of medical supplies were delivered, an increase of over 200 percent. Some 22.3 million doses of vaccine were received to cover over four million children and nearly one million pregnant women. The CPA increased the Iraqi health care budget from $13 million in 2002 to $211 million in the second half of 2003, over a 1,500 percent increase in spending.[90] Most importantly, Iraqis noticed the improvement in health care. A majority of Iraqis polled in August said the CPA was doing "a very/fairly good job" at providing medical care.[91] In another survey, 43 percent rated the availability of quality health care to be "very or somewhat good" as opposed to the 28 percent who rated it negatively.[92]

In late November 2003, the Ministry of Health developed an ambitious transition plan that called for a full transition to Iraqi man-agement by January 2004, with the CPA senior advisor assuming a liaison role to provide technical assistance until the transition to sover-eignty. To accomplish this goal, the plan listed four critical actions to be achieved over the next 90 days: formulation of the permanent Min-istry of Health structure and staff; "legitimization" of the Minister of Health; establishment of a Program Management Office to coordinate refurbishment efforts; and the completion of the Ministry of Health strategic planning initiative.[93] Although the state of the various Iraqi

---

[89] "Ministry of Health Update for Ambassador Bremer," August 3, 2003; Coalition Provi-sional Authority, "Strategic Plan 60-Day Report," August 1, 2003.

[90] Memo from Bremer to Jeffrey, September 5, 2003.

[91] Department of State, "Iraqis Offer Dim Evaluation of Reconstruction Effort Thus Far."

[92] "Baghdad Living Conditions 6 Months after the War: Preliminary Results from Gallup Survey."

[93] James Haveman to Assistant Secretary of Defense for Health Affairs, November 26, 2003; memo from James Haveman to Paul Bremer, "Ministry of Health Transition Plan-ning," November 27, 2003.

ministry transition plans varied widely among the advisor teams, the Ministry of Health was assessed in March as "well placed for transition to sovereignty with a clear strategy and supporting management structure, a capable Minister and senior advisor team."[94] Consequently, on March 28, the Ministry of Health was the first to be returned to full Iraqi control.

Even as the Ministry of Health was leading the way in administrative capacity-building, it was exhibiting the performance problems endemic to many other Iraqi ministries. First, the health sector was not immune to the security challenges that plagued Iraq during CPA's tenure. The Tikrit hospital lost its staff due to rising violence, and a private hospital in Balad was reportedly being used to shelter terrorists and Saddam loyalists in November.[95] In May 2004, CPA started to receive alarming reports of a rise in kidnappings of physicians in Baghdad.[96]

A bigger immediate problem was continued systemic disruptions in the distribution of medicines throughout Iraq. In December, the CPA governorate coordinator in Ninewa reported medicine shortages in the North because of warehouse distribution problems from Baghdad.[97] In April, the ministry was forced to make a $5 million immediate purchase of basic medicines, syrups, and antibiotics from Iraqi companies and completed a fund transfer for the $10 million immediate purchase of critical medicines from international sources expected to arrive by the end of the month.[98] Yet shortages persisted, and in May a significant dearth of pharmaceuticals and medical supplies at the hos-

---

[94] See Memo from Giles Denham and Andrew Rathmell to Paul Bremer, "Ministry Transition: Post-sovereignty support requirements," March 11, 2004; and Memo from Giles Denham and Andrew Rathmell to Paul Bremer, "Ministry Transition—Final Report of Post-Sovereignty Support Plans," March 27, 2004.

[95] Memo from Emad Dhia to Paul Bremer, "Summary of November 8 IRDC Conference," November 14, 2003.

[96] *The Baghdad Mosquito*, May 14, 2004.

[97] Herro Mustafa to Paul Bremer, "Weekly GC Update—Ninewa," December 20, 2003.

[98] Memo from "Advisor for Health" to Paul Bremer, "Ministry of Health Update," April 11, 2004.

pital and clinic level and in government warehouses, especially in the south, was reported. When Ala'adin Alwan took over as the new Minister of Health in May 2004, the ministry staff told him that hospitals were out of stock of 40 percent of the 9,000 drugs deemed essential by the ministry. Of the 32 drugs used in public clinics for the management of chronic diseases, 26 were unavailable. These shortages were attributed to two causes: excessive utilization of free pharmaceuticals, and structural obstacles to effective distribution, especially the corrupt vestiges of Kimadia.[99]

Education

Over the 20 years prior to Iraq's liberation, government neglect, ideological manipulation, and the chaos of the war's aftermath had crippled Iraq's educational system. UNESCO estimated that some 25 percent of children ages 6–11 had dropped out of school. In rural areas, the dropout rates for girls approached 50 percent. A postwar UNESCO study of 4,044 secondary schools determined that approximately 9 percent of school buildings were structurally dangerous to students, 21 percent were badly in need of essential repairs, and 43 percent were somewhat in disrepair. Most schools and universities had closed due to the war and were not immune to the postwar looting that plagued Iraq. It was estimated that 80 percent of supplies and equipment in vocational and technical schools had either been looted or rendered unusable. The period of looting caused damage to roughly 3,000 schools, both vocational and academic.[100] Iraqi schools were also chronically overcrowded, sometimes with up to 180 students in a single classroom, with an average of one book for every six students.[101] As a CPA over-

---

[99] See Chandrasekaran, *Imperial Life in the Emerald City*, p. 219; "Strategic Plan 60 Day Report," August 1, 2003; and Memo from Richard Jones to Paul Bremer, "Response to Basra Weekly GC Update," May 23, 2004.

[100] Ministry of Education, "Presentation Package for Madrid Conference," October 21, 2003.

[101] Bremer, *My Year in Iraq*, p. 64.

view of the Ministry of Education summarized, Iraq's "national education system has collapsed. There is a need to build a modern education system from scratch."[102] Overall, the World Bank assessed Iraq's need for the education sector to be $4.8 billion.[103]

In response to a request by Bremer, the CPA's senior advisor to the Ministry of Education listed the top three policy priorities for the ministry as repairing the school infrastructure, capacity-building at the ministry, and curricular reform, including new textbooks.[104] Consequently, Bremer and his advisors set a goal of rehabilitating 1,000 schools and distributing more than a million kits for individual schools by September 30. The CPA also increased teachers' salaries from the equivalent of $3 to $150 per month and started to purge textbooks and curricula of the pervasive Ba'athist propaganda. This meant printing and distributing over five million books before schools reopened in October.[105] Toward this end, the CPA and UNICEF made a joint request that the UN humanitarian coordinator for Iraq support the printing of more than 66 million copies of new textbooks for 5.5 million Iraqi students, at a cost of $67 million.[106] Later, in its comprehensive strategic plan for Iraq, "Achieving the Vision to Restore Full Sovereignty to the Iraqi People," the CPA expanded on these goals and set out seven key tasks for improving the quality of and access to education.[107]

---

[102] Dorothy Mazaka, "Overview of the Ministry of Education," July 11, 2003.

[103] Ali Tulbah to Bremer Memo, October 18, 2003.

[104] Dorothy Mazaka to L. Paul Bremer, "Memo on Ministry of Education Policy Priorities," June 7, 2003.

[105] Bremer, *My Year in Iraq,* pp. 69, 115.

[106] Dorothy Mazaka and Carel de Rooy, "Letter to Mr. Ramiro Lopes da Silva," July 29, 2003.

[107] Specifically, the CPA set out to (1) revise, print, and distribute textbooks; (2) rehabilitate 1,000 priority primary and secondary schools; (3) reform the curriculum; (4) develop and implement vocational training programs linked to employment centers; (5) provide technical assistance for Ministry of Education capacity-building to improve the quality of education and lay the foundation for education reform; (6) introduce special programs for girls' education and reducing girls' illiteracy; and (7) develop special accelerated learning programs for secondary school and adolescent drop-outs. See Coalition Provisional Authority, "Achieving the Vision to Restore Full Sovereignty to the Iraqi People," Draft, September 25, 2003.

By the end of September, the CPA could point to some clear progress in education as Iraqi schools set to open. Over $62 million had been spent on Iraqi schools, allowing the rehabilitation of 1,061 schools. The CPA had distributed 7,000 "schools in a box" for primary schools and was in the process of providing 1.5 million school bags with school supplies for secondary students and distributing 100 percent of needed blackboards and teachers' desks for secondary schools.[108] Moreover, all schools had reopened by the start of the new school year.[109]

The CPA made less progress with regard to capacity-building and textbooks. The Ministry of Education was the largest public-sector employer in Iraq and, as of July 6, it still had not paid any salaries since the fall of Baghdad.[110] After the destruction of the ministry's headquarters in April, its staff had been occupying a series of temporary, unrenovated buildings, and in December still did not have adequate office space from which to conduct its business.[111] Moreover, the withdrawal of UN personnel from Iraq created problems with the distribution of textbooks, forcing teachers to use Saddam-era material in the fall and work around the Ba'athist content.[112]

In early December, Minister of Education Dr. Ala'adin Alwan wrote Bremer to outline his ministry's priorities for the CPA's final seven months. These were obtaining a ministry headquarters building immediately; rehabilitating at least 3,000 schools; developing primary school teacher training and continuing to implement secondary school teacher training to achieve full coverage; implementing a plan for capacity-building and training for the Ministry of Education staff; and conducting a national dialogue on education reform and beginning

---

[108] Undated Command's Comments on Bremer to CJTF-7 Memo, "Establishing and Implementing Governing Policies" August 8, 2003.

[109] Bremer, *My Year in Iraq,* p. 187.

[110] David Oliver, Jr. and John Rooney, "Pensions and Salaries: Status as at 6th July 2003," July 7, 2003.

[111] Leslye Arsht to Ambassador Bremer, "Memo on Ministry of Education Building," October 29, 2003.

[112] Memo from Leslye Arsht, Bill Evers to Ambassador Bremer, "Your Meeting with Dr. Alwan, Interim Minister of Education," October 2, 2003.

the curriculum revision process in consultation with the Governing Council. The CPA and the Education Ministry were largely successful in achieving these goals. By early March, the number of schools rehabilitated had risen to 2,500. The CPA coordinated the review and editing of 48 primary and secondary math and science textbook titles, and USAID and UNESCO printed and distributed more than 8.7 million textbooks, providing one book for every two students.[113] Senior advisors to the Ministry of Education noted that Dr. Alwan "has reached out to all the political parties and religious groups and successfully included them in the national discussion about curriculum reform."[114] CPA officials were also impressed with the ministry's progress toward transition, noting "The Ministry has a very capable Minister and a newly restructured senior team . . . and there will be no need for staffing in any successor organization to CPA."[115]

In the spring of 2004, amid the myriad concerns regarding the politicization of the de-Ba'athification process, the CPA governance team emphasized to the National De-Ba'athification Commission that Bremer considered it a priority to rehabilitate a large number of the teachers who had earlier been dismissed for being senior Ba'ath party members.[116] Subsequently, on April 4 Bremer wrote to Chalabi, stating: "Thousands of teachers—often nominal Ba'ath Party members but not criminals—remain out of work even while the Iraqi educational system suffers. . . . The highest priority must be given to accelerating the appeals process, particularly in the education sector."[117] Two weeks later, Bremer informed National Security Advisor Condoleezza Rice that the CPA would issue an order to establish an independent de-Ba'athification committee removed from the Iraqi political process, and that in his April 23 address to the Iraqi people he would announce:

---

[113] CPA, "Administrator's Weekly Report—Essential Services," March 5, 2004; and CPA Baghdad, "Countdown to Sovereignty: Rebuilding Iraq," March 2004.

[114] Memo from Governance Team to Paul Bremer, "Ministers," May 9, 2004.

[115] Memo from Denham and Rathmell to Bremer, March 27, 2004.

[116] Memo from Michael Ratney to Paul Bremer, "Sending a Message on De-Ba'athification," March 19, 2004.

[117] Letter from L. Paul Bremer to Ahmad Al-Chalabi, April 4, 2004.

"The Commission and I have decided that the committee's first priority will be facilitating a return to productive work of primary and secondary teachers . . . who lost their jobs because they were formerly of the rank of *firqah* in the Ba'ath Party."[118] Bremer was careful to clarify to the Governing Council that "De-Ba'athification was and remains the right policy for Iraq," but he instructed the Minister of Education to ensure that those teachers whose appeals had already been heard and approved were returned to the Ministry of Education payroll; that those who opted for pensions begin receiving them as soon as possible; and that local de-Ba'athification review committees were constituted as necessary and pending appeals from former *firqah*-level teachers were heard within 20 days. To his own staff he wrote, "This needs <u>daily</u> follow up," on the memo covering a letter to Dr. Alwan.[119] Bremer's efforts appeared to have the intended effect. Between his April 23 address to the Iraqi people and June 6, approximately 5,000 teachers and administrators originally de-Ba'athified were cleared for reinstatement, and the CPA representative in the troubled Al Anbar Province reported that all of Anbar's de-Ba'athified primary and secondary school teachers were back at work by the second week of June.[120]

Bremer's decision to relax de-Ba'athification in the education sector was based more on a political calculation than any real shortage of teachers. Saddam had dismissed more teachers for not being Ba'athist than Bremer had for the obverse, and many of these proved ready and eager to come back. Further, the Ba'athists had badly politicized the Iraqi educational system. In October 2003, the UN–World Bank Joint Needs Assessment had condemned the politicization of the Iraqi education system under Saddam, which influenced everything

---

[118] Memo from Paul Bremer to Condoleezza Rice, "Independent De-Ba'athification Committee," April 19, 2004.

[119] Memo from Paul Bremer to Governing Council, "Clarification of De-Ba'athification Policy," April 25, 2004; letter from L. Paul Bremer to Dr. Ala'addeen Al Alwan, April 26, 2004; memo from Michael Ratney to Paul Bremer, "Letters on De-Ba'athification," April 26, 2004.

[120] See memo from Meghan O'Sullivan to Paul Bremer, "De-Ba'athification and the Transition," June 6, 2004; Stuart Jones to Paul Bremer, "Weekly GC Update—Al Anbar," June 10, 2004.

from curriculum to teaching staff to admissions policies. In December, CPA-South Regional Coordinator Hilary Synnot reported that the curricula were distorted for political purposes and that senior teaching staff were required to be members of the Ba'ath Party and were often active supporters.[121] In fact, at the same time that Bremer's staff was preparing the letter to Chalabi decrying de-Ba'athification in the education sector, the CPA was touting this measure as an achievement in other documents, declaring: "The politicized education system was dismantled. More than 12,000 headmasters, headmistresses, and teachers, who were former Ba'ath Party members, were dismissed with a process for appeal at the local and national levels."[122] Moreover, when Chalabi asked the CPA in December to propose draft regulations to implement de-Ba'athification, it provided him with a draft based largely on the de-Ba'athification procedures established by the Ministry of Education.[123]

The CPA representative in Najaf reported that the Director General for Education there said 2,553 former teachers had been fired under Saddam for political reasons and estimated that de-Ba'athification would result in the dismissal of only 200 teachers, suggesting that rehiring teachers fired by the former regime could easily make up any shortfall created by de-Ba'athification.[124] In Muthanna Province, the CPA representative reported his contacts as saying that for every low-ranking former Ba'athist teacher who lost his job there were many more Shi'ite teachers whose careers the Ba'athists had cut short because of their refusal to join the party.[125] Even in Anbar Province, where eventually 2,000 teachers would seek reinstatement, CPA officials were

---

[121] See "Section 2207 Report to Congress," December 29 Draft; email from Sir Hilary Synnot to Raad Alkadiri, "South Iraq: Report on Education," December 25, 2003.

[122] CPA, "Countdown to Sovereignty," March 23, 2004.

[123] Memo from Candace Putnam to Paul Bremer, "De-Ba'athification Update," December 29, 2003.

[124] CPA Office of the Ministry of Education Memo for the Record, "Unpaid Salaries for Unauthorized Spring-Summer Hires," November 25, 2003.

[125] James Soriano to Paul Bremer, "Weekly GC Update—Al Muthanna," May 18, 2004.

reporting that the schools, in their absence, were "good."[126] Whereas CPA put the number of teachers and administrators de-Ba'athified at 12,000, the Ministry of Education said that approximately 20,000 teachers had been fired by the previous regime and that "With the exception of some southern provinces, there is currently no real teacher shortage in Iraq."

Bremer's effort to reinstate former Ba'athist teachers was in response to pressures from Sunni leaders to rehabilitate respected members of their community and reflected Bremer's desire to bring this community into the political process. He also felt that once Saddam had been captured, the need to reassure the Iraqi people of a return of the Ba'athist regime was lessened. The exclusion of these ex-Ba'athist teachers may or may not have significantly affected the performance of the Iraqi education sector, but it did antagonize influential elements of the Sunni community. The CPA's primary complaint about Chalabi's administration of the National De-Ba'athification Commission was that it was more enthusiastic about rooting out residual Ba'athists in the public sector than the national reconciliation aspects of de-Ba'athification. Bremer's reversal on the dismissed teachers was an effort to right that balance.

This gesture naturally encountered immediate Shi'ite pushback. That community's leaders interpreted Bremer's rollback of de-Ba'athification as proof that Washington was pandering to Iraq's Sunni population. "He insists the policy wasn't changed, but why else would he televise the announcement?" an Iraqi asked a CPA official the next day. Mowaffak Al-Rubaie warned that the CPA's National Reconciliation campaign was incorrectly perceived by Shi'ites as a reintegration of Ba'athist criminals and the marginalization of the Shi'ites. Thus the decision to soften de-Ba'athification risked trading the goodwill of Iraq's 14 million Shi'ites and six million Kurds for the sake of Iraq's six million Sunni Arabs. Bremer's staff was aware of this risk when the decision to reinstate the de-Ba'athified teachers was made. On April 15, O'Sullivan and Martinez wrote to Bremer: "Returning Ba'athists and senior military officials . . . could undermine much of what we have succeeded in doing since Iraq's liberation. . . . The worst case scenario is

---

[126]See Memo from Dhia to Bremer, November 14, 2003.

that we lose the Sunnis over Fallujah and lose the Shi'a over our efforts to win back the Sunnis."[127]

Local Government

In contrast to the heavy and early emphasis placed on restoring basic services in Iraq, the CPA was slow to promote the development of local governance. In May, the Marine battalion responsible for administering Najaf sought to hold an election to replace a CIA-installed mayor who was unpopular with the city's residents and receiving negative press in the Western media. However, a day before the registration process was formally to begin, CPA instructed the Marines to cancel the election.[128] The CPA feared that elections "could create a legitimate counter authority to the CPA, *making its ability to govern more difficult.*"[129] Bremer subsequently wrote to the commander of the coalition forces: "In order to insure a consistent application of election policies and procedures, I request that you advise all subordinate military personnel that they should no longer initiate any election activity, including the calling of elections for any office whatsoever."[130] Instead, the CPA favored using appointed authorities when creating town councils or, if appointments were not deemed possible or desirable, using consultations and caucuses as a fallback.[131]

In early July, an independent study led by former Deputy Secretary of Defense John Hamre recommended expanding Iraqi ownership of the rebuilding process at the national, provincial, and local levels. Officials within the CPA also recognized the importance of local gover-

---

[127]Memo from Meghan O'Sullivan and Roman Martinez to L. Paul Bremer, "Points on NSC Memo," April 15, 2004.

[128]Gordon and Trainor, *Cobra II*, pp. 490–491; Packer, *The Assassins' Gate*, p. 189.

[129]Scott Carpenter to Ambassador L. Paul Bremer, "Memo on Interim Local Selection Processes," May 20, 2003. Emphasis added.

[130]L. Paul Bremer to Commander, Coalition Forces, "Memo on Election Administration," June 11, 2003.

[131]Memo from Carpenter to Bremer, May 20, 2003.

nance, listing "empowering local advisory councils" as a 30-day objective for the strategic goal of improving provision of public services at the local level.[132] Yet, although more than 600 neighborhood, city, district, and provincial councils had been established in Iraq, they lacked any formal authority or money and enjoyed questionable legitimacy.

In early August, the CPA began circulating a "Draft Memo on Establishing and Implementing Governing Policies" to spell out the authority of the local councils. However, this draft was almost universally panned by the military commanders throughout Iraq as being too vague or too restrictive of local government. The 1st Armored Division commander, Major General Martin Dempsey, warned that local councils would languish and dissolve without clear responsibilities and authorities. Major General Ray Odierno of the 4th Infantry Division noted that by saying what local governments *can't* do rather than what they *can* do, the order risked creating impotent local organizations. The 1st Marine Expeditionary Force's commanding general, Lieutenant General James Conway, objected that the draft centralized too many powers, such as the ability to hire and fire provincial ministry officials, in Baghdad rather than enabling local leadership.[133] At the August 15 regional coordination meeting, CPA regional officers made the similar case that local councils should have a say in local appointments, noting that the Baghdad ministries sometimes replaced experienced workers or protected corrupt people.[134] These concerns had not been addressed as of the September 29 regional coordination meeting, at which the participants noted that lack of clarity regarding the relationships among various national, regional, and local Iraqi governing bodies was causing confusion on the ground. They also warned that

---

[132] John Hamre, Frederick Barton, Bathsheba Crocker, Johanna Mendelson-Forman, and Robert Orr, *Iraq's Post-Conflict Reconstruction: A Field Review and Recommendations,* July 17, 2003; memo from Ryan Crocker to Paul Bremer, "Ensuring a firm foundation for political transition over the next 18 months," July 17, 2003.

[133] Undated Command's Comments on Bremer to CJTF-7 Memo, "Establishing and Implementing Governing Policies," August 8, 2003.

[134] See memo from Julie Chappell to Paul Bremer, "Regional Co-ordination Meeting," August 16, 2003.

where ambiguities existed about authorities in local areas, other groups would seek to fill the vacuum, often to the CPA's detriment.[135]

In October, the CPA reported to Congress concerning local councils: "The existence of these governing bodies provides the people of Iraq with a direct and meaningful say in their community's affairs for the first time in 35 years. The innovative decentralization processes initiated in each of the governorates in Iraq since April 2003 has brought a new sense of local ownership and prioritization for the delivery of services."[136] Yet this was not apparent to the Iraqi officials working with the CPA, who bemoaned the lack of clarity regarding their authority. The mayor of Tikrit told CPA officials that the government structure was confusing, as "I find that many of my department heads report directly to governorate, or even to ministries in Baghdad."[137] The deputy governor of Salah ad Din echoed this complaint, stating that the first priority for the Governing Council and CPA should be to clarify the relationship between the central government and local government, focusing on decentralization of powers.[138]

A more frequent lament of local Iraqi officials was that the lack of resources delegated to local governmental bodies left them unable to solve problems and gain the trust of their constituents. "You give us no responsibility," a sheikh on the Maysan Provincial Council told the CPA's governorate coordinator. "The people come to us demanding things and we cannot deliver."[139] On two separate occasions in August and September, a group of six Iraqi governors from south-central Iraq wrote Bremer to request that CPA invigorate the role of local administrations and give them the authority to accomplish their work and

---

[135]Memo from Julie Chappell to Paul Bremer, "Regional Coordination Meeting," September 29, 2003.

[136]Coalition Provisional Authority, "Update Report on United States Strategy for Relief and Reconstruction in Iraq," October 14, 2003, pp. 10–11.

[137]CPA Headquarters Cable, "The Mayor of Tikrit: Confusion, Complaints, and the Coming of Ramadan," September 21, 2003.

[138]William R. Stewart, "Salah ad Din Governorate Weekly Situation Report," October 15, 2003.

[139]Stewart, *The Prince of the Marshes*, p. 98.

that CPA grant them "the administrative and financial powers required to handle the affairs with [a] view to restore security and stability."[140] The Najaf Governing Council threatened to resign en masse in September if long-standing requests for resources were not addressed.[141] Similarly, a CPA visit to the outlying areas of Baghdad Province found that local leaders' most common request was for local control of local budgets.[142]

Despite the CPA's pronouncements about local ownership, public opinion polling showed that a majority of Iraqis did not think a local advisory council actually existed for their neighborhood.[143] Regional CPA officials were generally sympathetic to the Iraqi requests for greater authority over reconstruction funds. Hank Bassford, CPA governorate coordinator for Baghdad, told Bremer, "They have the staff and understand the problems but have no operating budgets."[144] Similarly, CPA South-Central Region's July 22 situation report warned: "Local governance is at a standstill due to the inability to draw funds for basic services at the local level."[145]

In mid-October, CPA's Office of Policy Planning suggested that developing local governance be made a planning priority in the next 90 days, noting: "As we empower Iraqis and reduce the military presence, we need an interim policy framework for local governance."[146] In

---

[140] See Iskander Jawad Witwit, Abdul Kareem Birgis, Ne'ama Sultan, Haider Mahdi, Hazim Al-Sha'lan, and Akram Al-Yasri, "Working Paper of the Governors," August 21, 2003; and "Governors Working Paper," September 17, 2003.

[141] Robert Ford, email to Mike Gfoeller, "Report for Najaf, September 11," September 11, 2003.

[142] Memo from Hank Bassford to Paul Bremer, "Welcome Home Brief for Ambassador Bremer," September 29, 2003.

[143] Iraqi Center for Research and Strategic Studies, "Results of Public Opinion Poll #3," October 23, 2003.

[144] Memo from Bassford to Bremer, September 29, 2003.

[145] Email from Timothy Krawczel to Ministry of Finance, "Frozen Local Government Operating Budgets," July 24, 2003.

[146] Memo from Office of Policy Planning to Paul Bremer, "Priorities for the next 90 Days of the Strategic Plan," October 16, 2003.

early November, CPA's governance team produced a series of memos on reforming provincial institutions, defining the relative authorities of Iraq's central and provincial governments, and developing a broader strategy on local governance. These memos acknowledged that local government institutions can "provide superb training for citizens in participatory, democratic politics," and set an objective of not merely clarifying lines of authority, but "empower[ing] local bodies and mak[ing] them responsible to local needs. . . . When possible, we should seek to strengthen local institutions and encourage the development of local power centers." Ultimately, the CPA governance staff recommended increasing the Iraqi sense of involvement in their local political structure through caucuses to "refresh" the provincial councils, and empowering the provincial bodies by increasing the resources at their disposal.[147]

After the November establishment of an accelerated timeline for the restoration of sovereignty, the CPA governance team reconsidered the local governance strategy. Rather than addressing the numerous complaints about ambiguity and impotence surrounding local governing councils, the governance team recommended scaling back any ambitions to empower local bodies. Bremer agreed, stating later, "When the TAL [the Transitional Administrative Law, or interim constitution] was being negotiated, I felt the power of the local or Provincial councils shouldn't be decided by the United States." Consequently, CPA Baghdad withdrew its previous proposals to give the provincial ministry offices increased powers and no longer advocated transferring to the governorate and/or local bodies' greater powers to collect local funds or taxes.[148] Bremer conveyed this decision to the CPA regional coordinators, governorate coordinators, and CJTF-7 commanders in

---

[147] See memo from Meghan O'Sullivan and Roman Martinez to Paul Bremer, "Reforming Provincial Institutions," November 2, 2003; memo from Meghan O'Sullivan and Roman Martinez to Paul Bremer, "The Relative Authorities of Central and Provincial Government," November 2, 2003; and memo from Meghan O'Sullivan, Roman Martinez, and Julie Chappell to Paul Bremer, "Strategy on Local Governance," November 3, 2003.

[148] Memo from Meghan O'Sullivan, Roman Martinez, and Julie Chappell to Paul Bremer, "Post-November 15 Local Government Plan," November 25, 2003; author interview with L. Paul Bremer, November 15, 2007.

a memo on November 27. The next day, he told participants at the Regional Coordinators and Commanders Conference that because the TAL would deal with issues of federalism and decentralization, the previous proposals increasing the responsibilities and authority of the provincial level were being withdrawn. Instead, $1 million would be allocated to each governorate to facilitate the work of the local councils and increase public participation and sense of responsibility.[149]

This new funding did little to mitigate the frustrations felt by local Iraqi leaders. Najafis were angered by constant interference from the Baghdad ministries and sought a decentralized government structure.[150] The Baghdad City Council continually expressed its desire for more authority and complained of being limited to political recommendations.[151] The governor of Basra complained to Bremer that delegations from Iran, Kuwait, and the United Arab Emirates had visited Basra looking for economic opportunities and making proposals for projects related to the oil industry and refrigerated storage, but that the Basra administration did not have the authority to make any deals.[152] In January, the CPA representative in Diyala reported that provincial and municipal officials there wanted the Iraqi budgetary process switched from a Baghdad-centric system to one where more responsibility rested at provincial and municipal levels.[153] After visiting all the neighboring provinces, the British coordinator for Basra Province noted a common theme wherever he went was complaints by Iraqi officials about the inadequacy of their operating budgets.[154] The Dhi Qar

---

[149] L. Paul Bremer to CPA Regional Coordinators, CPA Governorate Coordinators, and CJTF-7 Commanders, "Local Government Plan, Post-November 15," November 27, 2003; "Executive Summary of the Minutes from the Coordinators and Commanders Conference, 28 November 2003," November 28, 2003.

[150] Robert Ford to Paul Bremer, "Weekly GC Update—Najaf," November 22, 2003.

[151] "Read Ahead for Bremer Visit to Baghdad City Council," December 1, 2003.

[152] Hilary Synnot to Paul Bremer, "Notes from December 19 Bremer Meeting with Basra Governor Wael Abd al-Latif," December 23, 2003.

[153] Ed Messmer to Paul Bremer, "Weekly GC Update—Diyala," January 3, 2004.

[154] Email from Patrick Nixon to Paul Bremer, "Impressions of the South," February 18, 2004.

provincial council complained to a CPA governance official, "We are just talking and meeting with no authority," and on February 16, the chairman of the Wasit Provincial Council threatened to resign because the council "is just a talking shop" lacking authority.[155] Similarly, the chairman of the Zafaraniya Neighborhood Advisory Council (NAC) in Baghdad complained to a visiting journalist that "Our authority as the NAC is still shaky. . . . People don't trust us. They come up to us and ask for something, and we can't do anything for them."[156]

The CPA's regional officials sympathized with their Iraqi counterparts and advocated for more decentralization. In December, the governorate coordinator for Qadisiyah Province stressed, "We need to include the refreshed local and provincial councils in decision-making of the activity people care most about."[157] From Wasit, Mark Etherington argued in January: "The key here is that Councils must take responsibility for helping to tackle" complex problems such as unemployment, and that "this ownership will erode the hold of political parties in the minds of the poor and disadvantaged. Please allocate us funds." A month later, he lamented: "It is particularly unfortunate that the formal codification of council powers has not been issued, if only because it has eroded the confidence of council members at a time of particular vulnerability. This has retarded our ability to use the province council to effect reform."[158] Similarly, the CPA representative in Kirkuk stated: "If we put the responsibility for dealing with local political issues on the local administration, and back that up with real authority and resources, they are capable of managing their differences locally."[159]

In addition to not wanting to prejudge results of the TAL negotiations, CPA Baghdad was not inclined to give the local councils greater

---

[155] Diamond, *Squandered Victory*, p. 207; "Wasit Weekly Situation Report," February 20, 2004.

[156] Packer, *The Assassins' Gate*, pp. 296–297.

[157] Henry Ensher to Paul Bremer, "Weekly GC Update—Qadisiyah," December 11, 2003.

[158] Mark Etherington to Paul Bremer, "Weekly GC Update—Wasit," January 16, 2004; "Situation Report: Wasit, 19–25 February 2004," February 25, 2004.

[159] Paul Harvey to Paul Bremer, "Weekly GC Update—Kirkuk," March 27, 2004.

influence over reconstruction funds for logistical reasons. Iraq had no ability to transfer funds electronically, so all payments had to be made in cash. And, as at least one governorate coordinator noted, transporting and paying out significant amounts of cash in a "less-than-permissive security environment" was a dangerous undertaking.[160] Yet the CPA never placed enough emphasis on empowering local councils to ameliorate this problem. This had the unfortunate effect of leaving provincial and district councils unpaid for long periods of time. CPA representatives in Diyala, Najaf, Basra, Kirkuk, and Anbar all reported at various times that the Iraqi councils they worked with had not been paid their stipends in several months, leading the representative in Anbar to conclude "The system for paying salaries to provincial and municipal councils is broken."[161]

Finally, on April 6, 2004, Bremer signed CPA Order Number 71, which prescribed the respective authorities and responsibilities of the governorate, municipal, and local elements of government. Provincial governing councils could elect—and with a two-thirds vote remove—governors and deputy governors; approve or veto directors general and local officials in senior positions by a majority vote and remove them with a two-thirds vote; select chiefs of police to three-year terms by majority vote from nominees determined to be qualified by the Ministry of Interior; and amend specific local projects in the annual ministry budget plan on a two-thirds vote. Provincial and local government entities would generate and retain taxes, fees, and similar revenues in accordance with existing Iraqi law.[162] The implementation of this order was put on hold until the Governing Council could be persuaded that the governorates were ready for the devolution of such powers. Once implementation began, the provincial councils were generally pleased

---

[160]Author interview with L. Paul Bremer, November 15, 2007; Edward Messmer to Paul Bremer, "Weekly GC Update—Diyala," March 27, 2004.

[161] See Stuart Jones to Paul Bremer, "Weekly GC Update—Anbar," March 25, 2004; Tom Rosenberger to Paul Bremer, "Weekly GC Update—Diyala," November 21, 2003; memo from William J. Olson to Paul Bremer, "Executive Summary of Governorate Coordinator and Regional Coordinator SITREPs," February 28, 2004; Harvey to Bremer, "Weekly GC Update," March 27, 2004.

[162]Memo from OGC to Paul Bremer, "Local Government Powers," April 6, 2004.

with their new authority, even if it had taken nearly a year for the CPA to live up to its promise to expand local responsibility of the delivery of public services.[163]

Conclusion

The CPA was able to restore most basic public services to near prewar levels—and in some cases well above them—despite the lack of planning, the destruction of much Iraqi infrastructure by war, sanctions, mismanagement and looting, and the steadily deteriorating security situation. This was no mean achievement.

The CPA's efforts to boost power-generating capacity to well above prewar levels were probably misplaced. Under Saddam, Iraq had provided its population with as much electricity as regional states at a comparable level of development, such as Jordan. Jordan did not suffer chronic blackouts, whereas Iraq did. Iraq's problem was not limited supply but excess demand, driven by the fact that Iraq was not meaningfully charging customers for electricity they used. The CPA exacerbated this problem by choosing to forgo collection of even those minimal charges for electricity made under the old regime. Then, when the CPA reduced most import tariff barriers, an otherwise sound move, the result was a greatly expanded importation of white goods, driving demand up further.

The de-Ba'athification decision did not originate with Bremer and was approved by his superiors in the Pentagon. Unlike the decree to abolish the army, the decision on de-Ba'athification was discussed with other agencies, although its exact nature and extent may not have been thoroughly reviewed outside the Department of Defense. If officials in other agencies were surprised by the scope of the decree, the fault was not with Bremer, who was told, the day before leaving for Iraq, that the measure had been approved and was ready for promulgation. He

---

[163] Memo from Zaid Zaid to Paul Bremer, "Read Ahead for Iraqi Governing Council Meeting: April 14th," April 13, 2004; Herro Mustafa to Paul Bremer, "Weekly GC Update—Ninewa," May 2, 2004.

instituted this measure without substantive change. As with the army decree, the Pentagon had not kept Garner and his team apprised of the evolving policy discussion in Washington, however. The result, in both cases, was to embitter the transition from ORHA to the CPA.

There seems little doubt that both de-Ba'athification and the decree abolishing the army could have profited from further review. But the expanding chaos in Iraq and the sense of drift occasioned by uncertainty over who would be governing the country created a strong incentive to move quickly. The administration had dispatched Bremer with instructions to take a firmer hand, and Bremer felt both these measures would reassure the bulk of the population that Saddam's dictatorship was truly over and not destined to return, and they probably did. The result was a marginalized and angry Sunni minority but a largely quiescent Shi'ite and enthusiastically supportive Kurdish majority. This was not the worst possible outcome, although it was far from the best.

It is difficult to judge how seriously de-Ba'athification impacted the performance of the Iraqi government. Presumably some of the dismissed officials were competent, although the performance of those who remained might raise doubts in that regard. Ali Allawi, a moderate and secular Shi'ite leader who became Minister of Finance under the CPA, estimates in his memoirs that "up to ten thousand individuals" were initially removed by the order in the first three months of the CPA, a number that is probably less than the number of jobs vacated in the United States at the local, state, and federal levels whenever one party gives way to another. Iraq is, of course, a much smaller country, but one with a larger proportion of civil servants. According to Allawi, "De-Ba'athification in the early days of the CPA proceeded in a generally straightforward way. The vast majority of individuals caught in the first round of dismissals were those who could clearly be identified in the higher levels of the Party ranks, and the case against them was clearcut. It was only after the process had been transferred to Iraqi control, with the formation of the Supreme Council on de-Ba'athification, that the application of the process became more capricious."[164]

---

[164] Allawi, *The Occupation of Iraq*, pp. 149–150.

Given the intense hostility of much of the population toward the Ba'ath party, some purge was inevitable as well as justified. It was also natural for the newly emergent political forces to want to displace Sadaam's supporters in favor of their own. The CPA's formula was probably a good deal less punitive than the Governing Council, or any representative Iraqi government, would have applied under the then existing circumstances. Bremer has acknowledged that it was a mistake to have turned the implementation of this decree over to self-interested Iraqi politicians who had every incentive to free up more government jobs for their supporters, and little interest in rehabilitating even the most innocent former Ba'athists. It is worth noting that Bremer made this decision on the advice of his most senior political experts and later reversed it, in part, over their warnings.

The CPA charted an uncertain course with respect to local government. This was a missed opportunity, but it was hardly unusual. Most experts recommend beginning the process of democratizing an authoritarian society at the grass roots, first holding local elections to allow a new generation of leaders to emerge, then proceeding to national elections only when civil society, free media, and nonsectarian political parties have had time to get organized. In practice, this almost never happens. In most post-conflict environments, the international community has little political presence beyond the capital. Frequently, great urgency is attached to forming a national government, the powers of local governments are seldom well established, and societies emerging from conflict are often prey to serious centrifugal forces that could be exacerbated by the premature empowerment of local officials.

All these conditions applied in Iraq. As a result of its chronic shortage of personnel, particularly experienced mid-level Arabic-speaking officers, the CPA was never able to station more than a handful of officials in each Iraqi province—often no more than one or two. For their part, Iraqis had no modern experience with federalism and considerable skepticism regarding it. By contrast the CPA was under great pressure, first from Iraqis and then from Washington, to constitute and transfer sovereignty to a national government. And finally, the fragmentation of Iraq into three or more warring states was an ever-present danger that might have been advanced by empowering local governments before establishing a national one.

For all these reasons, Bremer's caution in empowering local governments before reestablishing Iraq's central authority is understandable, if regrettable. Once negotiation of an interim constitution began, Bremer was right not to preempt Iraqi decisions on the shape of local governance. But municipal and provincial elections in the summer or fall of 2003 would have allowed new indigenous leadership to emerge, particularly in the Sunni areas where so many of the elites had gone underground.

CHAPTER SIX

Promoting the Rule of Law

Describing the condition of the Iraqi justice system, Clint Williamson, the CPA's responsible senior advisor, reported to Bremer that it "was in a state of almost total devastation at the end of April. Most ministry buildings had suffered extensive damage from looting, and as a result were non-functional." The Ministry of Justice in Baghdad "was a burned out shell from which all of the furniture, equipment, and records had been stolen. Of 18 courthouses in Baghdad, 12 were gutted. Approximately 75 percent of the remaining estimated 110 courthouses in Iraq were destroyed as well." While a large contingent of ministry employees continued to report for work, there was little to do since they had no offices and no furniture—not even paper to write on. "In short," Williamson concluded, "the justice system was completely shut down."[1] From the start, the CPA was hamstrung by the absence of prewar planning for judicial reform, infrastructure that had been eviscerated because of the war and subsequent looting, and a justice system that was in tatters because of Saddam's neglect and abuses. The CPA itself experienced chronic staffing shortages that constrained its ability to rebuild the justice system and deal with such issues as corruption. Finally, the Abu Ghraib scandal and other detainee challenges badly damaged American bona fides in this area.

The Iraqi legal system was based on an amalgam of criminal codes from Europe (especially France), Egypt, and Syria. The codes

---

[1] Informational Memo from Clint Williamson to Ambassador L. Paul Bremer, "Subject: End of Mission Report," June 20, 2003.

were largely secular, though some accommodations were made for traditional tribal laws. Personal status laws were drawn largely from *sharia* (Islamic law). But the Ba'athist regime, which took power in 1968, undermined the very concept of a rule of law and used the justice system to extend and consolidate its authority. Judges had to be approved by the Ba'ath Party. The intelligence, security, and military services had their own special courts, which tried and sentenced thousands of Iraqis with little regard to due process—indeed, little regard to formal codes of law. The Iraqi judiciary had been complicit in the crimes of the state, so reform of the justice system was a key element in dismantling the Ba'athist state.[2]

Williamson had been a career federal prosecutor and had also served as a trial attorney at the International Criminal Tribunal for Yugoslavia, where he worked on cases for such prominent war criminals as Serbian president Slobodan Milosevic and the notorious paramilitary leader Zeljo Raznatovic, better known as "Arkan." He envisioned a "light footprint" approach to the justice system. In a memo to Bremer in May 2003, Williamson stated that "pursuant to our discussion yesterday, and in line with the parameters you suggested," the international presence would be small since "the Iraqis themselves are to staff the courts and the central ministry and are to have responsibility for all operations."[3]

The CPA's justice footprint included a senior advisor to the Iraqi Ministry of Justice, initially Williamson, who acted as the de facto minister until an Iraqi minister was put in place later in the year. The CPA staff also included a chief operational advisor and financial officer and international advisors in several Iraq Ministry of Justice offices, the Central Criminal Court, and regional offices.[4] Several U.S. government agencies were engaged in these efforts. The U.S. Department of Justice (DOJ) worked on judicial and prison reform, and was sup-

---

2   Allawi, *The Occupation of Iraq*, pp. 159–160.

3   Action Memo from Clint Williamson to Ambassador L. Paul Bremer, "Subject: CPA Justice Presence," May 25, 2003.

4   Action Memo from Clint Williamson to Presidential Envoy Ambassador L. Paul Bremer, "Subject: CPA Justice Presence," June 16, 2003.

ported by U.S. Army civil affairs and judge advocate general officers. Many of the advisors in the field were contractors, with a leavening of detailees from agencies including the U.S. Marshals Service. A significant amount of the work with Iraqi prisons and courts was undertaken by the U.S. and coalition military.[5]

Despite the intended light footprint, staffing problems plagued CPA's rule-of-law effort even more than in many other areas of CPA activity. In August 2003, two justice advisors, Gary DeLand and Lane McCotter, departed Iraq for the United States, leaving no other DOJ prison advisors in the country. This left a significant vacuum, DeLand pointed out, because "there are currently no replacements in the pipeline." He continued that the names and phone numbers of two possible replacements were provided to the U.S. Department of Justice in July, but "neither of those two men had been contacted by DOJ."[6] Donald Campbell, a judge on the Superior Court of New Jersey and a major general in the U.S. Army Reserves, sent a distressed email to CPA chief of staff Patrick Kennedy noting CPA was unable to meet its goals for the prison system because of staffing and resource problems. "There is no funding for the 80+ people," Campbell noted. This left him in the untenable situation of telling "the only capable expert I now have to spend two days writing the funding requirements . . . instead of getting the two prisons (with over 1,000 beds) open next week."[7]

Staffing problems plagued other aspects of the rule-of-law effort. Michael Dittoe, who had been an assistant U.S. attorney in the Southern District of Florida and a legal advisor in Kosovo before coming to Iraq, told Deputy Secretary of Defense Paul Wolfowitz that "there is a critical need for [international] prosecutors/lawyers to be deployed in

---

[5] Seth G. Jones, Jeremy M. Wilson, Andrew Rathmell, and K. Jack Riley, *Establishing Law and Order After Conflict* (Santa Monica, Calif.: RAND, 2005), pp. 136–145.

[6] Pre-Departure Briefing Memo from Gary W. DeLand to Major General Donald Campbell, "Subject: Replacements for Departing Prison Department Advisors," August 18, 2003.

[7] Email from Donald Campbell to Patrick Kennedy, "Subject: CPA Justice Sector Staffing," August 9, 2003.

the field to act as monitors and mentors to the judges and lawyers."[8] This was, he noted, the greatest threat to the Iraqi justice system.

"Of all the U.S. agencies," McManaway later charged, "the Department of Justice's performance in supporting the USG effort in Iraq was the worst by any measure."[9] Given the generally poor performance of agencies across the board in this regard, this is quite an indictment.

The deteriorating security situation contributed to the staffing challenges. In an urgent memo to Bremer in 2004, for example, Edward Schmults, CPA senior advisor to the Ministry of Justice, expressed serious concerns with the U.S. Department of Justice's International Criminal Investigative Training Assistance Program (ICITAP). Schmults, who had been a deputy attorney general in the U.S. Department of Justice, noted that "59 of the 71 ICITAP-contracted civilian prisons advisors deployed to Iraq have been locked down in their hotel by the Department of Justice because of security concerns, unable to perform their assigned mission." Schmults continued that the "lockdown, if it persists, will logically lead to the failure of that mission." The lockdown left no teams of correctional advisors at most prison sites. The one major exception was at Abu Ghraib, where rioting among the prisoners in April 2004 had caused a breakdown in law and order. Schmults reported to Bremer two weeks after the CPA team arrived in Abu Ghraib,

> The decision to redeploy this team was in direct response to an urgent plea from the US Army Military Police Brigade Commander overseeing the military operation at Abu Ghraib that we reengage our functions at that critical facility so that his MPs could quit killing inmates. While we were awaiting the military escort to the prison, the MPs shot and killed two more inmates at the prison. None have been killed since the arrival of the con-

---

[8]   Email from Michael J. Dittoe to Jessica LeCroy, "Subject: Back Brief Dep. Sec. DeF," October 26, 2003.

[9]   Author interview with Clayton McManaway, July 22, 2008.

tracted civilian advisors at Abu Ghraib, and the violence among the inmates has diminished.[10]

Establishing the Judiciary

In May and June 2003, Clint Williamson worked furiously to jump-start the Ministry of Justice and reopen courthouses. U.S. officials helped consolidate all criminal court functions in Baghdad into two courthouses. They initiated an emergency payment of $20 for Baghdad-based Ministry of Justice employees on May 10, and disbursed April salaries and $30 emergency payments to the same employees on June 1. A team of mostly U.S. Department of Justice officials conducted an in-depth nationwide assessment of the court system.[11]

The CPA tackled the problem of judicial inheritance from the Ba'ath regime through a series of orders between April and June 2003. It established a committee to vet judges, suspended the application of most of the 1969 Ba'athist penal code, and banned torture. CPA Order Number 7 suspended capital punishment and stated, "in each case where the death penalty is the only available penalty prescribed for an offense, the court may substitute the lesser penalty of life imprisonment, or such other lesser penalty as provided for in the Penal Code."[12] CPA Order Number 15 created a Judicial Review Committee to investigate the competence and suitability of judges and prosecutors, and to remove those deemed unfit for office.[13] This committee paid particular attention to vetting judges and prosecutors for past corruption, Ba'athist links, and complicity in the former regime's atrocities. But

---

[10] Info Memo from Edward C. Schmults to the Administrator, "Subject: Prisons Advisor Lockdown Crisis Headed to Mission Failure," April 27, 2004.

[11] Info Memo from Clint Williamson to Ambassador L. Paul Bremer, "Subject: End of Mission Report," June 20, 2003.

[12] Coalition Provisional Authority, "Order Number 7: Penal Code," June 9, 2003.

[13] Coalition Provisional Authority, "Order Number 15: Establishment of the Judicial Review Committee," June 23, 2003.

the process was slow and tedious, and by September only 20 percent of Iraqi judges and prosecutors had been reviewed.[14]

In a June 2003 public notice, the CPA exempted itself and foreign military forces, civilian government officials, and contractors from coming under the jurisdiction of Iraqi laws. Some senior Iraqi officials complained that "most Iraqis could not miss the irony that the CPA had replaced Saddam's rule by decree with another form of arbitrary authority, albeit apparently sanctioned by international law and mostly benign in its intent."[15] Given the state of Iraqi jurisprudence, exempting the intervening authorities and their agents from local jurisdiction was neither unusual nor inappropriate. Serious problems eventually emerged, however, after the demise of the CPA, due to the failure of the U.S. government to put in place any effective alternative form of jurisdiction covering the tens of thousands of civilian contractors granted immunity under the CPA ordinance. This is a problem with most UN- as well as U.S.-led interventions, but the scale of contractor presence in Iraq and the reliance on contractors to perform security functions made it a much more serious issue in this instance.

Commenting on the state of judicial reform, Judge Daniel Rubini, senior advisor to the Ministry of Justice, remarked, "We can continue pasting together feathers and hoping for a duck, but it will not likely fly." He continued that the "CPA Ministry of Justice has been chronically understaffed from the first, to the point of serious mission impairment," and the negative implications were dire. CPA's anti-corruption effort stalled because of the paucity of international staff, and staff deficiencies depleted CPA's rule-of-law credibility as "we abandon missions in the face of deteriorating conditions."[16] As Rubini explained in another email, he had "no announced replacements" for those in

---

[14] Memo from Donald F. Campbell to the Administrator, "Subject: Your Meeting Tomorrow with Mr. Hashim Abdel Rahman al-Shibli, Interim Minister of Justice," September 7, 2003.

[15] Allawi, *The Occupation of Iraq*, p. 160.

[16] Judge Daniel L. Rubini, "Coalition Provisional Authority Ministry of Justice: Prioritized Objectives (Excluding Prisons), Amman Justice Sector Conference" (Baghdad: Coalition Provisional Authority, January 2004).

his office departing imminently. "For all existing vacancies, I have requested by-name about 20 personnel starting beginning November, I have info as to one arrival in mid January but I have no information through the system as to status of any other of my requests. Of course this continues to impact on mission accomplishment."[17]

Rubini was not the only one who called attention to this problem. Colonel Scott Norwood, the CPA's military advisor, highlighted "the DOJ train wreck" on justice staffing.[18] A memo from CPA legal advisor Gary DeLand described the problem in excruciating detail:

> When I was recruited for this mission approximately five months ago, I was told there would be six advisors who would be assigned three regions of Iraq, each region supported by staff. The mission was to include three months of assessment resources, followed by the development of a master plan, creation of a Department of Corrections, and remodeling and building of prisons and detention centers. Even before the first advisors were on the ground the mission changed to one of standing up facilities. In early May, the initial complement in Iraq was only five persons; Bill Irvine, Team Leader, from the U.K.; three Americans (O. Lane McCotter, Larry DuBois, and Terry Stewart); and one Canadian. I was due to deploy with this group; however, the DOJ had a problem obtaining my passport in a timely manner. Ken Grant, an accountant from the UK, was the only support staff person provided.
>
> Prior to my arrival, the Canadian quit and went home due to the aforementioned change in the team's mission. During my first couple of weeks after my arrival in Baghdad, two of the three original American team members, Larry DuBois and Terry Stewart, quit and went back to the USA. That left only Bill Irvine (who primarily handles administrative matters) and Americans McCotter and me (who work primarily in the field) to stand up

---

[17] Email from Daniel Rubini to Scott Norwood, Matthew Waxman, Carl Tierney, Michael Dittoe, Homer Cox, Ralph Sabatino, Lance Borman, and Bruce Fein, "Subject: Revised Justice Sector Presence," December 30, 2003.

[18] Email from Scott Norwood to Steve Smith, "Subject: Revised Justice Sector Presence," December 30, 2003.

detention facilities and prisons, develop a national training program, and handle logistics and a myriad of other functions. In late July, Bill Irvine departed to enjoy three weeks leave and Lane McCotter returned to the U.S. for two or three weeks due to a death in the family and to try to recruit some American corrections officials to come to Iraq. For over two weeks, I have been the only advisor working in the Department of Prisons. As bad as that may sound it is about to get worse—much worse.[19]

Despite these difficulties, by September of 2003, the CPA judged that roughly 90 percent of courts across Iraq were up and running, though many were still in poor condition.[20] The backlog of cases increased the number of suspects who were let go or languished in pretrial detention facilities. In one instance, the British frigate *HMS Sutherland* apprehended a Panamanian-registered tanker carrying at least 1,100 metric tons of gas oil near Um Qasr, heading for international waters. As one CPA memo to Bremer noted: "There have been further arrests and impounding of vessels . . . but current practice is to hand detainees over to the local police and court system—most are released without adequate punishment or more importantly and the deterrence not proportional to the crime." At the end of November, Judge Rubini reported that there had only been 20 criminal convictions in Baghdad since the criminal courts reopened in May, and only 80 trials nationwide.[21]

Yet by April 2004, Chief Judge Medhat, the president of the Council of Judges and chief judge of the Court of Cassation, reported that the Iraqi judiciary had adjudicated 3,037 cases, an all-time Iraqi record.[22] There were also some improvements in detainment practices. "Jordan had sent roughly 100 interrogators to Iraq," noted Fred Smith,

[19] Pre-Departure Briefing Memo from Gary W. DeLand to Major General Donald Campbell, "Subject: Replacements for Departing Prison Department Advisors," August 18, 2003.

[20] Memo from Paul Bremer to Reuben Jeffrey, "Subject: Material for Speeches," September 3, 2003.

[21] Memo from Daniel L. Rubini to Ambassador Richard Jones, Nov. 29, 2003.

[22] Info Memo from Edward C. Schmults to the Administrator, "Subject: Record Case Volume for Iraqi Judiciary," May 16, 2004.

deputy director of CPA's Washington office and then deputy senior director of national security affairs. "But they didn't stay for long. When I asked why, I found out that they couldn't use their methods on detainees."[23]

In June 2003, the CPA helped establish a Central Criminal Court consisting of specially vetted judges and prosecutors to try high-profile cases of national importance.[24] Located in Baghdad, it consisted of an investigative court, a trial court, and an appellate court. Bremer was anxious to get this court moving. In a handwritten note to CPA advisor Donald Campbell, for example, he pleaded: "can't we get the first case in front of the CCC by end of July?"[25] The court, he hoped, would demonstrate that Iraqis were taking control of judicial functions. The first Central Criminal Court trial was held on August 25 and, over the next several months, CPA forwarded a range of cases to the court.[26] One defendant was Sabbah Nouri Ibrahim Al Salani, the office manager of the Minister of Finance, who was arrested on charges that he illegally detained and attempted to extract false confessions from bank employees through threats and intimidation. The investigation uncovered evidence that Sabbah was part of a wider conspiracy initiated by other individuals within the Ministry of Finance and the Iraqi National Congress to illegally obtain government property intended to benefit members of the Iraqi National Congress.[27] Other cases sent to the Central Criminal Court involved arms smuggling using a truck marked with the Red Crescent emblem; oil smuggling by a merchant vessel called Navstar I; and crimes allegedly committed by Abu Munim, the

---

[23] Author interview with Fred Smith, July 23, 2008.

[24] Coalition Provisional Authority, "Order Number 13: The Central Criminal Court of Iraq," June 18, 2003. Order Number 13 was later amended in June 2004.

[25] Bremer's comments were in the margins of Info Memo from Donald F. Campbell to Presidential Envoy L. Paul Bremer, "Subject: Update on Judicial Review Committee and Central Criminal Court of Iraq," June 27, 2003.

[26] Memo from Paul Bremer to Reuben Jeffrey, "Subject: Material for Speeches," September 3, 2003.

[27] Action Memo from Office of General Counsel to the Administrator, "Subject: Referral of Alleged Conspiracy to Commit Fraud to the Central Criminal Court of Iraq," April 10, 2004.

former governor of Najaf.[28] Douglas Brand, the chief police advisor for Iraq, noted that the Central Criminal Court worked fairly well, including coordination between Iraqi police and judges.[29]

But in a frank assessment, Judge Rubini acknowledged that "while much progress in the courts has been made, the entire justice system is fragile and could unravel in the near term. The courts are currently functional and capable of handling the common crimes including felonies." He continued, however, that the courts "are overwhelmed and intimidated from handling certain 'big fish' cases involving matters such as organized crime, political figures, police, and corruption." Indeed, corruption remained a major problem, Rubini noted, and "pervades every level of government and trusted positions in the private sector. Corruption is perhaps the deadliest poison-pill that threatens the social contract of the country."[30] As the CPA prepared to hand over authority to the Iraqis in 2004, there were also deep concerns about the capacity of senior justice officials. One CPA assessment lamented that Minister of Justice Hashim Abdel Rahman al-Shibli was "weak" and did not "have the energy to make a minister," and consequently recommended removing him. The same was true for Abd Al-Basit Turki, the Minister of Human Rights, who CPA accused of having "no experience in the business" and for having "repeatedly attacked CPA."[31]

War Crimes and Crimes Against Humanity

"Why did President Bush decide to overthrow Saddam Hussein?" asked Douglas Feith, Under Secretary of Defense for Policy, in his memoirs, answering that "it was to end a range of threats. No other contemporary leader—and few in history—had a record of aggression to match

---

[28] Cable from Headquarters Coalition Provisional Authority to SECTATE WASHDC and SECDEF WASHDC, "Subject: Central Criminal Court of Iraq," November 3, 2003.

[29] Author interview with Douglas Brand, January 16, 2009.

[30] Rubini, "Coalition Provisional Authority Ministry of Justice."

[31] Coalition Provisional Authority, "Iraqi Ministries," May 10, 2004.

Saddam's. He had started major wars of conquest. He had brutalized his citizens and killed them in enormous numbers."[32]

On March 25, 2003, Rumsfeld appointed the Secretary of the Army as the executive agent for investigating war crimes and for securing and preserving evidence of atrocities. The Army assigned this mission to the 3rd Military Police Group, which directed and coordinated the collection of evidence at suspected mass grave sites. Completed investigative reports were then forwarded to CPA's Office of Human Rights and Transitional Justice, which was responsible for establishing a public awareness campaign, conducting forensic assessment and exhumation, training Iraqis, and eventually transferring responsibility to them.[33]

The CPA's primary role was helping develop a record of atrocities committed by Saddam Hussein's regime, creating an Iraqi-run National Archive, and providing access to documents for criminal investigations.[34] Bremer reported to Secretary Rumsfeld in a memo in May that work had begun quickly and "some of the confirmed mass grave sites have been temporarily secured by the military to exploit for evidence of war crimes."[35] By July, CPA had identified 102 possible mass graves throughout Iraq.[36] In a memo to Secretary of State Colin Powell, Bremer noted that there were 1.3 million people missing in Iraq due to execution, wars, and defection, and the CPA "estimated that 300,000 of them were in mass graves."[37]

---

[32] Feith, *War and Decision*, p. 181.

[33] Memo from Jaymie Durnan to Secretary Rumsfeld, "Subject: Papers Requested in Preparation for Ambassador Bremer's Visit," July 18, 2003.

[34] Coalition Provisional Authority, "Human Rights Documentation Efforts," September 19, 2003.

[35] Action Memo from L. Paul Bremer to Secretary of Defense, "RE: Mass Graves," May 20, 2003.

[36] U.S. Department of Defense, *Iraq: The Path to Democracy* (Washington, D.C.: Office of the Secretary of Defense for Legislative Affairs, July 23, 2003). Estimates on mass graves were CPA estimates.

[37] Info Memo from L. Paul Bremer to Secretary Powell, "Subject: Background on Mass Graves in Iraq," September 13, 2003.

CPA's plan to deal with mass graves consisted of five steps: (1) public awareness and preparation, involving the initiation of a campaign through Iraqi media outlets to explain to the public the necessity of preserving and preparing sites for forensic examination; (2) conducting preliminary site assessments and prioritizing sites for a complete and thorough assessment; (3) a full forensic examination; (4) improving the capacity of Iraqis to take over the investigations through training; (5) transferring the results of forensic examination to Iraqi authorities for continued criminal and identification procedures.[38] The CPA's Office of Human Rights and Transitional Justice established a Mass Grave Database, which recorded the sites of mass graves. Examples included the 1988 Anfal campaign against Kurds in the north; the March 1988 use of sarin and VX on the town of Halabja; and the 1991 massacre of Shi'ites in the south, including in Hillah and Basra.[39] By the fall of 2003, the deteriorating security situation began to seriously disrupt the CPA's rule-of-law efforts. Judge Rubini noted in a memo to Bremer that "judges and prosecutors will undoubtedly become targets as they mete justice to terrorist and former regime loyalists." Security was provided by DynCorp for the central criminal court of Iraq, and CPA pushed through efforts to erect barriers around courthouses in Baghdad as protection from car bombs.[40] In addition, an increasing focus on military operations diverted attention away from criminal justice issues. In an action memo to Bremer, Ministry of Justice advisor Don Campbell and the Commanding General of CJTF-7 argued that "a considerable amount of intelligence is available to military agencies but it is not being used to support the criminal justice process."[41]

---

[38] Coalition Provisional Authority, "Mass Grave Action Plan" (Baghdad: Coalition Provisional Authority, July 2003).

[39] Info Memo from Sandy Hodgkinson and Dave Hodgkinson through Scott Carpenter to the Administrator, "Subject: Update of Human Rights and Transitional Justice Activities," September 19, 2003.

[40] Info Memo from Daniel L. Rubini to the Administrator, "Subject: Status Update on Projects in Supplemental Budget Request and Needs Assessment," October 9, 2003.

[41] Action Memo from Commanding General CJTF-7 and Senior Advisor, Ministry of Justice, to the Administrator, "Subject: The Investigation and Prosecution of Terrorism and Organized Crime in Iraq," September 21, 2003.

Capturing Saddam

"Ladies and gentlemen, we got him," Bremer announced to a stunned crowd of Western, Iraqi, and other Arab journalists packed into the CPA conference center in Baghdad. Several Iraqi journalists rose to their feet and cheered wildly, almost uncontrollably.

"Saddam Hussein was captured Saturday, December 13, at about 8:30 pm local time in a cellar in the town of ad-Duar, which is some 15 kilometers south of Tikrit," he continued. "This is a great day in your history. For decades hundreds of thousands of you suffered at the hands of this cruel man. For decades, Saddam Hussein divided your citizens against each other. For decades he threatened and attacked your neighbors. Those days are over forever."[42]

The hunt for Saddam had been relentless. It was primarily the mission of the U.S. military, which was responsible for his eventual capture. The CPA had announced a reward for those who could provide information leading to his arrest, basing its approach on New York City's Crime Stoppers Program. It promoted a public awareness campaign through leaflet drops, posters, and television and radio ads; established a call line for Iraqis to discreetly provide information to the CPA; and created a system to analyze, screen, and respond to reported information. "We cannot and must not let any thing slip through the cracks," police advisor Bernie Kerik had told Bremer.[43]

Once in U.S. custody, Saddam was initially handled as a prisoner of war, though CPA documents made explicit that he "will have rights that ensure a fair trial with respect to the various crimes that he is alleged to have committed."[44] Work to develop a special tribunal started at a frenetic pace. Part of the groundwork had already been laid. In September 2003, Salem Chalabi and members of the Office of

---

[42] Speech by Paul Bremer on the capture of Saddam Hussein, December 14, 2003. In his memoirs, Sanchez relates that Rumsfeld ordered that he, not Bremer, should announce the capture of Saddam. Bremer was informed of this instruction but chose to ignore it.

[43] Email from Bernard B. Kerik to Paul Bremer, "Subject: Reward for Saddam," July 4, 2003.

[44] Coalition Provisional Authority, "Additional Legal Considerations," December 15, 2003.

Human Rights and Transitional Justice had submitted to the Governing Council a draft statute to establish an Iraqi Special Tribunal (IST) to try members of the former regime for war crimes and other international offenses. Bremer told Secretary Rumsfeld that the tribunal's purpose was to "prosecute senior regime officials for war crimes and human rights atrocities in Iraq." One of the key sticking points was whether the tribunal would award the death penalty. Several members of the Governing Council wanted to lift the suspension of the death penalty, but this was opposed by some in the CPA and was vehemently opposed by the United Kingdom.[45] The tribunal was officially announced by the Iraqi Governing Council on December 10. The CPA's earlier order suspending the death penalty remained in effect, and its future application would be a matter for the future Iraqi government to decide.

The creation of the tribunal was of interest to several countries in the region, such as Kuwait, which sought to expand its scope. In December 2003, the Kuwait chargé d'affaires, Ahmad Razouqi, delivered a note to Secretary of State Colin Powell requesting that the tribunal extend its authority to include atrocities committed in Kuwait. "The Government of Kuwait supports this initiative," the note stated, "provided that it includes within the jurisdiction of that Special Tribunal, the prosecution of crimes committed in Kuwait and against Kuwaiti citizens, and other detainees removed from Kuwait during the period of illegal occupation."[46]

---

[45] Memo from Ambassador Clayton McManaway to the Governing Council, "Subject: Iraqi Special Tribunal," October 29, 2003; Info Memo from Office of the General Counsel to the Administrator, "Subject: Status of Iraqi Special Tribunal (IST)," October 29, 2003; email from Jessica LeCroy to Paul Bremer, "Subject: Interim Response to Taskers re Tribunal, CCC, and Reconciliation," October 28, 2003.

[46] Memo from Pierre-Richard Prosper to Ambassador L. Paul Bremer, "Subject: Kuwaiti Note Verbale to Secretary Powell on Iraqi War Crimes," December 5, 2003; Letter from Dr. Mohammed S. Al-Sabah to the Hon. Colin Powell, November 2, 2003. Prosper was the U.S. Ambassador-at-Large for War Crimes Issues.

There was disagreement over which U.S. agency should lead efforts to create the tribunal.[47] After receiving word that some in Washington favored giving the Department of Justice the leading role, a group of senior CPA personnel wrote to Bremer noting that "it is our opinion that this decision will greatly impede our preparation for Iraqi prosecutions before the IST." They argued that DOJ attorneys had "little experience in international Crimes Against Humanity and War Crimes cases," unlike the Departments of Defense and State, and that the Department of Justice "has been unable to fully staff MOJ needs" in Iraq.[48] In a memo to Attorney General John Ashcroft, Bremer emphasized "the need to integrate your DOJ personnel into the CPA's existing structure" and expressed concern that the Department of Justice would focus too much on directly prosecuting war criminals rather than letting Iraqis do it. Indeed, Bremer was somewhat perturbed about the possibility of a separate DOJ investigative and prosecutorial unit reporting directly to Washington outside of the boundaries of the CPA, which would duplicate efforts, increase bureaucracy, and delay initial trials. "Iraqis will erroneously interpret such delays," Bremer warned Ashcroft, as due "to political factors, including Saddam Hussein's status as a POW."[49]

Concerns also arose about the capacity of Iraqis. In a memo to Bremer, Judge Daniel Rubini, the senior advisor to the Ministry of Justice, argued that "while Iraqis will be used wherever possible, top administrative positions require a measure of security and expertise that is not available on the local market."[50] Rumsfeld complained

---

[47] See, for example, the Draft National Security Presidential Directive / NSPD-####, "Subject: United States Assistance to the Iraqi Special Tribunal," February 2004. The effort was to sort out the responsibilities of U.S. government agencies.

[48] Action Memo from Sandy Hodgkinson, David Hodgkinson, Scott Carpenter, and Scott Castle to the Administrator, "Subject: Designation of Lead Agency for Supporting Iraqi Special Tribunal," January 28, 2004. Also see Action Memo from Sandy Hodgkinson, David Hodgkinson, and Scott Castle to the Administrator, "Subject: Supporting Iraqi Special Tribunal," February 4, 2004.

[49] Memo from L. Paul Bremer III to Attorney General Ashcroft, "Subject: Iraqi Special Tribunal—DOJ Proposed Strategy," February 5, 2004.

[50] Action Memo from Daniel L. Rubini to the Administrator, "Subject: CPA Ministry of Justice Staffing Needs," January 13, 2004.

that the Iraqi courts were taking too long to prosecute individuals who attacked coalition forces. "It is important that people who attack Coalition forces in Iraq—or who are caught in possession of Manpads, RPGs or other special category weapons—are prosecuted and that the trials are prompt and fair," he wrote to Bremer and John Abizaid. Rumsfeld favored having U.S. tribunals try those individuals, since "I am concerned that the [Iraqi] process seems to be moving so slowly."[51] Yet Bremer pushed back, arguing that with regard to cases in the Central Criminal Court of Iraq involving crimes against coalition forces, it was important to ensure an expeditious and fair process. "At the same time," Bremer cautioned, "we must take care to not give the impression that we are in any way interfering with the independent judiciary we all have worked so hard to achieve."[52]

When the CPA handed over authority on June 28, the Iraqi Special Tribunal was still being organized. The first proceedings before the tribunal began in October 2005, when Saddam Hussein and seven other defendants were tried for allegations of crimes against humanity. Saddam was later sentenced to death by hanging on November 5, 2006, and was executed on December 30.

Handling Detainees

CPA senior advisor Donald Campbell worked with Hashim Abdel Rahman al-Shibli, interim Minister of Justice, to develop a comprehensive plan to build, guard, and run the Iraqi prison facilities in compliance with international standards. There were approximately 2,000 prison employees on the prison payroll, though, as Campbell admitted, "no member of the CPA advisory staff knows for certain." Former employees needed to be carefully screened and new guards and other

---

[51] Memo from Donald Rumsfeld to Amb L. Paul Bremer and General John Abizaid, "Subject: Prosecuting Iraqis for Security Offenses Against Coalition," October 20, 2003.

[52] Memo from Ambassador Bremer to Secretary of Defense, "Subject: Detainee Operations in Iraq," May 12, 2004.

personnel hired. In addition, security needed to be improved for judges at home and in their courtrooms.[53]

In July 2003, for example, CPA officials had pressed the U.S. military to establish a preliminary tracking system in order to be able to "provide relatives and loved ones of detainees with vital information, updated weekly." The benefits of this approach, Campbell noted, would help in "allaying fears of a second edition of Saddam's secret courts and prisons, and calming unduly nervous families."[54] In January 2004, roughly 6,500 civilians were being held by coalition forces as security internees. "Of these," one CPA cable noted, "approximately 1,500 are interned because they possess information of value to the Coalition and are being exploited for intelligence purposes. The remaining 5,000 are interned as imminent threats to security because they were involved in anti-Coalition or anti-state activity." The problem, as the cable noted, was that "there is insufficient evidence to support prosecution in any legitimate form."[55] In May 2004, Edward Schmults sent a memo to Bremer on the release of 359 prisoners, many of whom were at Abu Ghraib, because "there was little or no evidence" of wrongdoing, "including 160 as to whom no file existed at all."[56]

The CPA had reservations about transferring high-value detainees (HVDs) into Iraqi custody. As CPA's Office of General Counsel explained to Bremer, "If HVDs were to be turned over to the Governing Council . . . the Coalition would lose control over the prosecution of these individuals. This may adversely affect intelligence gathering, as the Coalition would not be able to use prosecution as a threat (or

---

[53] Memo from Donald F. Campbell to the Administrator, "Subject: Your Meeting Tomorrow with Mr. Hashim Abdel Rahman al-Shibli, Interim Minister of Justice," September 7, 2003.

[54] Action Memo from Donald F. Campbell to the Administrator, "Subject Detainee Tracking System," July 31, 2003.

[55] Cable from HEADQUARTERS COALITION PROVISIONAL AUTHORITY to SECDEF WASHDC, "Subject: Update on Central Criminal Court of Iraq (CCCI) and Thoughts on Supplementing the CCCI Process," January 11, 2004.

[56] Memo from Edward C. Schmults to the Administrator, "Subject: Release of 359 Prisoners," May 20, 2004. Also see Info Memo from Lt Col Coacher to Senior Advisor—MOJ, "Subject: Release List for Persons at Abu Graib and Tesferat Rusafa," May 19, 2004.

refraining from prosecution as an incentive) to gain cooperation." The office also argued that "the Coalition may not be able to gain access to HVDs in order to interview them."[57]

Human rights groups expressed concerns with the handling of detainees. Amnesty International sent an assessment to Bremer about detention practices and the treatment of prisoners in custody, including at Abu Ghraib. It noted that there were serious "reports of torture or ill-treatment by Coalition Forces not confined to criminal suspects" involving prolonged sleep deprivation, prolonged hooding, and prolonged restraint in painful positions. "Such treatment," it concluded, "would amount to 'torture or inhuman treatment' prohibited by the Fourth Geneva Convention and by international human rights law."[58]

Internal CPA documents showed similar concerns. In a memo to Bremer, Mark Kennon, the governorate coordinator for Salah ad Din, noted that "the issue of detainees hits home more sharply here than most places because of the large number of young people detained here last winter. The removing of such people from their homes in the middle of the night," he argued, "and the resulting long incarcerations without access to lawyers or family visits, or even the release of the most basic information is more and more often being compared to practices of the former regime."[59] By May 2004, coalition forces were holding 7,805 security internees, 2,055 criminal internees (755 of whom had been referred to the Central Criminal Court and 512 to Iraqi Common Court), and 92 high-value detainees.[60]

Bremer was also concerned about how Iraqi families were being informed about the status of relatives being held by the U.S. "We're posting it on the Internet" was the response he received in the course

---

[57] Info Memo from the Office of General Counsel to the Administrator, "Subject: Transfer of High Value Detainees (HVDs) to Iraqi Authorities," July 17, 2003.

[58] Amnesty International, *Iraq: Memorandum on Concerns Relating to Law and Order* (London: Amnesty International, July 2003), p. 11.

[59] Info Memo from Mark Kennon to the Administrator, "Subject: Your Visit to Tikrit April 24, 2004," April 23, 2004.

[60] Coalition Provisional Authority, "Current Situation: Detention, Interrogation and Legal Process," May 14, 2004.

of one staff meeting. How many Iraqis, he wondered pointedly, had access to the Internet?[61]

One of the most persistent challenges was a lack of investigative and intelligence resources, which resulted in many people being held unnecessarily. In April, the Review and Appeal Board recommended releasing more than 65 percent of the detainees whose cases had come before it, which meant, as one CPA memo to Bremer noted, "that many people who should be let go are spending long periods in detention without reason." There were at least two concerns about this development. One was that the United States was holding four out of five detainees for several months without adequate cause; a second was that the United States was letting dangerous individuals back on the street because the case files had insufficient information or were incorrectly filled out. An action memo to Bremer noted that the "information management technology used by CJTF-7 to track detainees is completely inadequate to the task."[62]

Releasing dangerous individuals could have a negative impact on security as CPA assessments acknowledged, but so did holding innocent people for extended periods because it undermined local support for coalition efforts. As one assessment concluded, "the manner in which detention operations are being conducted is undermining our strategic aims, in other words, our tactics are at odds with our strategy."[63] It was a lose-lose situation—and it was about to get worse.

Abu Ghraib

Although the CPA had no responsibility for the actual operation of the already notorious Abu Ghraib prison, it fell to Bremer to decide whether it should continue to be used as a detention facility. CPA's internal memos reveal a lively and contentious debate on this issue.

---

[61] Author interview with Fred Smith, December 14, 2003.

[62] Action Memo from Dobie MacArthur to the Administrator, "Subject: Detention Operations Recommendations," April 3, 2004.

[63] Action Memo from Dobie MacArthur to the Administrator, April 3, 2004.

American officials were acutely aware of Abu Ghraib's grisly symbolism, but many argued that the massive looting of other prison facilities left few good alternatives. Donald Campbell told Bremer in July 2003, "a maximum-security prison is urgently needed. Only Abu Ghraib prison could safely house an appreciable number of high-security detainees within three years." Campbell recommended the reconstruction of maximum-security cell blocks at Abu Ghraib "despite its grim reputation," and suggested that "a memorial should be located at Abu Ghraib as soon as possible, with appropriate notice to the public." He contended that "all agree that Abu Ghraib's deservedly horrid reputation counsels against perpetuation of its use as a prison any longer than operational necessity demands," but "the only alternatives we have available are even less palatable."[64]

Most other senior CPA officials agreed. As one memo noted, it was "the only prison site that has been assessed by the prisons assessment team that is capable of housing maximum security, dangerous inmates that can be refurbished in a reasonable period of time and in a cost efficient manner."[65] Walter Slocombe, Bremer's senior advisor for defense and security affairs, also recommended that CPA "immediately fund the reconstruction at Abu Ghraib prison at a cost of $1.76 million," but to "direct, as part of that effort, that the 'Death House' be fenced off from the remainder of the facility." The death house was situated in the northeastern corner of the maximum-security block, and was where Saddam's regime had ruthlessly tortured prisoners.[66] One of the most dreaded areas was a tri-level room. The upper level consisted

---

[64] Action Memo from Donald F. Campbell to Presidential Envoy L. Paul Bremer, "Subject: Maximum-Security Prison," July 1, 2003.

[65] Memo from Office of Reconstruction and Humanitarian Assistance to Office of the Under Secretary of Defense (Comptroller), "Subject: Funding Request ($ in Thousands)," June 2003.

[66] Action Memo from Walt Slocombe to Presidential Envoy L. Paul Bremer, "Subject Final Endorsement of MOJ Funding Request for Abu Ghraib Prison," June 18, 2003. There were a number of ideas of what to do with the part of the prison that had been used for torture, including destroying it and building a memorial to those who were executed. Memo from Lane McCotter to Colonel Greg Garner, "Subject: Memorial for Abu Ghraib Prison," June 18, 2003.

of two hanging brackets attached to the ceiling with a lever-operated trap door, and the person executed would fall to the lowest level once the lever was pulled.

Bremer sent a memo to Secretary Rumsfeld in June 2003 noting that "after a review of alternatives, I have decided to reopen Abu Ghraib Prison" with the provision that a museum be established in the section of the prison that had been used for the most heinous crimes. "We will make a model facility of the refurbished structure, and that is my answer to the political question, what does this facility symbolize?—the rule of law and humane treatment for even the most heinous offenders of it."[67] The CPA created a Human Rights Ministry office at Abu Ghraib to "provide transparency and information, and counter Iraqi fears" that Abu Ghraib was "still being used as a center of disappearance, torture and execution," and to act "as a showcase for the high standards of detention that the U.S. military provides."[68]

One of the most disturbing reports received by the CPA about Abu Ghraib had to do not with torture allegations but rather with the persistent radicalization of detainees there. A number of detainees were becoming significantly *more* radicalized after spending time at Abu Ghraib, thanks to an extremist network operating there. A memo to Bremer provided a telling example and described the plight of one prisoner. Coalition forces had arrested the man, and he was sent to Abu Ghraib, "where a group of devoted Muslims adopted, clothed, fed, and protected him from other prisoners. Because this group lived separately from the rest of the prison population," the cable continued, "they provided a nurturing environment that allowed them to brainwash potential recruits. When the young man was transferred to the facility in al Qassim, another group of extremists were waiting for him. After two months, they had successfully converted the young man and given him a list of contacts on the outside to pursue when he was released. A few

---

[67] Memo from Paul Bremer to Secretary Rumsfeld, "Subject: Abu Ghraib Prison," June 19, 2003.

[68] Action Memo from Sandy Hodgkinson to the Administrator, "Subject: Abu Ghraib Human Rights Ministry Office," January 14, 2004.

days later," the cable concluded, "the young man had found himself in a car bombing plot that nearly succeeded."[69]

Abu Ghraib burst onto the headlines in January 2004. Bremer was in Washington, and Dan Senor informed him of the news. "Apparently some MPs guarding detainees forced them to engage in homosexual acts," he reported. "They made one of them crawl around on the ground with a dog's leash around his neck. There may have also been women involved, whether our women MPs or women detainees isn't clear." Recognizing the explosiveness of the issue, Senor remarked that "it was likely to rip through the Arab press in a toxic way, and we needed to get out in front of it."[70]

Bremer announced that the United States would conduct a full investigation. Eventually the Department of Defense launched a series of inquiries led by Major General Anthony Taguba, former Secretary of Defense James Schlesinger, and others. Bremer also directed the immediate reform of detainee policy "to promote fair, speedy, and transparent procedures for dealing with detainees and reviewing their status without compromising the overall security of the Iraqi population."[71] This included the release of several high-profile Sunnis who had been in custody for an extended period without clear-cut charges. He also extended the operation of the Review and Appeal Board and Conditional Release Program to review the status of detainees.

Bremer presented to Secretary Rumsfeld proposals designed to enhance Iraqi participation in coalition detention operations. Rumsfeld approved several of them: creating an Iraqi ombudsman to review and assess complaints; placing Iraqi police observers at brigade-level detention collection centers to facilitate earlier movement of criminal

---

[69] Memo from Jessica LeCroy to the Administrator, "Subject: Your Question on Suicide-Bomber in Arbil," March 6, 2004. Attached to the memo was a draft cable, written by Kris Keele. CPA Baghdad to SECDEF WASHDC SECSTATE WASHDC NSC WASHDC CJCS CDR USCENTCOM AMEMBASSY ROME USDA WASH DC, "Subject: CPA 0703: March 4 Meeting with KRG Minister of Interior Karim Sinjari," March 6, 2004.

[70] Author interview with Dan Senor, October 31, 2003. Also see Bremer, *My Year in Iraq*, pp. 280–281.

[71] Memo from L. Paul Bremer III to Commanding General, Combined Joint Task Force Seven, "Subject: Detainee Policy Reform," April 15, 2004.

suspects into the Iraqi criminal justice system; and establishing a contingent of Iraqi law enforcement personnel at Abu Ghraib to coordinate the movement of reclassified detainees back to appropriate criminal jurisdictions. Rumsfeld also pushed for a greater unity of effort for detainee operations.[72] "Of particular importance," Bremer told Ricardo Sanchez, "is the initiative to aggressively publicize progress toward the Iraqi prosecution of high value detainees before the Iraqi Special Tribunal."[73]

The scandal was a bombshell for members of Iraq's Governing Council. In meetings with U.S. officials, they denounced the abuses, requested information about what the U.S. military was doing to correct the abuses, and decided to "form a committee by Iraq Judges and public prosecutors to investigate the anti-humanitarian violations committed by those in charge of Abu Ghraib prison."[74] In addition, the Governing Council proposed that Abu Ghraib "be razed to the ground and a housing complex be built on the property to house those who were imprisoned or tortured by the past regime."[75]

Some in the U.S. military felt that prisoner abuse at Abu Ghraib was being blown out of proportion. Sanchez later criticized the media for unparalleled "speculation and lies about what really occurred at Abu Ghraib."[76] Senior U.S. policymakers nevertheless understood the strategic implications for the United States and its image abroad. In preparing President Bush for a phone call with Governing Council President Sheikh Ghazi al-Yawr, National Security Advisor Condoleezza Rice cautioned that the Abu Ghraib scandal had "heightened feelings

---

[72] Memo from Donald Rumsfeld to Jerry Bremer and Gen John Abizaid, "Subject: Detainee Operations in Iraq," April 30, 2004.

[73] Memo from L. Paul Bremer III to Lieutenant General Ricardo Sanchez, "Subject: Enhancing Iraqi Participation in Detention Operations," May 18, 2004.

[74] Letter from Muhyi K. Al Kateeb, Secretary General of the Governing Council, to Ambassador Paul Bremer, May 13, 2004.

[75] Memo from Lydia Khalil to the Administrator, "Subject Readahead for May 12 GC-CPA Meeting," May 12, 2004.

[76] Sanchez, *Wiser in Battle*, p. 375.

of anxiety among Iraqis."[77] CPA officials were deeply concerned about the impact on their efforts in Iraq. "The Abu Ghraib prison problem cannot, in my view, continue to meander along only an investigative and legal path," noted Edward Schmults, CPA's senior advisor to the Ministry of Justice, since CPA was "overwhelmed" by negative publicity.[78] These concerns triggered an effort to close Abu Ghraib down immediately, and the CPA examined the feasibility of two 4,000-bed quick-build facilities: one for criminals at Rashid Air Base in eastern Baghdad; and the other for security internees at Camp Bucca near Um Qasr.[79]

CPA officials nevertheless had continued concerns about closing Abu Ghraib. "If we had to close Abu Ghraib now," Schmults noted, "the [Iraqi Corrections Service] would have to move approx. 2000 (of whom approx. 200–300 will be released in the near future). We have no place to move them to." The only option would be to erect tent structures in a fenced area, which would take at least 90 days if the tents and other items were available. "A prison to replace Abu Ghraib," Schmults concluded, "would cost approximately $150–200 million and would take about 2 years to build."[80] President Bush ultimately decided to keep Abu Ghraib open, at least temporarily. As Bremer explained to Sanchez, "In accordance with policy direction from President Bush, MNC-I will continue to operate Abu Ghraib until a new maximum security facility is completed. CPA will not construct 'quick-build' facilities in Baghdad or at Camp Bucca."[81]

---

[77] Memo from Condoleezza Rice to POTUS, "Subject: Telephone Call With Shaykh Ghazi Al-Yawr of Iraq," May 17, 2004.

[78] Memo from Edward C. Schmults to the Deputy Administrator/Chief Policy Advisor, "Subject: Compensation for Abu Ghraib Abuse Victims," May 8, 2004.

[79] Info memo from Edward C. Schmults to the Administrator, "Subject: Abu Ghraib Destruction Bullet-Point Timeline," May 23, 2004; Action Memo from Edward C. Schmults to the Administrator, "Subject: Abu Ghraib Draw-Down and Destruction Plan," May 26, 2004.

[80] Email from Edward Schmults to Executive Secretary and Brian McCormack, "Subject: Urgent Tasker: Follow Up—Abu Ghraib," May 12, 2004.

[81] Memo from L. Paul Bremer III to Commanding General, MNF-I, "Subject: Abu Ghraib," May 27, 2004.

Fighting Corruption

The CPA set combating corruption as one of the top priorities for the Ministry of Justice.[82] Senior CPA officials had no illusions about what they were up against. In a meeting with Attorney General John Ashcroft, Bremer lamented that "corruption is pervasive in Iraq."[83] He told the Governing Council shortly thereafter that "corruption inevitably wards off investment and foreign aid, impedes economic growth, and becomes a mechanism for extortion from the people. It is in the interest of Iraq to develop a robust, credible anti-corruption program."[84] Judge Daniel Rubini wrote a scathing memo to Bremer noting that "governmental corruption is one of the biggest immediate threats to the integrity of restoring faith in the new Iraq. It is a cancer destroying the government."[85] The World Bank and Transparency International both ranked Iraq as one of the most corrupt countries in the world.[86]

The CPA established an Office of Anti-Corruption and Integrity in Government. Bremer explained to the Governing Council that the mission of the office was to investigate, detect, and aid prosecution of corruption at all levels of government nationwide; monitor financial disclosure requirements; propose additional anti-corruption legislation; and conduct public awareness programs. The anti corruption campaign was composed of several components:

---

[82] Coalition Provisional Authority, *Towards Transition in Iraq: Building Sustainability* (Baghdad: Coalition Provisional Authority, December 2003), Annex A, p. xix.

[83] Coalition Provisional Authority, "Talking Points for Amb. Bremer to Discuss with Mr. Ashcroft re Office of Anti-Corruption and Integrity in Government," December 1, 2003.

[84] Info Memo from the Administrator to the Governing Council, "Subject: Proposed Agenda for the December 3 Meeting," December 2, 2003.

[85] Info Memo from Daniel L. Rubini to the Administrator, "Subject: Background Info for Meeting," December 19, 2003.

[86] For the World Bank, see the variables for "rule of law" and "control of corruption" in Daniel Kaufmann, Aart Kraay, and Massimo Mastruzzi, *2007 Governance Matters VI: Governance Indicators for 1996–2006* (Washington, D.C.: World Bank, 2007). For Transparency International, see Transparency International, *Corruption Perceptions Index 2003* (Berlin: Transparency International, 2003), p. 5.

- an independently funded, Iraqi-run government body authorized by Governing Council approved legislation
- a chief that reported only to the highest authority (who was Bremer as long as the CPA existed)
- national jurisdiction to investigate, gather information, subpoena, infiltrate, and prosecute government agencies, employees, or private parties
- financial analysts, auditors, investigators, and equipment for sophisticated investigations.[87]

Members of the Governing Council were generally supportive of the CPA's anti-corruption program, although, as Meghan O'Sullivan pointed out in a memo to Bremer, "it may not be clear to them immediately" that there might be direct "implications for them personally."[88]

The CPA established independent inspectors general in all Iraqi ministries. These individuals were responsible for audit and investigations relating to the programs and operations of their ministries. "Through their insistence on transparency and honesty in government," Bremer told his senior CPA advisors, the inspectors general "will ultimately be accountable to elected representatives" and "will be an important building block in the transition to democracy."[89] By March, most ministries had inspectors general in operation.[90] As expected, resistance emerged.

On May 9, 2004, Samir al Semadi, the Minister of Interior, sought to remove Nouri Gaber from his position as the ministry's Inspector General, citing vague personal observations and hearsay allegations. Bremer told the minister in a letter, "it is insufficient . . . to

---

[87] Info Memo from the Administrator to the Governing Council, "Subject: Proposal for an Office of Anti-Corruption and Integrity in Government," December 2, 2003.

[88] Info Memo from Meghan O'Sullivan to the Administrator, "Re: December 3 Meeting with the Governing Council," December 3, 2003.

[89] Action Memo from the Administrator to All CPA Senior Advisors, "Subject: Appointment of Inspectors General," December 20, 2003.

[90] Info Memo from Giles Denham to the Administrator, "Subject: Establishing Inspectors General Offices—Progress Report," March 18, 2004.

base a removal decision solely upon the observations and hearsay statements of a relevant minister. Some corroboration must be submitted."[91] Bremer had two concerns with the effort to remove Gaber. "My first problem was that I hadn't seen any real evidence about why he was removed," explained Bremer. "And the second was that the minister didn't have the authority to remove him. Only I had the authority to do it as the administrator."[92] But Samir blatantly ignored Bremer and replaced Gaber with an interim inspector general without providing further details.

Over the next few weeks, CPA officials became increasingly concerned over the Gaber ouster, especially since Gaber's office had uncovered a major border crossing corruption problem, which "someone out there did not like." The dismissal, CPA concluded, "appeared to be an entirely personal decision by the new Minister."[93] The situation had still not been resolved after a month, and the CPA's Office of General Counsel "concluded that there was insufficient cause to justify" Gaber's removal, partly since "US intelligence authorities stated formally they had no evidence to corroborate statements made by the then Minister." The case had become important in dealing with corruption, since all of Iraq's inspectors general throughout the ministries awaited the outcome of the case with notable anxiety.[94] With Bremer's direct involvement, Gaber was eventually reinstalled as the Ministry of Interior's Inspector General.[95]

---

[91] Memo from L. Paul Bremer III to Samir Al Semadi, "Subject: Removal of Inspector General," May 17, 2004. Also see Action Memo from Office of the General Counsel to the Administrator, "Subject: Removal of Inspector General by Minister of Interior," May 17, 2004, and Action Memo from David Kirk to the Administrator, "Subject: Removal of Inspector General by Minister of Interior," May 26, 2004.

[92] Author interview with L. Paul Bremer, August 12, 2008.

[93] Info Memo from David Kirk to the Administrator, "Subject: 'Emergency' Meeting of Inspectors General to Discuss Events at the Ministry of Interior," May 16, 2004.

[94] Action Memo from David Kirk to the Administrator, "Subject: Interior Minister—Inspector General," June 13, 2004. Also see Letter from L. Paul Bremer III to Falah al Nakib, Minister of the Interior, June 13, 2004.

[95] Author interview with L. Paul Bremer, August 12, 2008.

Corruption cases plagued the CPA. Don De Marino, a former U.S. Deputy Secretary of Commerce, noted that he had information "from reliable sources that members of the new Iraq Council were setting up offshore vehicles and trading companies using third parties in order to cash in on their new positions. No doubt," the memo concluded, "they are planning for the time when they will be picking the new ministers and in a position to receive their slice of the contracts and concessions to be awarded." De Marino suggested tasking the CIA to collect the financial details and then perhaps "to make an example of one of them," followed by establishing a financial disclosure policy for the Governing Council and the new ministers.[96]

The CPA referred an increasing slate of corruption cases to the Central Criminal Court of Iraq.[97] In November, Bremer sent a corruption case to the court involving Abbas Kunfuth, Iraqi Ambassador to the Russian Federation from September 2002 to July 2003. "On 29 July, 2003 two million dollars were stolen from the safe in the Iraqi Embassy in Moscow," the Office of General Counsel informed Bremer. "Several of the circumstances surrounding the robbery suggest that it was an inside job in which the Ambassador and people very close to him played roles." In addition, "all of the cash in this case was obtained through kickbacks in the UN oil for food program."[98] Another case involved a litany of bank tellers who were arrested, without a warrant or other judicial involvement, at the end of January 2004 by Iraqi police officers assigned to the Major Crimes Unit. In March, CPA Ministry of Finance employees brought forward allegations that Sabbah Nouri Ibrahim Al Salani, a Ministry of Finance official, was behind the actions, and the Ministry of Justice drafted a referral, which was signed by Bremer, referring the case to the Central Criminal Court of Iraq.

---

[96] Memo from Don De Marino to Jerry Jones, "Re: Feathering Nesting in the New Iraq," August 11, 2003. Also see memo from Jerry Jones to Reuben Jeffery, CC: Deputy Secretary Wolfowitz, August 12, 2003.

[97] Coalition Provisional Authority, "Order Number 13: The Central Criminal Court of Iraq," June 18, 2003. Order Number 13 was later amended in June 2004.

[98] Action Memo from Office of General Counsel to the Administrator, "Subject: Referral of Abbas Kunfuth Investigation to Central Criminal Court of Iraq," November 3, 2003.

On March 21, CPA issued a warrant for Sabbah's arrest, and he was subsequently arrested on March 24 and brought before the court.[99]

CPA documents strongly indicate that CPA and other U.S. government officials were acutely aware of the corruption that plagued Iraqi society and were serious in their efforts to counter it, despite the CPA's limited capacity and diminishing leverage. In addition to the establishment of inspectors general offices in Iraq's ministries and the referral of a range of cases to the Central Criminal Court, the CPA took other steps, including an increase in judges' pay. Edward Schmults, senior advisor to the Ministry of Justice, urged in a memo to Bremer, "that you approve a salary increase for judges and prosecutors in order to preserve judicial independence and discourage endemic bribery that formerly plagued the Iraqi judiciary." He argued that a good salary was a necessary, though not a sufficient, step to protect the judiciary from intimidation and decrease the temptation of corruption.[100]

Charges of CPA Financial Mismanagement

There were also growing allegations about corruption involving U.S. contractors. Responsibility for contracting in Iraq had been assigned to the Department of the Army at Rumsfeld's direction. In September 2003, Henry Waxman, the ranking minority member of the House Committee on Government Reform, sent a letter to Joshua Bolten, the Director of the Office of Management and Budget (OMB), noting that "a picture is now beginning to emerge of waste and gold-plating that is enriching Halliburton and Bechtel while costing the U.S. taxpayer millions and imperiling the goal of Iraqi reconstruction."[101] In October, the CPA sought—and received—approval to arrest Regard Yakou and Sabri Yakou. Regard Yakou was a naturalized U.S. citizen and

---

[99] Info Memo from Ed Schmults to the Administrator, "Subject: Chronology of Ministry of Finance Investigation," May 22, 2004.

[100] Action Memo from Edward C. Schmults to the Administrator, "Subject: The Judicial Salary Crisis," Marcy 29, 2004.

[101] Letter from Henry Waxman to Joshua Bolten, September 26, 2003.

Sabir Yakou was a British citizen with permanent resident alien status in the United States. Both were charged with violations of the Arms Export Control Act based on their brokering the sale of patrol boats to Iraq in violation of 22 United States Code Section 2778.[102]

From May 2003 to the transfer of sovereignty in June 2004, the CPA was responsible for managing, spending, and accounting for over $20.9 billion in Iraqi funds, of which it disbursed nearly $14.3 billion through transfers to Iraqi ministries and CPA-administered reconstruction projects. The remaining third was transferred to the government of Iraq on June 28, 2004.

The CPA's management of these funds eventually came under significant scrutiny. During the debates over the $18.4 billion reconstruction supplemental appropriation for Iraq, congressional Democrats insisted on creating an inspector general to oversee the CPA. The position of Special Inspector General for Iraqi Reconstruction (SIGIR) was negotiated, and Stuart Bowen was appointed to the position by the Bush administration.[103] Although Bowen was reputed to be a staunch supporter of the administration, he quickly became a severe critic of the CPA's reconstruction effort and financial management practices.

Subsequent to the demise of the CPA, Bowen accused it of losing track of $8 billion in Iraqi government funds. Specifically, the SIGIR noted that although CPA Resolution Number 2, signed by Bremer on June 15, 2003, required the CPA to hire an independent certified public accounting firm, no contract was awarded until October, when North Star Consultants was awarded a $1.4 million contract to review controls over the Development Fund for Iraq (DFI). However, the firm

---

[102] Action Memo from Michael Dittoe to Ambassador Patrick Kennedy, "Subject: Authorization for Arrest of United States Nationals," October 10, 2003.

[103] The Bush administration requested that the Defense Department have the right to classify and prevent the public release of any SIGIR reports deemed to endanger the national security, a provision which ran counter to the traditional use of inspectors general as public and independent watchdogs. However, Bowen later reported that neither the Pentagon nor the White House ever attempted to interfere in his publishing a report despite the often damaging nature of his accusations. See T. Christian Miller, *Blood Money: Wasted Billions, Lost Lives, and Corporate Greed in Iraq* (New York: Little, Brown, and Company, 2006), pp. 196–197.

had no certified public accountant and failed to perform the review.[104] SIGIR argued that the CPA had also failed to observe its own requirements regarding expenditures of Iraqi funds. Regulations for managing and overseeing the expenditures of Iraqi money were laid down in CPA Memorandum Number 4, which applied to contracts and grants made by the CPA. However, this regulation explicitly *did not* apply to "Iraqi Ministries and governmental agencies executing contracts or grants to fulfill requirement approved through the national budget process" as long as the "contracting procedure of the Ministry or agency is adequate to ensure the transparent use and management of Iraqi funds."[105] Nevertheless, SIGIR found that the CPA transferred DFI funds to Iraqi ministries without assurance the monies were properly used or accounted for. Oversight of Iraqi funds was "burdened by severe inefficiencies and poor management," and the CPA did not give "clear guidance" regarding the "responsibilities, procedures, and controls for disbursing DFI funds for the Iraqi national budget."[106] Consequently, the SIGIR concluded the following:

- The CPA did not exercise adequate responsibility over DFI funds provided to Iraqi ministries.
- CPA officials did not review budget execution at Iraqi ministries even though external assessments indicated budget and financial control systems required strengthening.
- The CPA did not implement adequate controls to ensure DFI funds were properly used for salaries of Iraqi employees.[107]

---

[104]SIGIR Audit 05-004, "Oversight of Funds Provided to Iraqi Ministries through the National Budget Process," January 30, 2005, pp. 8–9.

[105]Coalition Provisional Authority, Memorandum Number 4: "Contract and Grant Procedures Applicable to Vested and Seized Iraqi Property and the Development Fund for Iraq: Implementation of Regulation Number 3, Program Review Board," August 19, 2003.

[106]SIGIR Audit 05-004, pp. 6, 10.

[107]SIGIR Audit 05-004, pp. 6–7.

A SIGIR review of ten disbursements made by the CPA Comptroller's office between October 2003 and June 2004 disclosed the following:

- None of the ten disbursements—ranging between approximately $120 and $900 million—included documentation, such as budget spending plans to support the amounts provided to the Iraqi ministries.
- Six disbursements were made without CPA/OMB allocation memoranda.
- Two disbursements totaling approximately $616 million were not supported by disbursement vouchers.
- An improper approximately $120 million disbursement was made in May 2004 because of miscommunication between CPA/OMB and CPA Comptroller's office.[108]

Bowen's investigators alleged that the CPA could not account for $8.8 billion of Iraqi funds provided to the Iraqi ministries. Subsequently, Bowen testified before the House Committee on Oversight and Government Reform that this figure was too low and that he believed "the lack of accountability and transparency extended to the entire $20 billion expended by the CPA."[109]

The State Department initially tried to block the report, arguing that Bowen's mandate did not extend to investigating the use of Iraqi money, and it did manage to convince Bowen to delay the report's release until after the January 2005 Iraqi elections.[110] Bremer strongly rejected the SIGIR's criticism. He wrote to Bowen, "This draft report does not meet the standards Americans have come to expect of the Inspector General." He accused Bowen of not giving the CPA credit

---

[108] SIGIR Audit 05-004, p. 8.

[109] Stuart Bowen, Testimony before the House Oversight and Government Reform Committee, February 6, 2007. It should be noted that Bowen was careful never to say the money had been stolen by Americans or definitely wasted, only that nobody in the CPA could say with any certainty what happened to the funds.

[110] Miller, *Blood Money*, p. 198.

for its efforts to ensure accountability. He also noted the SIGIR auditors "presume that the coalition could achieve a standard of budgetary transparency and execution which even peaceful Western nations would have trouble meeting within a year, especially in the midst of a war. Given the situation the CPA found in Iraq at Liberation, this is an unrealistic standard."[111]

SIGIR's facts seem well grounded, but its expectations regarding what the CPA, or any occupying authority, could achieve in immediate improvements of the indigenous government's financial accountability were unrealistic. The CPA could and did account for the Iraqi funds that it had released to Iraqi ministries but, not surprisingly, had more difficulty accounting for exactly how those ministries expended the funds thereafter. Yet schools had to be opened, civil servants paid, hospitals staffed, garbage collected, and sewage treated. There was no alternative to using Iraqi institutions for this purpose and no possibility of installing new mechanisms throughout the Iraqi government for financial accountability overnight. SIGIR is correct that the CPA violated its own accounting rules and fell far short of standard best practices in how it disbursed Iraqi funds. Clearly, there were also some corrupt American contractors (i.e. the "Bloom/Stein Conspiracy" which defrauded the CPA of more than $8.6 million in DFI funds) that took advantage of the lax accounting practices.[112] However, the CPA's defenders are equally justified in pointing out that it never had enough personnel with the requisite contracting and accounting experience to monitor the disbursal patterns of the entire Iraqi government. Bremer is also correct to note that it would have been highly destabilizing to wait to pay Iraqi civil servants "until we had fully modern pay records," which likely did not exist in Iraq anyway, and that later "it would have been dangerous for security . . . to stop paying armed young men."[113] The quick disbursal of funds from April to June 2004 to tamp down the spread of the growing insurgency is equally justifiable and was, if

---

[111] SIGIR Audit 05-004, pp. 33, 36.

[112] See Miller, *Blood Money*, pp. 198–201.

[113] SIGIR Audit 05-004, pp. 34, 35.

anything, a belated response that should have been undertaken even earlier.

The bulk of the SIGIR reports deal with charges of mismanagement and corruption that occurred after the CPA's demise, suggesting that the problems uncovered were by no means unique to that institution but rather were indicative of a lack of capacity within the U.S. government as a whole to handle a reconstruction project of this magnitude under the violent and chaotic conditions then applying in Iraq. Given that the CPA, a hastily improvised institution with only limited capacity to impose internal—let alone external—financial controls, was handing billions of dollars in cash in an almost bank-less society, it is indeed remarkable how few charges of malfeasance involving CPA staff have emerged. It is also important to note that no prior effort at postconflict stabilization has ever been subjected to the degree of scrutiny accorded to the operation in Iraq. The SIGIR model has recently been extended to Afghanistan and, one hopes, to future American nation-building missions. It will be important, nevertheless, to recognize the limitations under which any postconflict authority operates and the preeminent need to establish security and restore basic government services under conditions of incipient or actual chaos, in which the usual standards for financial accountability can be met only at the cost of immense human suffering and forgone opportunity.

Oil for Food

The UN-administered Oil-for-Food (OFF) Program began in 1996. It followed several years of UN sanctions, including a ban on all oil exports. The United Nations and Iraq eventually reached an agreement that permitted limited oil sales by Iraq. The revenues, which were deposited in UN-controlled accounts, were supposed to be used to import food and medicine. The UN Security Council, including the United States, agreed to allow the Iraqi government rather than the UN to select the recipients of its oil sales. This opened up an avenue for massive kickbacks, thereby allowing Saddam's regime to divert a portion of the sales from their intended humanitarian ends to other purposes. Individuals in a number of countries, including France, Russia, and the United States, became involved in such schemes. This triggered a lengthy UN investigation. The U.S. administration and Congress, still unhappy about the UN's failure to endorse the invasion of Iraq, became quite critical of the UN's failure to curb this abuse. Largely lost in the ensuing outcry was the fact that the United States had knowingly approved the arrangements that gave rise to the abuse, that no U.S. or UN money had been lost as a result, and that—with comparatively minor exceptions—UN officials had not been involved in the corruption.

"CPA is aware of interest at the United Nations in pursuing a full investigation arising from allegations of bribes, kickbacks, and corruption in the OFF program," one CPA memo noted. "We welcome full UN involvement and recommend that the UN designate individuals to join with CPA, and possibly Iraqi nationals, to safeguard and inventory records at key ministries."[114]

The CPA and the U.S. military uncovered a number of oil-for-food abuses. In a memo to Secretary Rumsfeld, Pentagon Comptroller Dov Zakheim outlined a range of cases. In one case, for example, he told Rumsfeld that there were "a total of five Oil-for-Food contracts (totaling $247 million) to provide Mercedes Benz saloon cars and other similar brand luxury cars to Iraqi officials." Not only did these contracts appear overpriced, Zakheim told Rumsfeld, but they obviously had nothing to do with humanitarian relief. There was also a range of other questionable contracts for cigarette paper and gymnasiums. Even in cases where contracts appeared justified on humanitarian grounds, there were worrisome indications of contract overpricing.[115]

Former Speaker of the U.S. House of Representatives Newt Gingrich, then at the American Enterprise Institute, sent a fax to Secretary

---

[114] Action Memo from Scott Castle and James B. Warlick to the Administrator, "Subject: Allegations of Misconduct Regarding the Oil for Food Program—Next Steps," March 15, 2004.

[115] Action Memo from Dov S. Zakheim through Deputy Secretary of Defense to Secretary of Defense, "Subject: DoD Audit Support of Iraq Contracts," May 21, 2003; Info Memo from Dov S. Zakheim to Secretary of Defense, "Subject: UN Oil for Food (OFF) Contract Audits," July 18, 2003.

of State Powell outlining his concern with the enormity of the oil-for-food quagmire, which Powell then forwarded to Bremer. "It is vital to get ahead of this corruption scandal by appointing a special investigative task force both to help uncover past corruption and to root out current corruption," Gingrich warned. "This could explode this summer and fall and be very much to our disadvantage unless we get ahead of the curve and very loudly meet it head on."[116] But there was some disagreement within the CPA about how to handle this matter. In a memo to Bremer, James Warlick, who ran the Oil-for-Food Program and served as a counselor to Bremer, argued that finding the smoking gun by combing through Iraqi ministry records might yield nothing. "It's WMD—you know it's there, but you can't produce the evidence," he noted. Warlick continued by expressing concern that "CPA will be tarred with incompetence, or even cover-up," and recommended that CPA not take the lead.[117] CPA general counsel also argued that it was beyond the capacity of the CPA to conduct a full review of the Oil-for-Food Program and recommended that its main contribution should be to "safeguard materials to ensure that complete investigation can be undertaken at a future date."[118]

Bremer agreed with this recommendation. The CPA offered assistance to the UN Office of Internal Oversight Service investigation and directed all ministries to identify, safeguard, and inventory all relevant oil-for-food documents and to identify individuals with knowledge of abuses related to the oil-for-food program.[119] This was ultimately welcomed by senior administration officials. Secretary Rumsfeld wrote a note to Bremer stating, "Your recent pledge of assistance and cooperation to the UN office of Internal Oversight's investigation is an impor-

---

[116] Fax from Newt Gingrich to Secretary Colin Powell, March 10, 2004.

[117] Info Memo from James B. Warlick to the Administrator, "Subject: Dealing with Corruption," March 16, 2004.

[118] Action Memo from Scott Castle and James B. Warlick to the Administrator, "Subject: Allegations of Misconduct Regarding the Oil for Food Program—Next Steps," March 15, 2004.

[119] Memo from L. Paul Bremer III to Secretary Rumsfeld, "Subject: CPA Assistance in Oil-for-Food Program Investigations," March 27, 2003.

tant part of our efforts to support these investigative efforts."[120] Bremer also directed Iraqi ministries "to safeguard all information related to the Oil-for-Food Program, including contracts, amendments and annexes to contracts, and supporting materials." He ordered senior CPA ministry advisors to "assist the interim Iraqi ministers in identifying any current ministry officials who may have knowledge of misconduct arising from the administration of the OFF program."[121] Bremer decided to place the Iraqi Board of Supreme Audit, a respected Iraqi government body that had been in existence since the 1920s, in charge of overseeing an independent investigation. On May 12, 2004, the Board of Supreme Audit awarded a contract to Ernst & Young to assist the investigation. As Bremer explained to U.S. Representative Christopher Shays, a Republican from Connecticut who made multiple trips to Iraq, the Iraqi Board of Supreme Audit "is best prepared to conduct such an investigation, in view of its legal status as a separate and independent public agency; its expansive investigative authorities under Iraqi law; its experienced and trained staff of career public servants; and the likelihood that it will continue to function as Iraq's highest public audit organization following the transfer of governance authority to the Iraqi Interim Government."[122]

Coincidentally, Ahmad Chalabi was spearheading an effort to have the Governing Council conduct a separate oil-for-food audit using the consulting firm KPMG and the international law firm Freshfields Bruckhaus Deringer. The CPA, however, was opposed to having a political body, the Governing Council, conduct such an investigation.[123] Bremer's deputy, Richard Jones, bluntly told Chalabi in a

---

[120]Memo from Donald Rumsfeld to Ambassador L. Paul Bremer, "Subject: CPA Assistance in Oil-for-Food Program Investigations," March 26, 2004.

[121]Cable from CPA Baghdad to SECDEF WASHDC, SECSTATE WASHDC, WHITE HOUSE NSC WASHDC, CJCS WASHDC, CDR USCENTCOM MACDILL AFB FL, US MISSION USUN NEW YORK, "Subject: CPA 756: Oil for Food (OFF) Corruption Allegations—UN Assistance," March 15, 2004. Also see Action Memo from Scott Castle and James B. Warlick to the Administrator, "Subject: Allegations of Misconduct Regarding the Oil for Food Program—Next Steps," March 15, 2004.

[122]Letter from L. Paul Bremer III to the Honorable Christopher Shays, May 19, 2004.

[123]Bremer, *My Year in Iraq*, p. 315.

phone call that the CPA would approve only one inquiry, which would be directed by the Board of Supreme Audit. Bremer reiterated this to Chalabi shortly thereafter, noting that "CPA has approved only one investigation of the allegations—that overseen by the BSA." He continued by arguing that "the CPA will not authorize funding for other investigations into these allegations, and any such investigation could undermine the process already underway."[124]

Freedom of the Press

U.S. officials became increasingly agitated by U.S. and foreign press coverage for what they considered biased and anti-American reporting. CPA staffers believed the American media, for example, had wildly overstated the security concerns in Iraq, completely ignoring a range of positive reconstruction developments. One response adopted by CPA officials was to feed sympathetic information to tip the balance of media reporting toward more favorable coverage. "There are a handful of 'friendly' (conservative) columnists and television pundits that were supportive of our war effort, but are frustrated with the re-construction and are becoming critical," Dan Senor noted in a memo to Bremer. "We need to reach out to them." Examples included Charles Krauthammer, Tony Snow, William Kristol, Paul Gigot, and George Will. "The goal here," Senor argued, "is to arm our friends with the facts."[125]

"Rather than waiting for the press to take the initiative," Dan Senor wrote in a separate memo to Bremer, "we should use this month to choose and frame the next chapter." In 2002, he argued "the Administration let its guard down on message discipline on Iraq; repercussions were still felt in the early fall months during the lead-up to the U.N.

---

[124]Letter from L. Paul Bremer III to Dr. Ahmad Chalabi, May 12, 2004. Also see Action Memo from James B. Warlick to the Administrator, "Subject: Letter to Chalabi on OFF BSA Investigation/Audit," May 12, 2004; and Letter for E. Scott Castle to Claude Hankes-Drielsma, May 15, 2004.

[125]Memo from Dan Senor to Ambassador Bremer, "Re: Short-Term (D.C.-centric) Press Strategy," June 1, 2003.

debate."[126] Efforts to shape a more positive impression of events in Iraq proved of little avail, however, in the face of accumulating bad news. "The CPA was constantly struggling with the inability to launch an effective public diplomacy campaign," recalled CPA staffer Fred Smith. "Symptomatic of this challenge was the influence of Al Jazeera throughout the region and its ability to undercut the CPA's efforts."[127] (The name of this TV network, based in Doha, Qatar, means "the island," a reference to the Arabian Peninsula.) After the fall of the Ba'athist regime, a newly enfranchised Iraqi press emerged. Many Iraqis preferred news from foreign satellite channels, however, especially such pan-Arab channels as Al Jazeera and Al Arabiya. "There was a growing sentiment from the White House," noted NSC senior director Frank Miller, "that Al Jazeera can't defeat us. We recognized the importance of information."[128] The U.S. took a number of steps to counter Al Jazeera's influence, such as building a radio station in Jordan to beam news and information into Iraq.

CPA officials debated a range of legal measures to counter Al Jazeera's coverage. In a note to Bremer, Nabeel Khoury, CPA's spokesman for Arab media, laid out the CPA's concerns. First, Al Jazeera played unedited bin Laden tapes that incited "Muslims to take up arms against what he terms 'the crusader/Zionist coalition against the Muslim people.'" Second, Al Jazeera consistently interpreted reports of violence against coalition troops as "resistance," and always referred to coalition troops as occupying forces. Third, Al Jazeera regularly picked the "worst possible 'experts' such as the editor of the al-Quds al-Arabia daily out of London, to explain what's happening in Iraq." The result, Khoury bitterly complained, was that the comments were virulently anti-American. Fourth, and most disturbing, some Al Jazeera reporters were appearing on the scene soon after attacks against coalition forces, sometimes during the attack. "This has led to suspicions," Khoury told Bremer, "that they have had prior notification of an imminent attack

---

[126]Memo from Dan Senor to Amb. Bremer, "Re: A Plan for the Next Storyline," August 6, 2003.

[127]Author interview with Fred Smith, December 14, 2008.

[128]Author interview with Frank Miller, June 6, 2008.

and placed themselves in a position to tape it rather than alerting CPA authorities in order to save lives."[129]

In another memo to Bremer, Charles Heatly from the CPA Press Office commented that "the image projected by Jazeera is of the coalition acting imperiously, willfully harming (often killing) Iraqis, and facing popular resistance from ordinary Iraqi citizens," as well as making "factually inaccurate and insightful (sic) reports."[130]

What most concerned CPA and Iraqi officials was growing evidence that some journalists from Al Jazeera had prior information about the detonation of explosive devices in Iraq but did not disclose this information to Iraqi authorities. Their intention appeared to be obtaining exclusive news footage of the incident. According to CPA's General Counsel, Paragraph 219 of the Iraqi Penal Code provided "that persons who fail to notify public authorities of their awareness of the commission of such an offense under the internal security part of the Code" may be punished.[131]

Edwin Castle, CPA general counsel, provided Bremer several options, which were not mutually exclusive. The first was to close Al Jazeera down in Iraq and expel its non-Iraqi employees under paragraph 219 of the Iraqi Penal Code. This was likely to be politically explosive. The second was to take criminal action against the reporters involved. But Castle did not recommend that option because of concerns that the CPA did not have enough proof to convict them. The third option was to promulgate a CPA order clarifying that Iraq's criminal law compelled disclosure of information concerning criminal activities planned against the state. The order, Castle noted, should also broaden the criminal provision to ensure that all persons, including foreign media employees, were required under penalty of law to inform

---

[129]Email from Nabeel A. Khoury to Paul Bremer, "Subject: Taking on Al Jazeera," September 13, 2003.

[130] Memo from Charles Heatly to the Administrator, "Subject: Al Jazeera Television," November 11, 2003.

[131] Coalition Provisional Authority, "Addressing Media Conduct that Is Inimical to Security," September 14, 2003.

proper authorities of information regarding planned criminal activity against the CPA or coalition forces.[132]

But who should take the lead in responding to Al Jazeera, especially with the growing concerns that its reporters were being used as a tool of insurgents? Should it be the CPA or the U.S. military? "It was never entirely clear," said Dan Senor. "In some cases, it was the military and in many cases it was us at CPA."[133] In September, the CPA acquired affidavits "stating that the Al Jazeera reporters confirmed that they had advance knowledge of a bomb being placed in the area, and had come there seeking exclusive footage. They indicated they would be paid a bonus if they produced such an exclusive." This message continued: "It seems to me we have evidence of prior knowledge, and an economic benefit tied to an explosion—in this case, one that killed an American soldier." The author, Gary Thatcher, urged that the CPA shut down Al Jazeera's operations in Iraq until the CPA could secure clarification from Al Jazeera on its policies involving prior knowledge of violence and economic incentives to employees for "exclusive" photos of violence.[134]

The Iraqi Governing Council wanted to throw out Al Jazeera altogether.[135] On September 22, the Council agreed to issue Decision 48, which ordered the closure of Al Jazeera and Al Arabiya for one month for "repeated violations committed by the two networks in broadcasting voices that call for political violence in Iraq, at times to the point of blatant incitement to kill."[136] Governing Council members argued that Iraq was in a state of war and that the action against the networks was

---

[132]Email from Edwin Castle to Paul Bremer, "Subject: Al-Jazeera Options," September 14, 2003.

[133]Author interview with Dan Senor, October 31, 2008.

[134]Email from Gary Thatcher to Daniel Senor and Nabeel A. Khoury, "Subject: Taking on Al Jazeera," September 13, 2003. Also see Cable from CPA to SECDEF WASHDC, SEC-STATE WASHDC, NSC WASHDC, AMEMB DOHA, IRAQ COLLECTIVE, TREASURY WASHDC, "Subject: The CPA and Al Jazeera," September 30, 2003.

[135]Author interview with L. Paul Bremer, November 15, 2007; author interview with Dan Senor, October 31, 2008.

[136]Memo from Scott Carpenter to the Administrator, "Re: Daily Governing Council Sitrep," September 22, 2003.

an act of self-defense. The decision created some controversy because it was leaked to the media before the CPA had learned about it.[137] The Governing Council also imposed stiffer penalties against those involved in criminal and insurgent behavior:

- 25 years imprisonment for anyone convicted of sabotaging government resources
- 25 years imprisonment for anyone convicted of kidnapping
- 25 years imprisonment for anyone convicted of rape
- 25 years imprisonment for anyone convicted of carjacking
- 10 years imprisonment for theft or destruction of power sources, such as electricity, oil, or gas lines
- 10 years imprisonment for possession of heavy weapons, including machine guns, rifles with a caliber above 7.62 mm, rockets, grenades, or explosive devices.

"These penalties," the Governing Council stated, "are mandatory minimum sentences; the courts will have no discretion to impose lesser punishments."[138] In the midst of these discussions, CPA officials began to grow leery of acting too harshly against Al Jazeera, recognizing how this would be interpreted on the Iraqi street. State Department officials also expressed concern that harsh actions "will provoke a storm of protest from the media community" and "will be misinterpreted and used to undermine the credibility of our claims to be promoting democracy, free speech and open media in Iraq and the rest of the Arab world." In addition, expulsion might not be an effective way to remove anti-American coverage because networks could replace expelled journalists or hire Iraqis, who couldn't be legally expelled by CPA. In fact, the State Department argued that expelling journalists "may produce the opposite effect of stimulating greater anti-American coverage in

---

[137] Email from Scott Carpenter to Clayton McManaway, September 23, 2003.

[138] Proposed Governing Council Press Release, "Penalties for Crimes," September 10, 2003.

the regional and Arabic media or unexpected negative reaction from Iraqis."[139]

Despite these concerns, CPA officials began to exert pressure on Al Jazeera and other Arab media outlets. Rumsfeld sent an exasperated memo to Bremer and General John Abizaid noting that "our message isn't getting through" and "we have to do something about improving Al-Jazeera and Al Arabiya."[140] CPA chief of staff Patrick Kennedy subsequently sent a letter to Al Jazeera's board of directors, asking a series of pointed questions:

- What procedures will be followed when a staff member receives information regarding a pending attack, explosion or other unspecified incident which may result in injury or death to any person, including civilians, civil authorities or military personnel?
- What procedures will be followed when a staff member is instructed not to cross a security or police line?
- How will your staff be instructed to respond when questioned by military or law enforcement authorities at the scene of a violent attack or in the course of investigating the incident?
- What is your policy regarding any employee who attempts to deceitfully conceal evidence relevant to an ongoing investigation?
- Does your organization pay bonuses, cash awards or any other incentive for staff members who provide exclusive coverage of incidents involving violence or death?
- Have you ever made such payments to anyone in Iraq?
- What will your policy be in the future?[141]

There were some initial signs that CPA might be making progress. A CPA cable to Washington stated that "initial indications are

---

[139] Info Memo from the Office of General Counsel to the Administrator, "Subject: Authority to Expel Journalists from Iraq," April 24, 2004.

[140] Memo from Donald Rumsfeld to Jerry Bremer and Gen. John Abizaid, "Subject: Afghanistan and Iraq—the Message," September 16, 2003.

[141] Letter from Ambassador Patrick Kennedy to the Board of Directors of al-Jazirah, September 27, 2003.

that the [Al Jazeera] Network is taking CPA concerns seriously. The moves by CPA come against a wider backdrop of anger by the Governing Council about both Al Jazeera and Al Arabiya."[142] But this proved to be wishful thinking. Qatar's government had consistently rebuffed requests from senior Bush administration officials to change Al Jazeera's tone and content.[143] As Dan Senor noted in an email, U.S. government officials—including President Bush—had repeatedly expressed concerns to Al Jazeera's senior officials to no avail, including the chairman, Sheikh Hamad bin Thamer al-Thani, a cousin of Qatari Emir Sheikh Hamad bin Khalifa al-Thani. Al Jazeera refused to modify its editorial approach.[144] President Bush became convinced that Al Jazeera broadcasts were costing American lives and asked Bremer to shut it down. A leaked British account of the telephone conversation with Prime Minister Blair even had the President suggesting, presumably in exasperated jest, that the U.S. should bomb Al Jazeera headquarters in Qatar, a key American ally in the Gulf region.[145] None of this produced the desired effect. "In fact," Senor stated, "the Qatarese Foreign Minister explicitly told us that AJ is in tight competition with Abu Dhabi and many others. If the choice is between placating the Americans and maintaining market share, they 'must choose' the latter."[146]

By November, coalition forces had arrested 17 Al Jazeera reporters for misconduct, though most of the allegations, one CPA report

---

[142] Cable from CPA to SECDEF WASHDC, SECSTATE WASHDC, NSC WASHDC, AMEMB DOHA, IRAQ COLLECTIVE, TREASURY WASHDC, "Subject: The CPA and Al Jazeera," September 30, 2003.

[143] For example, Under Secretary of Defense Douglas Feith's comment to Crown Prince Jassim bin Hamad Khalifa al-Thani that the "U.S. government was displeased about the antagonism toward America—and sympathy for terrorists—that routinely aired on al-Jazeerah" went "nowhere." Feith, *War and Decision*, p. 95.

[144] Sanchez, *Wiser in Battle*, pp. 333, 352. On concerns about Al Jazeera's reporting during the Fallujah crisis, also see Bremer, *My Year in Iraq*, p. 328, and Allawi, *The Occupation of Iraq*, p. 276.

[145] "Paper Says Bush Talked of Bombing Arab TV Network," *Washington Post*, November 23, 2003.

[146] Email from Dan Senor to Nabeel A. Khoury, "Subject: Taking on Al Jazeera," September 13, 2003.

concluded, "have proven unsubstantiated, which Jazeera claims is a pattern of deliberate harassment."[147] In December 2003, the CPA initiated prosecution of three Al Jazeera employees. One was Salah Hassan, who coalition forces alleged "had prior knowledge of an Iraqi Civil Defense Force attack on 3 Nov 03" and was being held at Baghdad's Central Confinement Facility.[148] Another was Sattar Karain Kareem, who allegedly assisted insurgents in planning and conducting attacks on coalition forces. He was placed in Iraqi custody at Babil Province Prison and was convicted by Iraq's Central Criminal Court, as Iraqi Minister of Interior Nouri Badran proudly declared to Majed Khader, Al Jazeera's chief of the Baghdad Bureau.[149] The third was Shuaib Badir Daweesh, who was detained on November 13 for placing an improvised explosive device that killed and injured U.S. soldiers from the 4th Infantry Division.[150]

CPA complaints lingered into 2004, when the CPA general counsel recommended suspending Al Jazeera satellite transmissions for two months because of "repeated incidents of unprofessional and undisciplined reporting by Al Jazeera employees." He continued that "the resulting media coverage has been grossly inaccurate and has a destabilizing effect on security and civil order in Iraq."[151] CPA and U.S. military officials were also upset at Al Jazeera during the Fallujah crisis in the spring of 2004. Al Jazeera had a reporter embedded with insurgents inside the city, and, as Lieutenant General Sanchez acknowledged, "there was no way we were going to be able to get him out." Sanchez complained that Washington failed to coordinate strategic informa-

---

[147] Memo from Charles Heatly to the Administrator, "Subject: Al Jazeera Television," November 11, 2003.

[148] Email from Kathryn Sommerkamp to Michael Adler, Executive Secretary, "Subject: Update on Al-Jazeera Case—Salah Hassan," December 16, 2003.

[149] Letter from Minister of Interior Nouri Badran to Baghdad Bureau Chief Majed Khader, February 19, 2004.

[150] Email from Kathryn Sommerkamp to Michael Adler, "Subject: Al Jazeera Case Update," December 17, 2003.

[151] Action Memo from E. Scott Castle to the Administrator, "Subject: Suspension of Al Jazeera Media Activities," February 19, 2004.

tion activities to counter Al Jazeera. But CPA officials, including from CPA's Strategic Communications office, continued to believe that closing down the broadcaster's operations in Iraq would have significant negative consequences and intensify anti-American sentiments.[152]

Al Jazeera was not the only media problem. In March 2004, the CPA's Office of General Counsel recommended temporarily closing down the *Al Hawzah* newspaper, which was run by Muqtada al-Sadr's supporters, because of its incitement to violence. Bremer's advisors warned that although CPA had the legal authority to permanently close the newspaper down, a graduated response that included suspending *Al Hawzah's* operations for 60 days was better "so as not to appear heavy-handed with the press."[153] The CPA also ran into problems with Al Arabiya in December 2003, when that network restarted broadcasting in an apparent breach of the ban imposed by the Governing Council. In a letter to Al Arabiya, Iraqi Minister of Interior Nouri Badran alleged that Al Arabiya broadcast an audio tape purporting to be Saddam Hussein urging resistance against the occupation and threatening coalition forces. The audio tape and its rebroadcast, he noted, incited violence against individuals, which was a violation of CPA Order Number 14 on Prohibited Media Activity. "Accordingly," he concluded, "the Governing Council on November 24, 2003, directs the confiscation of all Alarabiya equipment used to uplink its satellite transmissions. In addition, Alarabiya is hereby prohibited from satellite transmission, to/from satellites, in Iraq until Alarabiya provides sufficient assurance about its behavior in the future."[154]

---

[152] Author interview with L. Paul Bremer, November 15, 2007.

[153] Action Memo from Office of General Counsel to the Administrator, "Subject: Response to Questions Regarding Temporary Closure of *Al-Hawzah* Newspaper," March 11, 2004. Also see Info Memo from Office of General Counsel to the Administrator, "Response to Questions Regarding Temporary Closure of *Al-Hawzah* Newspaper," March 13, 2004.

[154] Letter from Nouri al-Badran, Iraqi Minister of Interior, to Managing Director, Alarabiya, November 24, 2003 (English version). On CPA problems with Al Arabiya, also see Action Memo from Charles Heatly to the Administrator, "Subject: Al Arabiya," December 17, 2003.

Conclusion

Promoting the rule of law is invariably the most difficult aspect of democratic reform. In comparison, designing constitutions and holding elections can be comparatively easy. Unfortunately, this seems to have been the element of the CPA's work that received the least support from Washington in the form of qualified personnel and substantive backstopping from the most relevant agency, the Department of Justice. Not surprisingly, the CPA was able to make only limited progress in this area during its 14-month existence. Nevertheless, its efforts in this field, although hampered by lack of planning, acute personnel shortages, and the deteriorating security situation, were comprehensive and generally well conceived. If anything, the CPA's objectives were too ambitious, given the time and resources available. An impressive amount of progress was achieved in a short period: reopening the courts; establishing special tribunals; creating the basis for an independent judiciary; cleansing the legal statutes of abusive legislation; preparing the Iraqis to handle high-profile cases, such as the trials of Saddam and his principal henchmen; and putting in place barriers to official corruption. How enduring these reforms will be remains uncertain, but the overall level of achievement compares favorably to that of many earlier U.S.- and UN-led nation-building efforts.

Bremer resisted the introduction of international judges and prosecutors while arranging to provide substantial support to an Iraqi-led process that would eventually try Saddam and his main henchmen. The result was an Iraqi prosecution that engendered some criticism but was vastly more expeditious and inexpensive than international tribunals have proven to be elsewhere. Bremer acted with moderation and circumspection on such politically charged issues as the oil-for-food scandal and the behavior of Al Jazeera. In doing so, he resisted high-level pressures from Washington and avoided actions that might have abused his uniquely powerful position as the ultimate source of executive, legislative, and judicial power in Iraq.

CHAPTER SEVEN

Growing the Economy

Iraq had achieved middle-income status in the late 1970s, but Saddam Hussein's decision to invade Iran in September 1980 marked the beginning of a two-decade downward spiral in the country's economy. Iraq's port facilities were destroyed early in that conflict and Iraqi oil production came to a virtual standstill. Iraq borrowed on the international capital market to cushion the immediate revenue effects of this loss. By 1990, debt service payments were soaking up 55 percent of Iraq's oil revenue.

Mounting debt was one incentive for Iraq's invasion of Kuwait that year. The second war in a decade resulted in a crushing defeat for Iraq and UN sanctions that further crippled the Iraqi economy. No longer able to borrow abroad, the Iraqi government began to finance its operations by printing money. This caused inflation to rise from 6 percent in 1989 to nearly 500 percent in 1994 and 100,000 percent by the end of 2002.[1] Additionally, the policies of Saddam Hussein's highly centralized and corrupt regime led to major distortions of economic activity and the suppression of private-sector development on an unprecedented scale.

Although Iraq still possessed the world's second-largest usable oil reserves, the legacy of three wars and a decade of economic and political sanctions contributed to deep deterioration and degradation of the economy. Many of Iraq's human development indicators were

---

[1] Christopher Foote, William Black, Keith Crane, and Simon Gray, "Economic Policy and Prospects in Iraq," *Journal of Economic Perspectives*, Vol. 18, No. 3, Summer 2004, pp. 47–53; Bremer, *My Year in Iraq*, p. 67.

the lowest in the Middle East. As one visitor to Baghdad warned Secretary of Defense Rumsfeld and CPA Administrator Bremer, "The CPA is confronting a much more difficult problem than a traditional post-conflict reconstruction challenge. Iraq is also a completely failed economy. The CPA is confronting the equivalent of both a defeated Germany in 1945 and a failed Soviet Union in 1989."[2]

Macroeconomic Stabilization

In a May 2003 briefing for President Bush prepared by the CPA, "Revive the Economy" was listed as the first task of the postwar phase.[3] In a series of phone calls with congressional leaders, Bremer stated, "My first priority is the Iraqi economy and to get people back to work."[4] Similarly, Bremer told a *Washington Post* reporter, "If we don't get their economy right, no matter how fancy our political transformation, it won't work."[5]

The Iraqi government was effectively broke when the CPA took over.[6] The CPA nevertheless began with several sources of funding for its operations for the near term. The U.S. Congress had appropriated $2.4 billion before the invasion to provide humanitarian, rehabilitation, and reconstruction aid. The United States had also confiscated $1.7 billion in Iraqi property held within the United States, which was vested in the Department of the Treasury. U.S. forces in Iraq captured approximately $789 million in regime-owned cash, which was turned

---

[2] CPA Memo from John Hamre to Secretary of Defense and Administrator, "Preliminary observations based on my recent visit to Baghdad," July 2, 2003.

[3] See Coalition Provisional Authority, "Presidential Update" briefing slides, May 29, 2003. "Phase I—Stabilize Post-War Situation" was listed prior to "Phase II—Consolidate our Victory," and included "Establish Law and Order," "Restore Basic Services," and "Root out Saddamism" as the key objectives.

[4] Memo from Tom Korologos to Paul Bremer, "Talking Points for Frist, Daschle, Hastert, Pelosi Calls on June 3, 2003," June 3, 2003.

[5] Rajiv Chandrasekaran, *Imperial Life in the Emerald City*, p. 62.

[6] On the importance of generating revenues to support economic stabilization in the post-conflict period, see James Dobbins et al., *The Beginner's Guide to Nation-Building*, p. 162.

over to the CPA. In May 2003, Bremer established the Development Fund for Iraq at the Federal Reserve Bank of New York, into which $1 billion from the UN Oil-for-Food Program's escrow account was deposited. This money, too, was put at the CPA's disposal. The international community also offered $790 million in response to a UN Flash Appeal to meet urgent requirements in Iraq.[7]

Before the war, the Iraqi government claimed unemployment to be 50 percent. In the war's immediate aftermath, the CPA's Ministry of Labor Advisory Group informally estimated that 70 percent of Iraqis were unable to work because of the physical damage from the combat and the looting.[8] Bremer's top economic advisor, Peter McPherson, concluded: "Our most immediate need is to put money into the hands of ordinary Iraqis in order to prime the economic pump."[9]

The CPA began working to inject purchasing power into Iraq's economy almost immediately. In May, the World Food Program purchased the annual wheat and barley crop from Iraqi farmers. The CPA provided more than one million Iraqi civil servants with $20 emergency payments, and on May 24 began to make salary payments according to a simplified four-grade pay scale.[10] The administrators of Iraq's pension system had converted the entire pension database from a minicomputer to PC format prior to the war and stored these files at their homes before the looters descended on the Pension Administration building along the Tigris River in Baghdad.[11] Consequently, within ten weeks of the end of the conflict, in a country without proper communications or a functioning banking system, the CPA disbursed over

---

[7] See Scott Castle, "Iraq Relief and Reconstruction Funding Sources," May 5, 2003; and Dov S. Zakheim to Secretary Donald Rumsfeld Memo, "Summary of Financial Management in Iraq/Transparency Measures," July 8, 2003.

[8] Memo from Tom Foley to Paul Bremer, "Unemployment Data," November 14, 2003.

[9] Quoted in Bremer, *My Year in Iraq*, p. 67.

[10] David Oliver and John Rooney, "Pensions and Salaries: Status as of 6th July 2003," July 6, 2003.

[11] "Plan to Make Emergency Payments to Pension Recipients in Baghdad," unattributed, undated paper in CPA Archives. The authors note there were 254,000 pensioners in Baghdad alone, as well as 129,000 dependents of deceased pensioners.

$400 million to Iraq's 1.3 million public-sector workers and 1.1 million pensioners to pay off all arrears through June.

To begin stabilizing the dinar, on May 17 Bremer reestablished and made independent the Central Bank of Iraq. Additionally, the CPA began reopening Iraq's commercial banks, which had been shut down by the war. However, Iraq faced a potentially serious liquidity crisis due to a severe shortage of domestic currency. The emergency salary payments were made in U.S. dollars because there were insufficient Iraqi dinars on hand in Iraqi banks, and not enough could be printed in time to pay the May or June salaries.[12] One CPA advisor warned, "The financial position of the Iraqi banking system is uncertain. The ability of banks to meet a significant number of depositors' claims is uncertain."[13]

The state-owned Rafidain and Rasheed banks held a combined total of 119 billion dinars in cash; another 80 billion dinars were located at the Central Bank in Baghdad with another 40 billion at branches in the Basra and Mosul branches. The CPA estimated potential claims on the banks to be 332 billion dinars. Although a ratio of 239 billion in cash to meet deposit liabilities of 332 billion would be manageable under normal circumstances, cash balances in Iraqi banks had nearly halved in the CPA's first month. Potential net claims were rising rapidly as more branches of the Rafidain and Rasheed banks began reopening across the country.[14] Iraqis' confidence in their banking system was fragile and susceptible to a variety of shocks, and the Central Bank's role as the lender of last resort was compromised by its limited stock of dinars, many of which were flood damaged from postwar looting. The

---

[12] Memo from Peter McPherson to Paul Bremer, "Iraqi Salaries for May/June," June 7, 2003.

[13] Memo from Tony McDonald to Peter McPherson, "Initial Strategy for Reform of Iraq's Financial Sector," June 19, 2003.

[14] In fact, there were 747 billion dinars in private accounts, but only 50 percent of Rafidain and Rasheed branches were open in July. See memo from Jacob Neil to Clayton McManaway, "The Ticking Time Bomb: Managing Liquidity in the Banking System," July 17, 2003.

International Monetary Fund (IMF) described the situation as "a ticking time bomb."[15]

The CPA was faced with difficult trade-offs among the policy options available to maintain liquidity. Announcing a limit on withdrawals from Iraqi banks might send a signal to account holders that their money was not safe and thereby trigger the very run the CPA was hoping to avert. Placing a moratorium on withdrawals by Iraq's state-owned enterprises (SOEs) would protect the roughly 1.28 trillion dinars held in public accounts, but would also deny many SOEs access to working capital. Printing more dinars would risk increased inflation, and the only printing plates available were those with Saddam's image—and the CPA was emphasizing de-Ba'athification and the end of the old regime.[16]

In the end, the CPA took several steps to maintain liquidity. In early June, it decided to resume printing Saddam dinars. Only two denominations of the Saddam dinar had circulated widely in Iraq: the 250-dinar note (worth about 17 cents) and a 10,000-dinar note (worth about $6.50). Fears that the 10,000-dinar note could be widely counterfeited, along with its impracticality for everyday transactions, caused it to be traded at a 10–30 percent discount relative to the smaller note.[17] From June 9 to June 23, the Central Bank of Iraq printed 33 million 250-dinar notes, yet demand continued to rise. Some 750 million dinars in 250-dinar notes were produced each day until August 27, when the Central Bank of Iraq decided to halt its production because demand finally tapered off.[18] A moratorium on withdrawals by state

---

[15] Memo from Jacob Neil to Clayton McManaway, July 17, 2003. (The date of this memo is unclear, but it is relatively important because it relates to the memo from McDonald to McPherson in determining when certain corrective steps were taken. July 17 is the date it was logged into the CPA archives.)

[16] On the importance of stabilizing the currency in post-conflict situations, see Dobbins et al., *The Beginner's Guide to Nation-Building*, pp. 163–167.

[17] Foote et al., pp. 60–61.

[18] See "Senior Steering Group (SSG) Minutes," June 2, 2003; Coalition Provisional Authority, "Highlights for Secretary Rumsfeld," briefing slides, June 23, 2003; memo from Neil to McManaway, July 17, 2003; and U.S. Treasury Iraq Task Force, "Iraq: Report on Financial and Macroeconomic Issues," September 11, 2003.

companies was also enacted, while private businesses were allowed access to their accounts. (This moratorium was later made permanent in a controversial decision discussed below.)

The CPA achieved notable success in reducing inflation. The Iraqi Central Statistics Office calculated that the Consumer Price Index in the Baghdad area declined by roughly 3 percent from January to July 2003.[19] Although these early figures were probably unreliable, later surveys showed that from December 2003 to April 2004, prices declined each month except March.[20] Although monthly inflation data in Iraq have a distinct seasonal pattern, with seasonally low rates in the late spring and summer, the annual data reflected the downward trend in prices for fuel and basic food commodities. The April 2004 annual inflation rate was 19.6 percent, down from its peak of 47 percent in October 2003.[21] In the last month of its existence, CPA officials concluded that "Inflation appears for now to be under control in Iraq."[22]

As the CPA worked to maintain liquidity and stabilize Iraq's currency, its staff also scrambled to draft a budget for the Iraqi government for the remainder of 2003. The priorities for the budget, in the words of David Oliver, the CPA's senior financial officer, was "to avoid spending money on public companies which as a practical matter don't exist, provide resources to get the Iraqi economy working as fast as possible, and appropriately allocate the funds available."[23] For expediency's sake, the budget was prepared using the systems, forms, and formats that had traditionally been used for Iraqi budgets. Each Iraqi ministry and Kurdish equivalent prepared a budget with its coalition senior advisor. The proposed budget was then reviewed by appropriate officials from

---

[19] Memo from Marek Belka and Olin Wethington to Paul Bremer, "Information on Consumer Price Changes," November 30, 2003.

[20] Memo from Olin Wethington to Paul Bremer, "Briefing Memo for the Prime Minister Meeting: Consumer Price Trends," June 6, 2004.

[21] Memo from Olin Wethington to Paul Bremer, "Official Inflation Data: April 2004," May 23, 2004.

[22] Memo from Wethington to Bremer, June 6, 2004.

[23] Memo from David Oliver to Paul Bremer, "Budget Timeline and Process for July–December 2003 and January–December 2004," June 16, 2003.

the Ministries of Finance and Planning, briefed to USAID and UN representatives, and approved by the CPA Program Review Board in July.[24] This process revealed the limited capacity of Iraqi institutions to enact economic and financial policies according to best international practices. The Finance Ministry had controlled only 8 percent of the Iraqi budget under Saddam Hussein and was clearly overwhelmed by its new responsibilities.[25] While one Iraqi critic derided this process as "an amateur and unrealistic affair," former Deputy Secretary of Defense John Hamre, reporting to Rumsfeld on his visit to Iraq, acknowledged the budget process was rudimentary but called it "an incredible accomplishment this early in such a complex environment."[26]

Issuing New Currency

On July 7, Bremer announced the CPA's intention to issue new bank notes. This announcement followed extensive discussions between Treasury Department advisors, senior Iraqi officials at the Central Bank and Ministry of Finance, and the IMF. In these meetings, Iraqi officials stressed the fragile state of public confidence in the currency. The only plates available were for the 250-dinar note. The banks had no capacity for electronic transfers, and there were no checking accounts. Iraq was also being flooded with high-quality counterfeit notes. Acting Central Bank Governor Faleh Salman agreed that new notes would be well received by the population.[27]

The currency swap, scheduled to begin on October 15 and to continue through January 15, 2004, replaced existing "Saddam" and "Swiss" dinars (the currency in use in the Kurdish areas during Saddam's rule, so called because they were printed in Switzerland) for new

---

[24] "2003 Iraqi National Budget: July–December 2003," briefing slides, July 17, 2003.

[25] Bremer, *My Year in Iraq*, p. 110.

[26] Allawi, *The Occupation of Iraq*, p. 194; CPA memo from John Hamre to Secretary of Defense and Administrator, July 2, 2003.

[27] Author unattributed, "New Currency," June 23, 2003; memo from Peter McPherson to Paul Bremer, "Meeting with Acting Central Bank Governor, Faleh Salman," July 6, 2003.

Iraqi dinars, whose design would resemble Swiss dinars with some minor modifications. It was specifically decided not to print new Swiss dinars as the national currency. This would have required more than 80 percent of the population (in Iraq's center and south) to change their prices and wages. The Sunni and Shi'ite populations could also be expected to resist adopting the Kurdish region's money. Discussions with Iraqi leaders in both parts of the country led the CPA to set the conversion rate for the new currency at 1:1 for the Saddam dinar and at 1:150 for the Swiss dinar, a compromise between the market rate and the purchasing power parity rate for those two currencies. The number of denominations would be increased from two to six and would be printed in different colors to avoid confusion with the existing Swiss dinars.[28]

Under the direction of retired U.S. Brigadier General Hugh Tant, the Iraqi currency exchange required multiple aircraft, loaded with over six thousand tons of the new currency, flown to Iraq. Nearly 10,000 boxes of dinars were then transported to approximately 260 banks throughout the country. Although the first week of the currency exchange saw only light crowds as Iraqis adopted a wait-and-see attitude on lines and security, by the time the exchange was completed 4.62 trillion dinars were in circulation, 106 percent of the original estimate of demand.[29] Two months later, the New Iraqi Dinar had stabilized and appreciated against the U.S. dollar, a sign of domestic and international confidence in the new currency. "The whole operation, from design and contracting of the currency, to planning and distribution, to execution was performed flawlessly," recalled Fred Smith. "And the Iraqi people loved it. Getting rid of old Saddam dinars had a huge psychological effect."[30]

One reason for the success of the Iraqi currency exchange was that the CPA made strategic communications a central element of its

---

[28] See Foote et al, pp. 61–62, and memo from Paul Bremer to Governing Council, "Information Campaign for the Currency Exchange," September 15, 2003.

[29] Coalition Provisional Authority, "Administrator's Weekly Report: Economics, 10–16 January 2004," January 16, 2004

[30] Author interview with Fred Smith, Dec. 14, 2008.

implementation plan. In early September, the CPA began running public service announcements three times a day on the Iraqi Media Network and took out daily ads in Al Sabah. Daily news angles were pitched to Arab satellite channels, and advertising space was purchased in non-coalition newspapers. Two million informational handbills and 250,000 posters were distributed in the month before the exchange, and regional "help lines" were set up.[31] Officials from the CPA, the Central Bank, and the Iraqi Resources Development Council traveled to nearly all of Iraq's major cities to meet with local leaders to discuss details of the currency swap, answer questions, and encourage them to inform their communities about the swap to enhance public understanding.[32]

In cooperation with the Central Bank, the CPA worked to stabilize the new dinar through a transparent market-based process. Daily auctions were held to establish the exchange rate against the dollar, and the dinar was also freely traded internationally. The CPA also decontrolled interest rates within the country. Confidence was soon established, and the dinar began to appreciate, gaining 20-25 percent against the dollar through June of 2004.[33]

Reforming the Banks

The CPA's attempts to develop the Iraq private sector were consciously modeled on the lessons learned from Eastern Europe's postcommunist economic transformation. The CPA's strategic plan, *Achieving the Vision to Restore Full Sovereignty to the Iraqi People*, stated: "The unique recent experience of the Central and East European countries from

---

[31] Dan Senor, "New Iraqi Currency Roll-Out," Briefing for NSC Deputies, October 7, 2003.

[32] Senor, "New Iraqi Currency Roll-Out"; memo from Peter McPherson to Paul Bremer, "Iraqi Currency Swap—Notification to the Governing Council," September 3, 2003.

[33] Author interview with Olin Wethington, December 28, 2008.

their transformations should be helpful in this regard."[34] On September 21 and 22, CPA (along with the State Department and USAID) held a "Lessons Learned" conference in Baghdad to draw on the expertise of former ministers from Central and Eastern European countries on a range of issues relating to economies in transition. In November 2003, the CPA's Office of Economic Policy came under the direction of Marek Belka, an economist from Poland who had served as that country's deputy prime minister and finance minister during its transition from communism.

One CPA advisor argued that the experience of these postcommunist transitions, "illustrate that Iraq must move to a market system as quickly as possible by encouraging the growth of private firms."[35] However, CPA officials understood that many Iraqi private-sector businesses were unable to take advantage of opportunities to expand and create new jobs due to the lack of available credit. When Secretary of Commerce Donald Evans visited Baghdad, members of the Iraqi business community complained to him that access to credit was one of the primary hurdles for Iraqi companies. This primarily resulted from Iraqi banks' inability to deliver credit products of the type and volume needed.[36]

Iraqi banks had been cut off from international technology, standards, and business practices for 30 years. After the fall of the Ba'athist regime, the Iraqi financial sector was dominated by two state-owned commercial banks—the Rafidain and the Rasheed—and four state-owned specialized banks. All these institutions had been used as subsidy transfer organizations under Saddam Hussein and had been instructed as to which loans to fund. The CPA concluded that "Both management and employee banking skills [in the commercial banks] are very weak."

---

[34] Coalition Provisional Authority, *Achieving the Vision to Restore Full Sovereignty to the Iraqi People*, September 5, 2003.

[35] Memo from Peter McPherson to Paul Bremer, "Timelines for Iraq's Economic Recovery," July 12, 2003.

[36] See memo from Secretary of Commerce Donald Evans to President George W. Bush, "Recent Visit to Baghdad, Iraq, and Kabul, Afghanistan," October 24, 2003; and memo from Tom Foley and Michael Fleischer to Paul Bremer, "Private Sector Credit," December 5, 2003.

This was particularly true of the Rasheed. Whereas the Rafidain had once been one of the region's leading commercial banks, the Rasheed had only been founded in 1988, and Ba'ath Party influence was pervasive.[37] The information systems of the Central Bank and commercial banks were antiquated and—in some cases—nonfunctioning. The Rafidain and Rasheed bank branches had no interbank voice and data communications, and records were provided only once a month from branches to headquarters in Baghdad.[38] The World Bank also noted that the Central Bank's supervisory capacity was largely nonexistent, and the bank suffered from the absence of any supervisory legislation.[39] The CPA informed Congress that the rehabilitation of the Iraqi banking system "is vital to the nation's economic recovery."[40]

The CPA immediately began to enact a series of reforms to overhaul the Iraqi banking sector. To address the lack of human capital, the Central Bank was bolstered with personnel from the IMF, the Federal Reserve, and the central banks of Bahrain and the United Arab Emirates. On July 7, the CPA established the independence of the Central Bank. On September 18, the CPA issued the Banking Order (CPA Order Number 38), which established rules for a modern banking sector, set capital requirements, and provided a mechanism for dealing with troubled domestic banks. Because the Rafidain and Rasheed were seen as the only tools available to promptly deliver loans to small and medium businesses, they were encouraged to begin issuing commercial loans as soon as possible. According to one observer, "Bank staff cheered and hugged one another . . . when they learned that they will be able to operate on a commercial basis again."[41] By October, 280

---

[37] Memo from Peter McPherson and George Wolfe to Paul Bremer, "Restructuring Rafidain and Rasheed," October 4, 2003; and World Bank, "Iraq Update," Briefing to the Executive Directors, June 17, 2003.

[38] Under Secretary of the Treasury John Taylor, "Reconstruction of Iraq's Banking Sector," A Briefing Sponsored by the Bankers Association for Finance and Trade and the Arab Banks Association of North America, October 10, 2003.

[39] World Bank, "Iraq Update."

[40] "Section 2207 Report to Congress," December 29, 2003.

[41] Cited in Taylor, "Reconstruction of Iraq's Banking Sector."

branches were open across the country, and satellite voice and data systems for Rafidain, Rasheed, and the Central Bank were installed at 80 locations.

Although the reopening of branches throughout Iraq was a major achievement, these banks were not yet capable of providing loans in the volume needed to support economic growth and create jobs in Iraq. In December, Bremer informed President Bush that although 83 percent of state-owned bank branches were currently open, they had been unable to pump credit into the economy.[42] The CPA's Office of Private Sector Development (OPSD) estimated that private-sector businesses had an immediate need for $250 million in letters of credit and another $250 million in medium-term loans.[43] By the end of February 2004, Rafidain had approved 200 small- and medium-enterprise (SME) loans valued at $6.7 million, and the Rasheed had approved 126 SME loans valued at $2.1 million.[44] As two CPA advisors noted to Bremer, these loans were "not nearly as many or as fast as we need."[45]

The CPA initiated several programs to expand the availability of credit in Iraq. In July, Bremer approved the expansion of the microcredit lending program in southern Iraq to the rest of the country, making loans available to individuals who operate small businesses. In October, CHF International was awarded a $7 million grant for microfinance activities in Baghdad Governorate and ACDI/VOCA was given a $5 million grant for microfinance activities in the northern governorates.[46] In September, Under Secretary of the Treasury John Taylor told Bremer the International Finance Corporation was willing to set up a $100-million facility to support small and medium-sized

---

[42] Ambassador Paul Bremer to President Bush, "Draft Points for Message, 18 December," December 18, 2003.

[43] Memo from Foley and Fleischer to Bremer, "Private Sector Credit."

[44] U.S. Treasury Iraq Financial Task Force, "Iraq: Report on Financial and Macroeconomic Issues," February 27, 2004.

[45] Memo from Peter McPherson and George Wolfe to Paul Bremer, October 4, 2003.

[46] Memo from Peter McPherson and Charles Greenleaf to Paul Bremer, "Micro-Credit Lending Program," July 17, 2003; memo from Karen Walsh to Paul Bremer, "Status of Micro-Credit Program for CPA," September 26, 2003.

businesses in Iraq and get it operational as soon as possible. Bremer subsequently pledged $5 million of CPA funds in both 2003 and 2004 to support the IFC Small Business Facility. The CPA also requested $200 million in the 2003 supplemental budget for the Iraq-American Enterprise Fund, a private equity fund to invest in Iraqi businesses with substantial growth and job creation potential. The fund would provide investment capital and technical expertise to Iraqi businesses on a profit-oriented basis. It was specifically modeled on such East European funds as the highly successful Polish Fund.[47] This proposal was killed by the House Appropriations Subcommittee on Foreign Operations despite the CPA's pleas that the fund was intended to address "one of the key needs in building the economic fabric of Iraq."[48] In the end, the supplemental appropriation did provide another $46 million for micro- or SME lending.

To facilitate trade, Bremer approved a CPA order to establish the Trade Bank of Iraq. The charter and by-laws for the Trade Bank of Iraq were approved the weekend of November 8, and its first board of directors meeting was held on November 11. The next day, the Trade Bank's new president, Hussein al-Uzri, and the CPA signed the relevant agreements with an international consortium led by J.P. Morgan Chase to operate the bank. The Trade Bank of Iraq opened in December, and by mid-April it had issued 130 letters of credit totaling $581 million.[49]

The CPA pinned its hopes for revitalizing the Iraqi banking sector on the introduction of foreign banks into Iraq. Peter McPherson told Bremer that "International banks are keys to the economic development of Iraq," and would

---

[47] Author unattributed, "Talking Points on Iraq-American Enterprise Fund," September 30, 2003.

[48] See email from Dan Senor to Paul Bremer and David Oliver, "Re: Kolbe Meeting," October 6, 2003; and untitled email from Josh Bolten to Paul Bremer, October 21, 2003.

[49] See untitled memo from Peter McPherson to Paul Bremer, July 1, 2003; U.S. Treasury Iraq Financial Task Force, "Iraq: Report on Financial and Macroeconomic Issues," November 20, 2003; U.S. Treasury Iraq Financial Task Force, "Iraq: Report on Financial and Macroeconomic Issues," April 15, 2004.

- bring technology, know-how, and best practices
- bring new capital and attract funds from domestic sources
- increase efficiency and innovation in the banking sector through greater competition
- develop greater confidence in the banking system and assist investors and businesses from the banks' home countries to come into Iraq
- allow for greater access to international financing sources.[50]

CPA Order Number 38 allowed non-Iraqi banks to operate in Iraq as either a subsidiary with up to 100 percent ownership or as a branch, and would be treated under the law the same as domestic banks. The total number of banks majority-owned by foreigners would be limited to six for the first five years, although a foreign bank could own up to 50 percent of a domestic bank without that ownership counting toward the six. On January 31, Sinan Al-Shabibi, governor of the Central Bank of Iraq, announced that three banks (Hong Kong Shanghai Banking Corporation, National Bank of Kuwait, and Standard Chartered Bank) had been selected to proceed to the final stage of the foreign bank licensing process.

The immediate impact of these initiatives on the Iraqi banking sector was mixed. In December, the CPA reported to Congress that "The Iraq banking sector (state and private) is currently not functional."[51] In March, the Treasury Department's bank supervision team completed an assessment of the Central Bank's supervisory process and found it to be in substantial noncompliance with 25 assessment principles.[52] Although the Iraqi currency exchange had been a success, lack of confidence in Iraqi banks kept people from recirculating the new dinars. "People or businesses are apparently putting their money under the

---

[50] Memo from Peter McPherson to Paul Bremer, "International Bank Entry into Iraq," August 11, 2003.

[51] "Section 2207 Report to Congress," December 29, 2003.

[52] "Administrator's Weekly Report: Economics, March 20–26," March 26, 2004.

mattress based on this lack of confidence," noted one CPA official.[53] Even as he urged the CPA and Iraqi government to seek more opportunities to involve foreign banks and predicted that the Foreign Bank Entry Licensing Program "[w]ill make a significant contribution to Iraqi development over the long-term," Secretary of the Treasury John Snow conceded that "it would be a mistake to have high expectations for the near-term."[54] Consequently, CPA advisors were still complaining in March that Iraq's "[w]eak banking structure hinders economic advances of all kinds."[55]

Debt Relief

The CPA worked closely with the U.S. Treasury Department and the responsible Iraqi ministers on a uniquely successful effort to relieve the country of the tremendous debt burden left by Saddam Hussein. These negotiations were initially led by James Baker, a former U.S. Secretary of State and Treasury. Olin Wethington, who was McPherson's successor as Bremer's chief economic advisor, participated for the CPA. He continued to work on the issue from the U.S. Treasury Department after the CPA's demise until an agreement was reached in mid-November 2004 during a meeting of the Paris Club, an informal group of officials from 19 developed countries who offer such financial services as debt rescheduling and debt cancellations to ailing countries. The CPA insisted from the beginning that Iraqi officials should be involved and take ultimate responsibility for the outcome. The CPA helped the Iraqis prepare for the Paris Club negotiations and facilitated economic reform with the International Monetary Fund, which was

---

[53] Hugh Tant, "Note to Paul Bremer," January 5, 2004. However, between July 2003 and February 2004 the total deposits at Rafidain and Rasheed banks rose by 88 percent, from 2 trillion dinars to 3.9 trillion dinars. See memo from Olin Wethington to Paul Bremer, "Report on Bank Deposits in Iraq," April 4, 2008.

[54] Memo from Secretary of the Treasury John Snow to Ambassador Paul Bremer, "Accelerating Delivery of Credit to Iraq," November 7, 2003.

[55] Memo from Joan Wadelton to Paul Bremer, "Iraq Economic Overview Outline," March 23, 2004.

a precondition for Paris Club relief. The outcome was the largest debt relief package in history, resulting in 80 percent of Iraq's official debt being written off.

Although this process concluded after CPA was disbanded, its framework was established with the active involvement of CPA officials. It relieved the country of a huge burden, and provided a favorable prerequisite for future long-term growth.[56]

Promoting Foreign Investment

Just as CPA advisors hoped foreign licensing would reinvigorate the Iraqi banking system, they hoped that foreign direct investment (FDI) into Iraq would spur private sector development. Specifically, they believed that FDI would help finance new capital for Iraq's economic recovery and encourage the growth of domestic suppliers and transfer technology to Iraq. They also pointed out that among transitional economies, countries with high levels of FDI have also seen substantial growth of small and medium-sized private firms.[57] Thus, Bremer told a skeptical Iraqi Governing Council that "Foreign investment is . . . critical to the growth of the private sector."[58]

Several aspects of Iraqi commercial law impeded foreign investment in Iraq. Non-Arab foreign companies were precluded from owning stock in Iraqi-based companies. All non-Arab, nonresident foreign entities had to go through government-approved commercial agents to do business in Iraq, the corruption of which added prohibitive transaction costs. Moreover, the non-Arab foreign companies that were authorized to do business in Iraq had only a limited capacity to repatriate profits.[59] Consequently, it was clear to the CPA and other observers that a new

---

[56] Author interview with Olin Wethington, December 28, 2008.

[57] "Foreign Direct Investment in Iraq," Discussion Paper, August 12, 2003.

[58] Memo from Ambassador Paul Bremer to Governing Council, "Foreign Investment in Iraq," August 10, 2003.

[59] Email from Scott Castle to Jayme Durnan, "Re: Private Investment," July 21, 2003.

civil and commercial code would be needed to attract regional and international investment in Iraq's industries.[60]

In July, the CPA proposed an overhaul of the Iraqi commercial law impeding foreign investment. The initial draft of the law set a limit of 49 percent FDI, despite Iraqi experts' preference for a limit of 10–30 percent FDI. CPA advisors underlined the desirability of allowing a majority FDI ownership in specific cases, and by August had drafted a new proposal allowing for 100 percent foreign investment with the exception of prohibitions on foreign ownership in the area of oil and natural resources.[61] The new proposal received a mixed greeting from the Iraqi Governing Council. Although some members were enthusiastic, others were resistant because of political concerns that the law would allow foreigners to buy up Iraqi real estate.[62]

Former Minister of Trade Ali Allawi complained that the FDI order "was made with little discussion with the Governing Council," but the record shows that Bremer was greatly concerned with obtaining Iraqi buy-in.[63] He handwrote on a memo to Colin Powell in preparation for the Secretary of State's meeting with the Governing Council: "I have strongly urged quick action on changing Iraqi laws to permit foreign direct investment," and urged Powell to raise the issue.[64] Bremer wrote to the Governing Council stressing the benefits of FDI to Iraq, citing more jobs at good wages, higher investment, and technology transfer as reasons for the adoption of the order.[65] These assertions were supported by a visiting delegation from the World Bank, who stressed

---

[60] See, for example, Hamre et al., *Iraq's Post-Conflict Reconstruction.*

[61] Richard Bartelot, "Meeting on Investment Law and Policy, 8 July 2003," July 8, 2003; memo from Peter McPherson to Paul Bremer, "Foreign Investment Proposal," August 5, 2003.

[62] Ambassador Paul Bremer to Secretary Donald Rumsfeld, "Meetings with the Governing Council on FDI," August 13, 2003.

[63] Allawi, *The Occupation of Iraq*, p. 196.

[64] Memo from L. Paul Bremer to Secretary Powell, "Your Meeting at the Governing Council: Sunday, September 14, 1500–1600," September 12, 2003.

[65] Memo from Ambassador Paul Bremer to Governing Council, "Key Elements of a Foreign Direct Investment Order," August 18, 2003.

the importance of FDI in a meeting with the Governing Council's Finance and Economic Subcommittee on August 17, making clear that 100 percent FDI was the best solution.[66]

On September 15, the subcommittee agreed to the principle of unlimited FDI and offered two amendments to the proposed order. First, the Finance Ministry, not the CPA, would have the authority to draft regulations to govern the implementation of the law. Second, foreign investors were specifically not allowed to buy state-owned enterprises. The subcommittee suggested the order should be issued by the Governing Council in order to bolster investor confidence that the decision had been made by a body more permanent than the CPA. The Iraqis also suggested that $500 million of the proposed $20 billion supplemental request to Congress be set aside for loans to help domestic companies.[67] The next day the Governing Council approved the full economic stimulus package proposed by the CPA, which also included orders dealing with tariffs, taxes, and the banking regulations discussed previously. Although the order was expected to go forward without formal conditions, the Finance and Economics Subcommittee's most influential staffer, Salem Chalabi, told McPherson that the principle of unlimited FDI was approved on the condition of the emergence of an approximately $1 billion "credit umbrella" for small and medium-sized enterprises.[68] On September 19, Bremer signed CPA Order Number 39 into law.

As the FDI order was being debated by the Governing Council, McPherson and Scott Carpenter, the head of the CPA's governance team, recommended that Bremer "treat seriously the credit needs of the Iraqi people. If they get the sense you are not taking this seriously . . . things could fall apart."[69] Finance Minster Kamil Gailani unveiled the

---

[66] Email from Irfan Siddiq to Baghdad Governance, "World Bank meeting with GC Finance and Economic Subcommittee," August 17, 2003.

[67] CPA Headquarters Cable, "Economic Policy Discussions at the Governing Council," September 15, 2003.

[68] Memo from Scott Carpenter to Paul Bremer, "GC Approves Economic Stimulus Package," September 16, 2003.

[69] Memo from Scott Carpenter to Paul Bremer, September 16, 2003.

package of market reforms in a keynote speech before the Institute on International Finance on September 21. The Arab and international media immediately issued sensationalist reports playing up the "Iraq-for-sale" angle of the decision.[70] The Governing Council reacted to media reports of the speech about FDI by issuing a press release that suggested they were backing off from the FDI order. CPA advisors met with Ahmad Chalabi, then serving as acting president of the Governing Council, and obtained a new press release making clear that the Governing Council supported the FDI order and Minister Gailani.[71] However, the initiative in the strategic communications battle necessary to implement such a sweeping reform had already been ceded. Allawi described the backlash from the Iraqi business community as "fearsome," and months later the CPA conceded that FDI was a good policy lost to poor presentation.[72]

Some analysts have questioned the priority accorded the FDI law, given the unlikelihood of any substantial foreign investment as long as Iraq remained in the midst of civil war. A United States Institute of Peace Special Report on the CPA's experience with economic reconstruction in Iraq concludes,

> The CPA's conviction that foreign investment liberalization was so crucial that it had to be enacted immediately turned out to be a miscalculation. It engendered ill will and fed Iraqi suspicions needlessly, as foreign investor interest in Iraq was minimal. Foreign investors are not typically drawn to environments of ongoing violence that lack enforceable property and contract rights. Furthermore, few companies were willing to risk investing under an occupation authority whose laws might be rescinded by the Iraqi successor regime.[73]

---

[70] Allawi, *The Occupation of Iraq*, p. 197.

[71] Memo from George Wolfe to Paul Bremer, "Update on Recent Developments in the Office of Economic Developments," September 28, 2003.

[72] Allawi, *The Occupation of Iraq*, p. 198.

[73] Anne Ellen Henderson, "The Coalition Provisional Authority's Experience with Economic Reconstruction in Iraq," United States Institute for Peace Special Report 138 (April 2005), p. 11.

There was, nevertheless, evidence of interest in investing in Iraq after the overthrow of the Ba'athists. In May, representatives from the Crown Prince of Dubai's office told Defense Department Comptroller Dov Zakheim that they wanted to establish a commercial foothold in Iraq.[74] In June, a CPA advisor reported, "After spending the last few days at the Jordanian-American Business Association, it was apparent from attendees that thousands of businesses from across the world are ready and eager to invest in Iraq."[75] The CPA economic team reported it was beginning to hear from enough potential international investors that it was necessary to establish new offices in Washington, D.C., and Amman, Jordan, to deal with them in a systematic manner.[76] Even as violence was rising in Iraq in the fall of 2003, representatives of the Syrian business community continued to press for greater access to the Iraqi market, and the Basra delegation to an Iraqi reconstruction conference in Amman received a number of follow-up enquiries from British and other companies.[77]

Given that Arab foreign investors already had reasonably easy access to Iraq, it does seem unlikely that the proposed FDI order would have led to any early and substantial increase in investment. The Iraqi business community's opposition was based on the fear of competition and the loss of preferential treatment given favored firms under the old regime. Protectionist reactions of this sort were inevitable. Given the small immediate return such a measure was likely to yield, creating a new regime for foreign direct investment may not have been the best battle for the CPA to have undertaken during the limited time available to it. In the event, rising violence quelled whatever foreign investor interest there might have been for the next several years.

---

[74] Email from Dov Zakheim to Paul Bremer, "Dubai Offer of Assistance," May 16, 2003.

[75] Memo from Douglas Combs to Paul Bremer, "Commercial Service Office (as part of Business Development Council)," June 9, 2003.

[76] Untitled memo from Peter McPherson, Susan Hamrock, and Charles Greenleaf to Paul Bremer, July 4, 2003.

[77] USEMB Damascus Cable, "Syria-Iraq Oil and Electricity Agreements," September 4, 2003; Henry Hogger to Paul Bremer, "Weekly GC Update—Basra," December 15, 2003.

Reducing Subsidies

Subsidies ate up about half of Iraq's national budget, stifled the development of markets in food, power, and fuel, and caused persistent fuel and power shortages. The CPA sought to unburden the Iraqi budget from three costly economically inefficient subsidy programs: those for energy, food, and public enterprises. It succeeded only in the latter instance, and then only partially.

Energy

The official price for premium-grade gasoline in Iraq during the summer of 2004 was 50 dinars (or 3.2 cents) per liter, compared with more than $1 per liter in Turkey, $0.42 in Jordan, and $0.50 in Syria. This subsidy encouraged waste; distorted production decisions; and benefited corrupt officials, militias, political parties, and criminal enterprises through a thriving black market. The cost was estimated to be about $5 billion per year in lost state revenues.[78]

The consensus of economists—whether American, Iraqi, or international—was that liberalizing government control of prices was critical to economic development. In a July presentation to Ambassador Bremer, the World Bank warned: "Without a rapid liberalization of prices, Iraq will be hobbled in its effort to create a market economy."[79] The CPA's Energy Subsidy Reform Task Force "recommends energy prices be liberalized as soon as possible." Secretary Snow wrote Bremer to concur with the CPA's recommended approach for removing energy subsidies.[80] Accordingly, in October the CPA's top two economic priorities for the next quarter of the Strategic Plan were "Prepare [a] policy

---

[78] Foote et al., p. 67.

[79] Saumya Mitra, "Price Liberalization and Safety Net Protection," July 18, 2003.

[80] Memo from Tom Foley to Paul Bremer, "Recommendation for Liberalizing Energy Prices," November 16, 2003; memo from Secretary John Snow to Paul Bremer, "Response to Your Requests on Two Issues," November 6, 2003.

on increasing refined oil products prices" and "Prepare for electricity billing."[81]

The CPA's economic advisors recommended that energy subsidies should be reformed all at once rather than in a series of small steps and that this single-step liberalization must include a concurrent compensatory payment of cash to cover the increased cost of fuels, especially for consumers of kerosene and liquefied petroleum gas (LPG) and residential consumers of electricity. Since such a compensatory scheme likely could not be put into place until July 2004, in November the Energy Price Reform Task force recommended separating the energy price liberalization of fuels (other than kerosene/LPG) and nonresidential energy and initiating price liberalization of them sooner, possibly as early as January 1. Fuel subsidy reform should be implemented by having the Ministry of Finance impose a fee at the refinery equal to the regional parity price of crude oil plus the appropriate delivery costs. Electricity subsidy reform would be implemented by raising electricity to rational free-market rates. It was hoped that in addition to alleviating the immense burden on the Iraqi budget, the predicted increased energy rates would dampen demand and thereby reduce the blackouts and fuel shortages that were the cause of much Iraqi dissatisfaction with the CPA.[82]

Even as it was formulating these plans, the CPA was aware that price liberalization would be a very politically sensitive subject. The average Iraqi could face a direct increase in fuel costs of 10,000 dinars per month, and subsequent increases in food prices and services could have a similar effect, with the secondary increases disproportionately impacting the poor.[83] In fact, the CPA's economic advisors had rejected an earlier set of recommendations regarding liberalization made by the

---

[81] Memo from Office of Policy Planning to Paul Bremer, "Priorities for the Next 90 Days of the Strategic Plan," October 16, 2003.

[82] See memo from Tom Foley et al. to Paul Bremer, "Recommendation for Comprehensive Subsidy Reform," October 20, 2003; memo from Tom Foley to Paul Bremer, "Recommendation for Liberalizing Energy Prices," November 16, 2003.

[83] Author unattributed, "A Proposal to Reform Iraq's Subsidies on Oil Products," September 30, 2003.

Ministry of Finance because they were "driven mostly by financial and ideological factors and may ignore important political considerations."[84] By the fall, CPA advisors were advocating a more gradual approach, consisting of incremental steps toward reducing energy subsidies and monetizing the food distribution. Recognizing the public unrest that could result from a decision to implement a policy of full price liberalization, the CPA decided in December to postpone action on liberalizing energy prices. "The economic and commercial arguments in favor of reform are decisive," Bremer told President Bush. "However, these are outweighed by the political consequences of such a major sector at a time of insecurity and political debate."[85] Consequently, except for possible actions on gas price liberalization, the priority on energy subsidies shifted to ensuring "that the Iraqi leadership understand the importance of implementing structural reforms such as fuel price liberalization."[86]

Despite the CPA's trepidation, there appears to have already been a fair amount of support for such reforms within the Iraqi ministries. In January, Minister of Oil Ibrahim Bahr al-Uloum told CPA officials he had a task force that was preparing a comprehensive recommendation on energy subsidy reform. Both al-Uloum and Finance Minister Gailani agreed that a program of petroleum product price liberalization was necessary, but they disagreed on the specifics of such a program.[87] In April, the Iraqi interim ministers met with Bremer and debated the issue of subsidies. While recognizing that subsidizing

---

[84] Memo from Tom Foley to Paul Bremer, "Weekly Update and follow-up to issues raised at our meetings last week," September 7, 2003.

[85] Ambassador Paul Bremer to POTUS, "Draft Points for Message—December 18," December 18, 2003.

[86] Memo from Office of Policy, Planning, and Analysis to Paul Bremer, "CPA Priorities in 2004," December 18, 2003.

[87] Memo from Tom Foley to Paul Bremer, "Update for Meeting with Minister of Oil," January 29, 2004; memo from J. Michael Stinson to Paul Bremer, "Petroleum Product Price Liberalization," March 11, 2004. Specifically, al-Uloum favored increasing prices incrementally to reach parity with Kuwait after winter 2005, whereas Gailani favored a single large increase as soon as possible. See also memo from J. Michael Stinson to Paul Bremer, "Meeting with Minister of Oil—April 15," April 14, 2004.

basic services was a "sensitive issue" for many Iraqis, Minister of Planning and Development Mahdi al-Hafidh stated that it was necessary to reduce the burden of resource transfer programs. Agriculture Minister Abdul Amir al-Abood said he understood the value of supporting basic services, but he thought subsidizing the price of gasoline was unnecessary. Finally, Minister of Electricity Aiham al-Sammarae strongly opposed subsidies, noting that Iraq's electrical grid required $10 billion in expenditures against $10 million collected in revenues.[88]

The level of resistance these reforms engendered may not have been as severe as envisioned. One CPA economist noted that although virtually every stabilization program involves increases of controlled prices, "the vast majority of these increases are usually *not* accompanied by violence."[89] In a January 2004 public opinion survey, 30.4 percent of the Iraqis sampled said they were unwilling to pay more to avoid long gasoline lines—but 46.9 percent responded that they did not own cars.[90] This suggests that the bulk of subsidies were going to the rich, who owned more cars and had larger homes consuming more electricity than poor Iraqis.[91]

Food

Under the UN-administered Oil-for-Food Program, every Iraqi received a monthly food basket of staples. The CPA also considered monetizing this subsidy—that is, giving each Iraqi the cash equivalent of that same basket. CPA experts argued that this would provide a source of demand for Iraq's private sector, including, in particular, Iraqi farmers (most of the distributed food was bought abroad); expand the

---

[88] The only dissent came from the Minister for Displacement and Migration, Muhammed Jasem Khdeir, who commented that the government would pay a high political price for eliminating subsidies and recommended the current subsidy level be maintained to provide stability. See "Minutes of Interim Ministers with CPA Administrator," April 20, 2004.

[89] Memo from Keith Crane to Bill Block, Tom Foley, Tony McDonald, and George Wolfe, "Social and Political Consequences of Increasing Prices," October 14, 2003.

[90] Coalition Provisional Authority, "Iraqi Opinion in Selected Cities, 1–7 January 2004," January 22, 2004.

[91] Foote et al., p. 67.

consumption choices for Iraqis; and establish the precedent that Iraq's oil wealth belongs to its people rather than to the government.[92] In August, Bremer was firmly committed to this reform, handwriting on a briefing paper about the 2004 budget: "Restate that we *will* monetize" over a bullet point about the food baskets that suggested the current form of distribution may be maintained.[93] By November, CPA staffers had been tasked to design a program for distributing cash instead of coupons for the public distribution system for food. The cash system would replace the coupon system by July 1, by which time the Ministry of Trade enterprises that distributed the food baskets would need to be commercialized. Eventually, a Smart Card system would replace the cash distribution system. Finally, trials of the new system would be conducted in concert with the Iraqi Central Statistics Office as soon as possible, with the transition being made gradually to the extent possible as neighborhoods and cities were added to the program.[94]

However, as with the elimination of fuel price subsidies, the plan to monetize the Iraqi food basket program ran into insurmountable political and logistical realities. Converting every Iraqi family to a cash payment would require the opening of over three million bank accounts. Four hundred million dollars in additional cash would have to be distributed around the country each month, and three million people or more would have to modify their behavior in a significant way.[95] A poll conducted in late November and early December showed that almost 90 percent of Iraqis preferred the continuation of food baskets to a cash-based system.[96] The CPA's trade advisors warned that because food is so central to Iraqi society and cultural life and because the public food distribution system is one of the only public services

---

[92] Foote et al., p. 67.

[93] Author unattributed, "2004 Budget: Priorities and Processes," August 3, 2003.

[94] Memo from Keith Crane to Paul Bremer, "Decisions for Food Monetization Meeting," November 9, 2003.

[95] Memo from Tom Foley to Paul Bremer, "Transition of the food delivery system after monetization of the Food Basket," October 28, 2003.

[96] Memo from Don Hamilton to Paul Bremer, "Poll: 9 of 10 Prefer Food Basket to Cash," December 29, 2003.

still functioning throughout the country, any changes to the status quo would be politically charged.[97] CPA economic advisors warned that procedures for monetizing the food basket would have to be a "zero-defect system," and that, "If even small numbers of people don't receive their allotments, the transition to the new system is likely to be politically explosive."[98] And CPA's senior advisor for labor and social affairs cautioned Bremer that the plan to monetize the monthly rations had "the possibility of huge humanitarian problems."[99]

In September, the CPA's trade advisors recommended moving to targeted cash payments only "when the economic situation permits."[100] Although Minister of Trade Ali Allawi agreed the food ration should be phased out over time, in November his ministry had the 2004 ration cards printed and ready for distribution, indicating his concern that the benchmarks for monetization would not be in place before the proposed July 1 deadline.[101]

In the end, as with energy subsidies, Bremer backed off plans for monetizing the food basket. Bremer telephoned Olin Wethington in mid-December and told him to inform the responsible Iraqi ministers that the CPA would not back either measure. Wethington was shocked at this reversal, and delayed the notification until he could meet face-to-face with Bremer. During that meeting, which occurred a few days later, Bremer reaffirmed his instruction, noting that it had White House backing. Wethington immediately called the Finance Minister, Ali Allawi, whom he caught in his car on the way to a Governing Council meeting where he planned to argue for approval of the

---

[97] Memo from Judith Appleton to Paul Bremer, "CPA Trade Approach," September 8, 2003.

[98] Memo from Keith Crane to George Wolfe and Clayton McManaway, "Implementing the Monetization of the PDS for Food," October 20, 2003.

[99] Quoted in Chandrasakeran, *Imperial Life in the Emerald City*, p. 227.

[100] Memo from Appleton to Bremer, September 8, 2003.

[101] Memo from Susan Hamrock to Paul Bremer, "Continuation of Food Subsidies—Trade Minister Allawi's Comments in 'Al Bayyinah,'" November 23, 2003.

decision. Allawi was similarly surprised and unhappy with Bremer's decision.[102]

Bremer later explained that his pullback was based on weak support from the Iraqi leadership and a concern over public reaction at a time when the CPA's main emphasis had shifted from further reform to preparing the transfer of sovereignty. "After several weeks in late November, early December, failing to find adequate political support for proceeding, I told Condi Rice that we just couldn't go ahead," Bremer noted. "She strongly agreed."[103]

By January, the proposal to conduct food monetization trials had been dropped from the CPA Strategic Plan. And by March, the most the CPA could hope for was to get the Iraqi Governing Council to announce trials by late 2004.[104]

State-Owned Enterprises

The legacy of Iraq's statist economy included 189 Iraqi publicly owned companies, which the CPA inherited. These state-owned enterprises operated at low capacity, if at all. They were badly overstaffed and had antiquated plants and equipment. Many had been closed for a long time. Preliminary assessments established damages to their facilities from the post-liberation looting at more than $400 million.[105] Even before the war, the SOEs had survived only because they received $500 million in annual subsidies from the Iraqi government.

In the near term, the CPA reduced these direct subsidies to $245–295 million.[106] More controversially, the CPA placed a moratorium on all debts and receivables that existed between state entities before June 1, 2003, which totaled 1.2 trillion dinars (about $800 million), about

---

[102] Author interview with Olin Wethington, December 18, 2008.

[103] Author interview with L. Paul Bremer, December 28, 2008.

[104] Memo from Office of Policy, Planning, and Analysis (Andrew Rathmell) to the Executive Board, "Strategic Plan Update for January 2004," January 16, 2004; memo from Olin Wethington to Paul Bremer, "Thoughts on Large Economic Priorities to End June 2004," March 28, 2004.

[105] Chandrasekaran, *Imperial Life in the Emerald City*, p. 122.

[106] "Budget Briefing," briefing slides, June 30, 2003.

10 percent of the government's annual budget. Eventually all such debts would be cancelled, with the cash requirements of the ministries and SOEs to be covered through the budget process. This step was taken for several reasons. First, given the high levels of corruption within Iraqi society, there was a legitimate fear that SOE managers would misappropriate any funds to which they had access. Second, there was concern that the Iraqi banking system did not have enough cash for withdrawals if the accounts were unfrozen, and that even if there were enough cash, unfettered access to the SOE accounts would risk serious inflation. Finally, McPherson and Oliver cited the "doctrine of impossibility" with regard to Iraqi intragovernmental debt, noting that the records of who-owed-what-to-whom were missing, destroyed, or so convoluted that an honest accounting of these debts could never be made.[107]

There was significant dissent about this policy even within the CPA. Advisors to the Ministry of Finance argued against the wholesale cancellation of intragovernmental debt because it would be most beneficial to the SOEs that were most in arrears in payment for services. It would potentially undermine bank capitalization and substantially weaken the future ability to consolidate government cash balances because it would encourage enterprises to hide those balances. They also felt it would substantially weaken the future ability to issue bills. Most important, the advisors feared that debt cancellation would risk denying potentially viable firms access to the working capital they needed to restart operations. (The SOEs' working capital was composed entirely of cash deposits in the state-owned banks or accounts receivable from other SOEs and inventories). Instead, these advisors recommended announcing that accounts would remain frozen until January 1 and that the SOEs' positive balances in the frozen accounts would be written down by the amount of budgetary support the SOEs

---

[107] Memo from Peter McPherson to Paul Bremer, "Working Capital of State-Owned Enterprises," June 19, 2003; author unattributed, "Two Issues Regarding SOE Bank Balances," July 1, 2003; and memo from Peter McPherson to Paul Bremer, "Treasury Bills on Ministry of Finance Remain Payable," September 19, 2003.

planned to receive.[108] In sharing some of these concerns, Bremer asked McPherson, "Are we confident of the impact of such a step, particularly on enterprises we may want to save?"[109] Yet despite these objections, the decision to cancel all intragovernmental debt was eventually carried out.

In the long term, the CPA planned to "corporatize and privatize" the SOEs. Initially, the CPA aimed for small-scale privatization or leasing of competitive SOEs from August to October. Assessments of larger SOEs to determine their suitability for privatization were not foreseen until November 2003 to January 2004. In fact, OSPD was able to conduct assessments of the 153 SOEs not held in the oil, electricity, and finance ministries by October. Its analysis concluded the following:

- Twenty-six were not viable stand-alone businesses and should be retained by the government and reclassified as government agencies.
- Nine were not viable stand-alone businesses and should be consolidated into other SOEs.
- Four were viable stand-alone businesses but should not be privatized.
- Twenty-five were not viable stand-alone businesses and should be closed.
- Eighty-five were good candidates for privatization.

The OPSD's privatization plan called for the formation of an agency for implementing the privatization process, an employee transition plan that included vocational retraining and job placement programs for employees who did not remain with their current employer after privatization, and an education and communications plan to better inform all constituencies about the benefits and effects of privatization, as well as the details and effects of the privatization plan.[110]

---

[108] "Two Issues Regarding SOE Bank Balances," July 1, 2003.

[109] Memo from McPherson to Bremer, June 19, 2003.

[110] Memo from Tom Foley to Paul Bremer, "Privatization Memo," October 4, 2003.

The SOE transition plan called for 103,000 employees to be fired or retired in 2004 (65,000 through forced retirements, 38,000 through firings); 7,500 each in 2005 and 2006. In addition to the savings through reduced subsidies, the annual cost to the Iraqi government of paying SOE employees would be reduced by $88 million beginning in 2005, with combined savings over the three years from 2005 to 2007 of approximately $300 million.[111]

The first step of the privatization plan was to obtain the Governing Council's approval of the proposed privatization law by November 30. However, in contrast to the Governing Council's tentative support of the FDI law, the Iraqis resisted the privatization of the SOEs from the start. When the FDI law was being debated in September, the Economic and Finance Committee scrapped the provisions that allowed companies to buy shares of SOEs because they wanted privatization to be dealt with separately.[112] On October 29, Tom Foley, Director of the OPSD, briefed the Committee on the CPA's privatization plan. Ahmad Chalabi made a strong intervention against moving forward on privatization, arguing that speed in this area would create social and political problems and would increase accusations that the CPA and Governing Council was stripping Iraq of its assets. Chalabi acknowledged that this suspicion was unjustified, but since it was widely held there was no possibility of the Governing Council agreeing to the policy of privatization.[113] As an Iraqi advisor to the CPA observed, Chalabi was "very conscious of the public perception of the [Governing Council] and will not think that proceeding with privatization will help the [Governing Council]'s reputation and credibility."[114] Governing Council member Hamid Majid Moussa supported Chalabi, saying it was a time to heal social wounds, not exacerbate them, and that the

---

[111] Memo from Tom Foley to Paul Bremer, "SOE Transition Plan Costs," December 17, 2003.

[112] Memo from Irfan Siddiq to Paul Bremer, "Economic Policy Discussion at the Governing Council," September 14, 2003.

[113] Email from Irfan Siddiq to Jessica LeCroy, "FW: GC Finance and Economics Committee Meeting, 29/10/03," October 30, 2003.

[114] Memo from Irfan Siddiq to Paul Bremer, "Meeting with Chalabi," November 2, 2003.

more appropriate time for privatization would be under a stable, elected government. Other Iraqis (including Central Bank Governor Shabibi) acknowledged the economic logic underpinning the privatization plan but argued that time was needed to discuss the issue and to establish a real financial market in order to produce a sound valuation of public assets.[115] Subsequently, the Governing Council issued Decision Number 90, which was intended "to stop any plans or activities to privatize SOEs in order to conduct a measure study of the condition of these SOEs and to evaluate the social, economic and political obstacles linked to privatization."[116]

Although the CPA still continued to develop plans for privatization, without Governing Council cover they were clearly nonstarters amidst the rising violence in Iraq. In December, Bremer informed President Bush that because of the priority given to creating jobs, "we will not press forward on reform of the state owned enterprises."[117] The January update to the CPA's strategic plan emphasized that SOE disposal was now at the behest of the Iraqi ministries.[118] And in May, when the OPSD proposed the elements of a new SOE strategy, Bremer admitted, "I am very skeptical that much of this can be done at this point."[119] Thus, as with the efforts to liberalize fuel prices and monetize the public distribution system's food basket, the CPA's plan to privatize the Iraqi SOEs was undone by the reality of Iraqi politics.

Expanding Employment

In his book *Imperial Life in the Emerald City*, *Washington Post* correspondent Rajiv Chandrasekaran argues that Iraqi unemployment stemmed

---

[115] Email from Siddiq to LeCroy, October 30, 2003.

[116] Memo from Col. Richard Reynolds to Paul Bremer, "Financial impact from recent GC decisions," November 17, 2003.

[117] Bremer to President Bush, December 18, 2003.

[118] Memo from OPPA (Rathmell) to the Executive Board, January 16, 2004.

[119] Handwritten note on memo from Michael Fleischer to Paul Bremer, "SOE Action Plan," May 2, 2004.

from the CPA's overzealous efforts to promote the private sector in Iraq at the expense of government-funded job creation. He writes that the USAID and Treasury Department's key economic document "outlined no program to create jobs. The words *tax* and *privatize* were mentioned dozens more times than the word *employment*."[120] Chandrasekaran's account seriously underestimates the CPA's efforts to create jobs in Iraq. At his June 2, 2003, press conference, Bremer acknowledged that unemployment "is an enormous problem."[121] Two days later he briefed President Bush on the economy, saying, "Our most urgent problem is unemployment."[122] On June 7, Bremer announced the Iraq Construction Initiative, a $100 million fund to improve the infrastructure using Iraqi construction companies and providing employment to "tens of thousands" of workers immediately.[123] This initiative reflected McPherson's belief that "We can't rely on private-sector job creation through the end of this year and perhaps for a couple of years." Instead, he argued, job creation would have to come from infrastructure improvements.[124]

Additionally, CPA officials were concerned about the effects of pervasive unemployment on the deteriorating security situation in Iraq. As early as July 2003, the Center for Strategic and International Studies (CSIS) warned that idle hands must be put to work and basic economic services provided immediately to avoid exacerbating political and security problems.[125] In October, the CPA South-Central Regional Coordinator warned: "Terrorist and criminal organizations opposed to the Coalition are actively and successfully recruiting among the unem-

---

[120]Chandrasekaran, *Imperial Life in the Emerald City*, p. 116. Emphasis in the original.

[121]"Opening Statement for Ambassador Bremer Press Conference, 2 June 2003," June 2, 2003.

[122]Bremer, *My Year in Iraq*, p. 71.

[123]Peter McPherson to Paul Bremer, "Iraq Construction Initiative—Statement," June 7, 2003.

[124]Memo from Peter McPherson to Paul Bremer, "Timelines for Iraq's Economic Recovery," July 12, 2003.

[125]Hamre et al., *Iraq's Post-Conflict Reconstruction*.

ployed both in al-Anbar and in our Shi'ite provinces."[126] The top concern of Sunni tribal leaders from Anbar who met with CENTCOM commander General John Abizaid in November was jobs, and the governorate coordinator in Anbar warned that no campaign of political engagement would overcome discontent until CPA did more to meet basic needs. "We need to create approximately 30,000 targeted jobs in the next month that would provide continuous employment for one year."[127] The December Section 2207 Report to Congress observed that "High unemployment rates in Iraq are a persistent source of insecurity and instability for the country."[128] The Interim Minister for Agriculture told Bremer that if the unemployment problem was not solved soon the enemy forces would only grow stronger in numbers.[129] And in Diyala Province, unemployment was "brought up at nearly every meeting and is, according to interlocutors, the main reason behind anti-coalition activity and instability in Diyala."[130]

In August 2003, the CPA proposed to the Governing Council a "300,000 Jobs Project" which would employ that number of workers in skilled and unskilled jobs, divided into short-term (90 days) and long-term positions. Bremer asked the Governing Council to form a subcommittee to work with CPA on the details, and the Council was "uniformly pleased and thankful" for the proposal.[131] The program was given to USAID for execution, scaled down to 150,000 jobs and eventually to 100,000. On September 20, Bremer approved the initiation of

---

[126]Email from Mike Gfoeller to Paul Bremer, "Bad Decisions on Electricity Threaten Stability in al-Anbar and Shi'ite Heartland," October 5, 2003.

[127]Memo from General John Abizaid to Ambassador Paul Bremer and LTG Ricardo Sanchez, November 10, 2003; Keith W. Mines, CPA Governance Coordinator, Al Anbar Province, "Coalition Provisional Authority Trip Report: The Cornered Tiger: Iraq's Sunnis After Saddam," December 16, 2003.

[128]"Section 2207 Report to Congress," December 29, 2003.

[129]"Minutes of the Meeting of Interim Ministers with CPA Administrator," April 20, 2004.

[130]Memo from Ed Messmer to Paul Bremer, "'Diyala Comes to Baghdad,' Scene Setter 14 May, 2004," May 10, 2004.

[131]Email from Meghan O'Sullivan Group, "August 6 Readout on 300,000 Jobs Project: Bremer to Governing Council," August 6, 2003.

the 100,000 Jobs Program, to be administered locally by the Ministry of Public Works and the local councils. Pilot projects would be located in the capital cities of 11 governorates and would be phased in over a two-month period beginning September 27, at a total cost of $14.4 million. The program identified nine southern governorates, and Anbar and Salah ad Din provinces as target areas, allocating an even share of 9,000 workers to each of the governorates.[132] (Separately, the Baghdad Employment Program began on August 16, employing approximately 70,000 people in 144 public works projects.)[133] Although the initial roll-out of the program took time to gather momentum and appropriately staff and implement the individual projects, 50,000 workers had been hired within the first two months. By January, this number had risen to 76,600, as many governorates chose to expand at slower rates in order to stretch the programs out for a more sustained effect.[134] The Iraqi reaction to the program was positive. One governorate coordinator reported in November: "We have provided temporary jobs for 7,200 workers throughout Al-Muthanna. This is a popular program; we urge continued funding until after the CPA's phase out."[135] Similarly, the CPA representative in Kirkuk recommended: "An important short-term remedial measure would be a second phase of the very successful jobs creation program."[136] And in February, 20,000 applicants descended on a newly opened job registration in Maysan province.[137]

Although the CPA was clearly aware of the connection between jobs and security, job creation was not sufficiently tied in to a compre-

---

[132] Memo from Andrew Bearpark to Paul Bremer, "Approval Request for 100,000 Jobs Pilot Program," September 20, 2003.

[133] Memo from Hank Bassford to Paul Bremer, "Program Highlights, Implementation Concerns, Baghdad Central," September 7, 2003.

[134] USAID, "Toward a Cleaner and Brighter Iraq," briefing slides, December 1, 2003; Coalition Provisional Authority, "Administrator's Weekly Report: Economics, 17–23 January, 2004," January 23, 2004.

[135] James Soriano to Paul Bremer, "Weekly GC Update—Al Muthanna," November 27, 2003.

[136] Paul Harvey to Paul Bremer, "Weekly GC Update—Kirkuk," February 13, 2004.

[137] Molly Phee to Paul Bremer, "Weekly GC Update—Maysan," February 22, 2004.

hensive counterinsurgency strategy until after the April 2004 uprisings. Rather than directing jobs to the provinces and cities most susceptible to anti-coalition violence, the 100,000 Jobs Program distributed jobs equally among the governorates. The lessons learned from the initial stages of the program concluded that "allocations to eleven initial governorates should not have been equal but based on respective population percentages, as originally envisioned."[138] Consequently, on the eve of the April uprisings, the relatively secure cities of Nineveh (14.84 percent) and Basra (12.17 percent) were to get the largest portion of jobs under the second phase of the job creation program, as opposed to the more restive cities in Anbar (5.75 percent), Salah ad Din (5.63 percent), and Baghdad (8.76 percent).[139]

After the outbreaks of violence in Fallujah and Sadr City, Bremer recognized the need to explicitly link job creation to the geography of the insurgency. On April 22, Bremer issued a Memorandum for the Record stating: "It is my intent to focus CPA reconstruction efforts in the following six cities: Baghdad, al Baqubah, Mosul, Ramadi, Tikrit, and Fallujah." Two days later, Samarra was added as a seventh city, and the program was officially named the Accelerated Iraqi Reconstruction Program (AIRP).[140] The AIRP accelerated $172 million in reconstruction funds to rapidly and visibly improve the daily lives of the citizens in those strategically vital cities. By June 11, 30 projects were under way, directly employing 5,407 Iraqis.[141] However, by this time CPA was fighting a rearguard action against a mutually reinforcing downward spiral of joblessness and anti-coalition violence.

Figures on employment and job creation throughout the CPA period are highly unreliable. Several press articles in November 2003 identified CPA as the source of a 70 percent unemployment estimate,

---

[138] USAID, "Toward a Cleaner and Brighter Iraq," briefing slides, December 1, 2003.

[139] Memo from Andrew Bearpark to Paul Bremer, "Mosul: Economic Revitalization Proposals and Program Review Board," March 28, 2004.

[140] L. Paul Bremer, Memorandum For Record, "Accelerating Iraqi Reconstruction Programs," April 22, 2004.

[141] David Nash to Paul Bremer et al., "SITREP #47—Accelerated Iraqi Reconstruction Program," June 11, 2004.

but the OPSD came up with a much lower figure, estimating that out of eight million Iraqis seeking employment, 1.67 million (with an error rate of up to 10 percent) were unemployed.[142] While the CPA told Congress that estimates of Iraq's unemployment rate ranged from 40 percent to 60 percent in December, its internal estimates placed unemployment at 20–30 percent.[143] In a poll conducted in six Iraqi cities from January 1 to January 7, 16.7 percent of Iraqis reported they were unemployed. A similar poll conducted from January 18 to January 25 showed only 12 percent unemployed.[144] In February, the Ministry of Planning and Development completed a survey that found 28 percent unemployed.[145] In June, CPA economists were still estimating unemployment to be between 20 percent and 30 percent.[146] A United Nations Development Programme study showed unemployment at the end of the CPA's lifetime to be about 10 percent.

Although there was some expansion of employment during the CPA's tenure, this was not always apparent to the average Iraqi. Excluding the Kurdish region, almost every governorate reported that joblessness was a serious and persistent problem:

- From Kirkuk in January: "Unemployment remains a major concern for all those I meet."[147]
- From Najaf in March: "High unemployment and low salaries plague Najaf."[148]

---

[142] Memo from Tom Foley to Paul Bremer, "Unemployment Data," November 14, 2003.

[143] "Section 2207 Report to Congress," December 29, 2003; "Draft Point for Message," December 14, 2003.

[144] IIACSS polls, cited in memo from Gene E. Bigler to Paul Bremer, "Your question about January opinion data on unemployment," March 29, 2004.

[145] Memo from Andrew Goledzinowski to Exec Sec, "Ministry of Planning and Development Cooperation Unemployment Survey," February 8, 2004.

[146] Memo from Olin Wethington to Paul Bremer, "Prime Minister Briefing: Key Economic Issues," June 7, 2004.

[147] Paul Harvey to Paul Bremer, "Weekly GC Update—Kirkuk," January 30, 2004.

[148] Memo from Dean Pittman to Paul Bremer, "Trip to Najaf," March 22, 2004.

- From Samarra in March: "[Sunni leaders] are not seeing the benefits of programs conducted in their governorate."[149]
- From Anbar in April: "We suspect that Iraq's economic expansion is not reaching Al Anbar. Although the World Bank estimated Anbar's urban unemployment to be 16 percent in fall 2003 . . . it feels like the numbers are much worse. Certainly there are many heads of household and many young men without meaningful work."[150]
- From Basra in June: "Concerns remain about jobs, especially for the under qualified."[151]
- From Muthanna in June: "Overall economic conditions [in Muthanna] are bleak. Long lines of unemployed youths seeking jobs are often seen outside government office buildings."[152]

More important, perhaps, the unemployment statistics masked the problem of underemployment and poverty. The same Ministry of Planning survey that showed unemployment to be 28 percent reported that the underemployment rate was 21.6 percent.[153] Similarly, a February focus group with Baghdad residents revealed: "The biggest economic concern among the group was the part-time nature of the work in which they are engaging."[154]

Promoting Long-Term Development

Although the CPA possessed the means to finance its initial efforts to stabilize the Iraqi economy, it lacked the resources necessary to recon-

---

[149] Memo from Ambassador Christopher Ross to Paul Bremer, "Your Meeting with Sunni Leaders from Samarra," March 5, 2004.

[150] Stuart Jones to Paul Bremer, "Weekly GC Update—Anbar," April 2, 2004.

[151] Memo from Henry Hogger to Paul Bremer, "Basra Visit," June 17, 2004.

[152] Memo from James Soriano to Paul Bremer, "Al Muthanna Visit," June 17, 2004.

[153] Memo from Goledzinowski to Exec Sec, February 8, 2004.

[154] Memo from Don Hamilton to Paul Bremer, "Yesterday's Focus Group," February 18, 2003.

struct Iraq's shattered infrastructure and continue to provide essential services to the Iraqi people. On July 17, Bremer's chief financial advisor informed him that without the influx of additional funds, the CPA would go broke sometime before the first quarter of 2004.[155] Bremer instructed Oliver to poll the senior advisors and compile a list of their ministries' needs. Oliver gave the senior advisors a week to submit their wish lists; when he tallied their requests, the total came to $60 billion. Oliver began paring the number down to a figure that would have a realistic chance of making it through the U.S. Congress and eventually cut the number down to $35 billion. Bremer recognized that even this reduced figure was unrealistic, halved it to $18 billion, and asked Oliver to prepare a detailed request based on that number. Oliver apportioned the $18 billion among the ministries and told them to develop a comprehensive plan of how they intended to spend the money in their sector.[156]

On August 5, Bremer informed Secretary of Defense Rumsfeld that he wanted the White House to seek an $18 billion supplemental appropriation from Congress. He cited three reasons: First, he called attention to current estimates that CPA would run out of money in January. Second, the timing of the supplemental appropriation in September would enable the administration to leverage Congress and the donor nations meeting in Madrid against one another. Finally, Bremer wanted to create a "safety net" of economic activity during a period in which he anticipated introducing significant—and painful—reforms to the structure of the Iraqi economy.[157]

The administration approved the request, which eventually totaled $20.3 billion, and on September 7 President Bush announced it in a nationally televised address on Iraq. The CPA chose not to prioritize the items within the supplemental request for fear that Congress would only provide funds for the highest priorities. CPA advisors felt that the $20 billion request would barely begin to address Iraq's needs. "We did

---

[155] Bremer, *My Year in Iraq*, p. 109.

[156] Chandrasekaran, *Imperial Life in the Emerald City*, pp. 160–161.

[157] Memo from Ambassador L. Paul Bremer to Secretary of Defense Donald Rumsfeld, "Supplemental Strategy," August 5, 2003.

not pad anything," Oliver wrote to his colleagues on August 21. "In fact, I'm terribly worried about being under our needs."[158] However, with $5.7 billion slotted for electricity projects and another $4.2 billion requested to train and equip Iraq's army and police, CPA's priorities were not difficult to discern. As the insurgency in Iraq started to take shape, Congress spent the next two months debating the measure. Finally, it approved $18.4 billion for the supplemental appropriation on November 6. That night, Bremer predicted to a reporter that the reconstruction funds would begin flowing into Iraq within weeks and that "We're going to transform this place." [159]

One of the first steps the CPA had taken in preparing the supplemental request was to set up the institutional architecture for spending the money. On August 9, Bremer approved Oliver's proposal to have a plan for conducting a bid solicitation and evaluation process in place before submitting the request to Congress.[160] Consequently, just prior to the passage of the supplemental appropriation in November, the CPA established a project management office (PMO), headed by retired Admiral David Nash, to handle contracting. However, Bremer's vision of the supplemental funding transforming Iraq quickly foundered on the shoals of bureaucratic politics, Congressional oversight, contracting bottlenecks, and rising insecurity.

USAID felt inadequately consulted in developing the plan and was concerned that too little money had been allocated for capacity-building, democracy promotion, agriculture, and economic development. "Development is not building things. It's not engineering. It's institution building," argued USAID administrator Andrew Nat-

---

[158] Email from David Oliver Group, "Generic Iraq Supplemental Questions," August 21, 2003.

[159] Quoted in Chandrasekaran, *Imperial Life in the Emerald City*, p. 162.

[160] See email from David Oliver to Peter McPherson, Tom Korologos, and Patrick Kennedy, "How we best answer the question of whether the Supplemental can be executed," August 7, 2003; memo from David Oliver to Paul Bremer, "Preparing for the Supplemental," August 9, 2003.

sios.[161] As the PMO draft contracting procedures were being circulated in Baghdad, the USAID/Iraq mission director strongly objected to the proposed mechanism of routing all the supplemental funds through the PMO, claiming, "The memo purposely excludes USAID from the supplemental funding and implementation mix."[162] Later that same day (perhaps not coincidentally), OMB made it clear that channeling all funds directly to CPA was not an option because it would provoke strong objections "from both agencies in Washington and from the Hill."[163] Consequently, OMB issued guidance in December on the management of the supplemental funds that reduced the amount PMO controlled for construction from $12 billion to $2 billion and spread the remaining work among four different agencies. One CPA advisor warned Bremer that the impact of this decision "will be devastating on our ability to string together all of the reconstruction activities into a coherent plan."[164] Similarly, Bremer warned Secretary Rumsfeld that "The constraints imposed severely harms and delays the reconstruction efforts" and "will significantly reduce our ability to effectively and efficiently implement the supplemental."[165] The change from a centralized process through PMO to a diverse, multiagency management approach put all sector construction requests for proposals on hold, caused a rework of all major contracts, and cost at least a 60-day delay in awarding contracting, according to an estimate by the CPA.[166]

---

[161] Special Inspector General for Iraq Reconstruction, *Hard Lessons: The Iraq Reconstruction Experience,* Washington, D.C.: U.S. Government Printing Office, 2009, p. 100.

[162] Email from Lewis Lucke to Al Runnels, "FW: Final Draft Info Paper on Execution of the FY04 Supplemental for Iraq Relief Reconstruction Fund," October 21, 2003.

[163] Email from Deidre Lee to Mary Tompkey, "RE: FW: Execution of FY04 IRRF Supplemental v3," October 22, 2003, citing email from Robin Cleveland to Kathleen Peroff, "Re: FW: Execution of FY04 IRRF Supplemental v3," October 21, 2003.

[164] Memo from LTG Keith Kellogg to Ambassador Bremer, "Effects of Changes to Use of the Supplemental," December 16, 2003.

[165] Memo Ambassador Paul Bremer to Secretary of Defense Donald Rumsfeld, "Effect of the Changes to the Implementation of the Supplemental," December 16, 2003.

[166] Author unattributed, "Iraq Reconstruction: Supplemental Acquisition Strategy," briefing slides, December 15, 2003.

Additionally, because CPA had been unable to fully account for the $3.7 billion obligated for contracts in fiscal year 2003, Congress attached extensive oversight requirements for spending the supplemental funds. The legislation stipulated that Congress be informed no later than seven days before any contract of $5 million or more was awarded and that all contracts be awarded on the basis of full and open competition procedures. Section 2207 of the legislation directed OMB (in consultation with the CPA) to "submit to the Committees on Appropriations not later than January 5, 2004 & prior to the initial obligation of funds appropriated by this Act . . . a report on the proposed uses for which the obligation of funds is anticipated during the 3 month period from such date, including estimates by the CPA of the costs required to complete each such project." Bremer had warned Joshua Bolten, Director of OMB, that these "excessive reports" meant that the CPA "will need more people out here, not performing people, but merely record-keepers, unnecessarily exposed to the security situation, and straining our support resources."[167] Just providing the extensive information to the "record-keepers" risked reducing the effectiveness of the already strained CPA-Baghdad staff, most of whom were already working 18–20 hours a day, seven days a week.

As a result of these bureaucratic and congressional restrictions, one senior CPA staffer noted: "We generally felt that we had three people working on an issue with CPA and Baghdad, and we had 30 people back in Washington asking questions about what the three people were doing."[168] By January 8, the PMO was already a month behind its original contracting schedule and did not expect to award the initial contracts until early March.[169] At the time of the CPA's dissolution on June 28, less than 2 percent of the $18.4 billion had actually been disbursed. It thus took nearly a year from when Bremer had

---

[167] Email from Colonel Scott Norwood to Joshua S. Bolten, "FW: Revised Priorities for Supplemental," October 21, 2003.

[168] Rodney Bent, Interview with United States Institute of Peace, September 14, 2004.

[169] Bill Smith, "PMO/Supplemental Update," briefing slides for Commanders and Coordinators Conference, January 8, 2004.

first formulated the supplemental request until the new resources came significantly on line.

Conclusion

Real gross domestic product growth in Iraq for 2004, the first year after the CPA's arrival, was 46.5 percent.[170] This is the second-highest growth figure for a comparable period in any of the 22 post-conflict environments studied in previous RAND publications. It was exceeded only in Bosnia, and it is much higher than that registered in post–World War II Germany or Japan or any of the other U.S.- or UN-led post–Cold War nation-building endeavors.[171]

The CPA achieved these results by curbing inflation, issuing a new currency, working with the Central Bank to stabilize that currency through transparent daily auctions, reducing external tariffs, reforming the banking system, expanding liquidity, and stimulating consumer demand. These results were achieved without a big influx of U.S. or other external assistance. External funding began to flow only after the end of the CPA. The CPA also promoted, supported, and helped broker negotiations in the Paris Club, which resulted in the largest dept relief package in history.

Iraqis were nevertheless disappointed and dissatisfied with the state of their economy under the CPA. This was in large measure the

---

[170] International Monetary Fund, *Iraq: Third and Fourth Reviews Under the Stand-By Arrangement, Financing Assurances Review, and Requests for Extension of the Arrangement and for Waiver of Nonobservance of a Performance Criterion*, Country Report No. 07/115 (Washington, D.C.: International Monetary Fund, March 2007), p. 13.

[171] James Dobbins, John G. McGinn, Keith Crane, Seth G. Jones, Rollie Lal, Andrew Rathmell, Rachel M. Swanger, and Anga R. Timilsina, *America's Role in Nation- Building: From Germany to Iraq* (Santa Monica, Calif.: RAND, 2003); James Dobbins, Seth G. Jones, Keith Crane, Andrew Rathmell, Brett Steele, Richard Teltschik, and Anga R. Timilsina, *The UN's Role in Nation Building: From the Congo to Iraq* (Santa Monica, Calif.: RAND, 2005); and James Dobbins, Seth G. Jones, Keith Crane, Christopher S. Chivvis, Andrew Radin, F. Stephen Larrabee, Nora Bensahel, Brooke Stearns Lawson, and Benjamin W. Goldsmith, *Europe's Role in Nation Building: From the Balkans to the Congo* (Santa Monica, Calif.: RAND, 2008).

product of wildly unrealistic Iraqi expectations. But the rhetoric of the CPA and the Bush administration, which had emphasized the material improvements in Iraq that would flow from the occupation, contributed to the disappointment. For instance, on August 29, 2003, in a broadcast to the Iraqi people, Bremer said, "About one year from now, for the first time in history, every Iraqi in every city, town and village will have as much electricity as he or she can use and he will have it 24 hours a day, every single day."[172]

The Special Inspector General for Iraq has criticized the decision to discontinue support for Iraq's state-owned enterprises, arguing that the CPA did not appreciate the dependence of other sectors of the Iraqi economy on these industries and the negative effect the decision would have on employment. This criticism is off the mark. Almost all the enterprises were operating at a loss, and they continued to do so. They consumed substantial amounts of electricity—electricity that was needed for Iraqi households and small businesses, the latter of which are the main employers in Iraq. The SOEs produced shoddy products at high costs. These products were readily available on the international market at lower cost and higher quality. Regarding the impact on employment, the CPA continued to pay the salaries of all SOE employees, despite the fact that they had no work to do. Finally, these state-owned enterprises were important sources of local patronage and would have become, had they been resuscitated under state ownership, a channel for funding Sunni insurgents in Anbar Province, and Shi'ite militias elsewhere.

The administration later reversed its policy on SOEs, and sought in 2007–2008 to resuscitate a number of them, without success. There is little reason to expect that such efforts would have been more successful four years earlier.

However, the SIGIR's criticism of the CPA's decision to cancel all SOE debts and liabilities was justified. Shortly before that measure was announced, the state-owned fertilizer plant had made a multimillion-dollar delivery to a private client. The client got to keep the fertilizer for free. The CPA's main concern was pilferage of funds on the part of cor-

---

[172]L. Paul Bremer, Broadcast to the Iraqi People, August 29, 2003.

rupt SOE management, but a temporary freeze on accounts would have served equally well to address this problem. Sorting out the accounts of bankrupt companies is a standard activity and was a necessary step in Iraq. Canceling all debts and liabilities simply made this activity more difficult, while rendering it impossible for even commercially viable SOEs, assuming any existed, to survive.[173]

The CPA failed to make significant cuts in Iraq's comprehensive and vastly counterproductive system of energy and food subsidies. Given the deteriorating security situation and the distinct possibility that those cuts would generate significant further unrest, this may have been prudent—and was, in any case, an understandable choice, albeit one with high long-term costs.

Temporary employment-generating schemes are seldom a good choice for scarce public resources in post-conflict environments. Such efforts almost invariably produce only a limited and very fleeting impact, and CPA's efforts in this regard were no exception. If the CPA is to be criticized in this area, it is for putting too much money into temporary employment rather than too little.

During the CPA period and for some time afterwards, economic growth was still seen as an independent contributor to security rather than as a dependent component of an overall strategy whose main focus was protecting the population. In these circumstances, it seems likely that much of the money spent on short-term employment made little contribution either to short-term security or long-term economic growth. Combined with a well-considered counterinsurgency approach of "take, hold, build," large-scale job schemes might have made some sense, but the U.S. military was still several years away from adopting such a strategy.

Problems in executing the $20 billion reconstruction program approved by the Congress in November 2004 emerged principally after the demise of the CPA but were due, at least in part, to choices made under its authority. In particular, the large proportion of these funds allocated to electricity generation and other forms of heavy infrastructure was unwise. By the end of the CPA, per-capita kilowatt hours gen-

---

[173] Author interview with Keith Crane, December 1, 2008.

erated in Iraq were similar to those in other regional countries at a comparable or higher level of development. Iraq was experiencing chronic blackouts primarily because of excess demand resulting from the fact that the state was not charging consumers for the power they used, not just because of deficiencies in the electric power system. This problem was inherited, but the CPA exacerbated it by deciding not to collect even those minimal charges that the former regime had levied. Investment in this sector, beyond the emergency repairs that the CPA had successfully implemented, should have been conditioned on the elimination of the subsidy and the implementation of plans to maintain and eventually amortize the costs of new power plants. Some of the money originally designated for the heavy infrastructure sector was eventually reprogrammed for capacity-building within the Iraqi government, which should have had a higher priority from the beginning.

The CPA economic policy has been criticized as being naively ideological in its devotion to deregulation and free-market principles. Some of its staff and some of its policies fit this mold. Cell phone licenses, for instance, were awarded at no cost. The CPA chose to stop collecting even the minimal charges Saddam's regime had put on electricity consumption. (This, of course, may be characterized as a socialist, rather than free-market gesture.) The CPA's effort to create a Baghdad stock exchange was premature, given the state of the Iraqi private sector. The time and energy expended to liberalize restrictions on foreign direct investment may also have been misplaced, as were efforts to privatize a large number of state-owned enterprises. On the whole, however, the CPA's economic policies were consistent with established best practices in post-conflict environments and, if anything, too cautious when it came to cutting subsidies. The overall effectiveness of the CPA's economic policies is indicated by the very substantial economic growth achieved in the first year following its creation.

In his memoirs, Ali Allawi, the Iraqi Finance Minister at the time, contrasted the "rank amateurism and swaggering arrogance of too many of the CPA recruits" with the "professionalism of the established development and reconstruction agencies," that is to say the U.S. Treasury and USAID representatives who were responsible for the CPA's successful currency and financial sector reform, banking regulation and supervision, and Iraq's international economic relations, including efforts at dept relief.[174] "The economic policy of the CPA," according to Allawi, "was a blend of wild-eyed and hopelessly unrealistic radical reforms, supposedly to introduce a liberal market economy, and a sober, methodical attempt to get the main engines of Iraq's economy gradually functioning again."[175] Somewhat inconsistently, however, Allawi also criticized the CPA for abandoning fuel price liberalization and monetization of the food distribution systems, both of which might have reasonably been thought to fall in the first category of measures.

Due to the spreading civil war, Iraq was unable to sustain a high level of growth beyond 2004. The CPA was nevertheless able to put in place a number of reforms that should serve the country well over the long haul, assuming recent improvements in security can be sustained.

Despite substantial success in the economic area, it was a mistake for the United States to have premised so much of its appeal to the Iraqi people on an improvement in their material circumstances. These premises fed already exaggerated expectations that proved impossible to fulfill. It would have been better to have confined American promises to (1) liberating the Iraqi people, (2) protecting them, and (3) allowing them to choose their own government. Had these three promises been made and kept, the substantial and, for the most part, well-considered economic reforms put in place by the CPA would have paid larger, quicker, and more enduring dividends than did the massive American aid package introduced at the end of the CPA's tenure—much of which was dissipated in security costs and ill-considered, often uncompleted projects.

This is not to argue that the United States should not have helped rebuild Iraq, as it did, but rather that it should have put security first and employed economic assistance as a contributing element in a larger strategy focused on public safety and political reform. This, however, would have required a different military as well as economic strategy.

---

[174] Allawi, *The Occupation of Iraq*, p. 267.

[175] Allawi, *The Occupation of Iraq*, pp. 123–124.

CHAPTER EIGHT

Running the CPA

A Defense Department team sent by Rumsfeld in late 2003 to assess the CPA's personnel situation characterized it as "a pick-up organization in place to design and execute the most demanding transformation in U.S. history."[1] Most of the initial State Department personnel detailed to the CPA had been on short-term assignments, and most had committed to staying only through the selection of the Governing Council, which took place on July 13, 2003. Ryan Crocker, an Arabic-speaking diplomat who had spent his professional life in the Islamic world, then departed, leaving governance matters to other capable, but less experienced, officers. A number of American and British regional experts came and went over the following year, but Bremer naturally tended to rely most heavily on the few officers who had been with him the longest and were most familiar with the immediate local situation.

"I could tell by July and August," Bremer later recalled, "that we had an operational part of the CPA and a policy part. We needed someone to head both. Too much was coming directly to me. Even the President pointed out that I had over 20 direct reports. It was overwhelming."[2] Clay McManaway recalled another deficiency: "[T]here wasn't much of a planning effort in the early stages." Andrew Rathmell, who became the head of the CPA's Office of Policy Plans and Analysis, reflected on leaving Baghdad that the "effectiveness of

---

[1] DoD Personnel Assessment Team, "Report to the Secretary of Defense," February 11, 2004, p. 3.

[2] Author interview with L. Paul Bremer, August 12, 2008.

our political-military reconstruction effort in Iraq in the first postwar year was lessened by the fact that CPA did not deploy with a comprehensive, long-term plan. Neither did it start with effective policy planning and coordination mechanisms."[3] Not until the late summer and early fall was CPA able to bring on board a larger contingent of strategic planners. The initial lack of strategic planning was compounded by shifts in personnel, since most of CPA's initial staff came on short tours of duty. By the fall of 2003, there had been almost a total changeover among Bremer's senior staff.

Staffing Shortages

Severe shortages of trained, experienced personnel, or of any staff at all, plagued the CPA throughout its existence. A number of sectors, such as police and justice, were chronically undermanned. David Brannan, an advisor in the Interior Ministry, recalled: "When I arrived in Baghdad in December 2003 to work on the police effort, there were only six functioning computers and barely two dozen personnel for the entire Iraqi Ministry of Interior. We were woefully under-resourced, and this never changed through the rest of CPA's existence."[4]

This problem plagued the CPA more generally. In some cases, staffing problems were due to bureaucratic disputes. In a memo to Secretary of State Colin Powell in April 2004, Bremer complained that the State Department's Bureau of International Narcotics and Law Enforcement Affairs had undercut CPA efforts and authority to name Thom Hacker as head of the police training academy in Jordan. "The lack of action by State INL [Bureau of International Narcotics and Law Enforcement Affairs] has not only caused delays in various mission areas," Bremer charged, "but has needlessly put personnel in harm's way. I find this most troubling as INL actions have already delayed the deployment of qualified individuals who would have already been

---

[3] Draft Memo from Andrew Rathmell to the Incoming Chief of Mission, "Subject: Achieving Long-Term US goals in Iraq: Strategic Planning in the Mission," April 28, 2004.

[4] Author interview with David Brannan, July 20, 2008.

on the ground, and secondly, they have usurped CPA's authority and, more importantly, our responsibility to manage this program."[5]

Personnel shortages may have originally resulted from an underestimation of the scope of the CPA tasks and challenges, but these shortfalls persisted even after the magnitude of CPA's mission became clear. In early January 2004, Bremer expressed concern about governance team staffing resulting from a large number of departures in the previous month.[6] A week later, CPA Baghdad reported 675 open billets out of a total of 1,448, meaning that it was manned at a 53 percent level. In March, as the transition to Iraqi sovereignty began to loom larger, one of Bremer's aides lamented: "It has been increasingly clear that there is an informal if not formal 'stop' order on new hires for CPA in anticipation of transition" to an embassy, despite CPA's ongoing need for personnel to accomplish its mission.[7] That month, the inspector general reported that only 56 percent of CPA's billets had been filled, and Bremer admitted, "We always had that problem right to the end."[8]

In early February, Richard Jones proposed documenting cases in which personnel shortages were preventing CPA from meeting its strategic objectives.[9] Although there is no evidence in the archival record that such a study was ever made, if it had been done, it likely would have noted the detrimental effects those shortages had on the regional offices and CPA's contracting efforts. As of August 17, the CPA had regional officers in only four provinces.[10] The September 2003 CPA

---

[5] Memo from L. Paul Bremer to the Secretary of State, "Subject: Department of State International Narcotics and Law Enforcement (INL) Impact on IPS Training and Equipping," April 12, 2004.

[6] Memo from Dean Pittman to Paul Bremer, "Governance Staffing Update," January 5, 2004.

[7] Memo from William J. Olson to Paul Bremer, "Staffing Muddle," March 21, 2004.

[8] L. Paul Bremer, Interview with PBS Frontline: "The Lost Year in Iraq" June 26 and August 18, 2006.

[9] "Executive Board Weekly Updates on the CPA Strategic Plan," March 16, 2004.

[10] Memo from Tom Krajeski to Paul Bremer, "Update on Regional Officers," August 17, 2003.

Personnel Status report showed that CPA North had only 7 out of 78 slots filled; Baghdad Central had 12 out of 65; South-Central 6 out of 76; and CPA South had only 5 out of 77 positions filled. In other words, no regional headquarters was manned at even 20 percent, and overall the four regional offices barely met 10 percent of their staffing requirements.[11] As one governorate coordinator noted in his weekly report to Bremer, "Personnel shortages and slow deployment of [Governance Teams] continue to limit effectiveness."[12]

The lack of personnel experienced in contracting procedures took a severe toll on CPA's reconstruction efforts. In October, the Dhi Qar governorate coordinator lamented, "We are failing to turn projects into progress on the ground" partly because "as a team we lack key contract/ project skills."[13] The same month, the regional coordinator for CPA-Baghdad informed Bremer that his office still lacked any implementation officers, and that by the end of November the office responsible for reconstruction in Iraq's capital would have "no implementation officers for sewer, water, trash and budget."[14] And in February 2004, USAID's deputy to CPA explained the effect that a shortage of contract officer personnel was having on the effort to add generation to Iraq's electricity sector:

> The CPA commitment to increase staff to nine personnel has not yet occurred; four staff are presently available. Although existing staff are working full out, without added personnel it will take another two months to place all procurements. With this delay, parts will not be available in the spring to improve capacity, reliability, and availability of existing generation units for the summer of 2004.[15]

---

[11] CPA, "Personnel Status," September 21, 2003.

[12] Memo from Mines to Bremer, December 6, 2003.

[13] John Bourne, "Dhi Qar Weekly Situation Report," October 15, 2003.

[14] Memo from Hank Bassford to Paul Bremer, "Welcome Home Update," October 30, 2003.

[15] Memo from Chris Milligan to Paul Bremer, "Weekly Infrastructure Update," February 7, 2004.

On a trip to Washington, Bremer approached the acting secretary of the Army, Les Brownlee, and confessed, "I have no contracting expertise over here at all. I am going to be in deep trouble. Can you help me?"[16]

In the summer of 2003, Bremer set up a CPA office in Washington within the Defense Department, headed by Reuben Jeffery. Prior to joining the CPA, Jeffery had been special advisor to President Bush for Lower Manhattan Development, where he coordinated ongoing federal efforts in support of the recovery and redevelopment of lower Manhattan in the aftermath of September 11, 2001. "Jerry Bremer told me I had two missions," noted Fred Smith, who arrived in July 2003 to work as the deputy director of CPA's Washington office. "The first was to get people to Iraq as quickly as possible, since CPA was so understaffed. The second was to take pressure off CPA officials in Baghdad by dealing with a range of actors in Washington, such as the White House, Pentagon, State Department, Capitol Hill, media, Justice Department, and Treasury. He wanted us to take the Washington pressure off the CPA staff in Iraq."[17] The pressure on CPA from Washington could be intense, and the amount of information asked for was often overwhelming. CPA officials joked about the "8,000 mile screwdriver" from Washington to Baghdad, which tried to micromanage all aspects of the CPA.

CPA rear, as it was sometimes called, was headquartered on the fourth floor of the D ring in the Pentagon, although the bulk of the staff worked in an office building a few miles and minutes away in Rosslyn, Virginia. Throughout much of 2003, CPA rear had a staff of about 35 people. Some worked on personnel issues and others worked with various U.S. government agencies around Washington. "Our biggest problem was getting personnel," said Smith, whose staff tirelessly worked to find and recruit candidates, ensure they were interested in coming to Iraq, get the necessary clearances, handle logistical details, and get them to Baghdad. But recruitment remained a problem. Colin Powell and Richard Armitage helped by identifying 200 State Depart-

---

[16] Packer, *The Assassins' Gate*, pp. 242–243.

[17] Author interview with Fred Smith, July 23, 2003.

ment officers who would come to Iraq. As CPA rear quickly discovered, however, "most quickly got cold feet and told us their spouses or families didn't want them going to Iraq."

The Defense Department hired a number of people in their twenties to help staff the ministerial advisory teams. After some unfavorable news articles appeared containing critical interviews with former CPA staff members, new employees were closely vetted on the basis of party affiliation by the Department's White House Liaison Office (WHLO) under Jim O'Beirne. Internships with conservative think tanks or recommendations from Republican Party leaders became a common route into the CPA for many young people.[18] "There was a clear political saliva test," noted Smith. "One of the most important qualifications for candidates was whether the individual was a good Republican."[19] Those individuals personally requested by Bremer eventually made it through this sieve, and the often young and inexperienced staff produced through patronage channels proved generally willing to work hard and eager to learn.[20] The CPA was grateful for every person it could get, and O'Beirne's efforts at least produced warm bodies at a time when other elements of the U.S. government were failing to meet even that test. That the CPA had to depend on such a source is evidence, however, of the generally inadequate level of Washington support for what was, after all, the administration's lead national security priority.

On June 21, Bremer hand-delivered a memo to Secretary of State Powell in Amman desperately requesting more personnel. "It is clear that we will need a significant number of additional personnel for the CPA . . . . There will be a particular need for Arabic speaking officers to expand our outreach beyond Baghdad and the three regional offices." Bremer estimated that CPA would require approximately 60 additional

---

[18] For a stinging indictment of O'Beirne and the WHLO's activities, see Chandrasekaran, *Imperial Life in the Emerald City*, pp. 91–94; Ariana Eunjung Cha, "In Iraq, the Job Opportunity of a Lifetime," *Washington Post*, May 23, 2004; and Rajiv Chandrasekaran, "Iraq's Barbed Realities," *Washington Post*, October 17, 2004.

[19] Author interview with Fred Smith, July 23, 2003.

[20] Author interview with Patrick Kennedy, September 5, 2008.

Foreign Service officers and stressed: "Meeting this need will require extraordinary methods, including detailing Arabic speakers from positions that do not require that skill and dispersing the entire Tunis language school to Iraq for six months."[21] Despite Bremer's clarity about the urgency of the matter, the State Department did not cable diplomatic and consular posts to request volunteers until more than two months later.[22] Although Secretary of Defense Rumsfeld wrote Powell offering to expedite all State Department personnel awaiting a clearance to get to Baghdad to staff the CPA, as of October 11, the State Department was filling only 37 percent of its billets on the CPA governance teams. Even with the influx of new arrivals expected by the end of the year, this number would only reach 52 percent.[23] A CPA report in February 2004 stated, "Other U.S. Government departments are reluctant to send the requested number of people or their best people because they see CPA as a DoD project," and noted that DoD was also slow to deploy its best people to Baghdad.[24]

Journalist and author George Packer attributed the State Department's failure to send its first team in sufficient numbers to Iraq to resentment at the leading role given DoD in running the civilian aspects of the occupation. Packer writes that "during the life of the CPA, the State Department didn't send all its best people to Iraq, even after the Pentagon's influence waned and Bremer began to use his back channel to Powell more and more." Packer goes on to quote an anonymous State official's confession:

> We didn't do our best job to get things uncocked or to help. I watched [Near Eastern Affairs], for example, essentially say, 'Okay, you don't want us—[F] you.' And then from there on out

---

[21] "SECSTATE Issues," June 21, 2003.

[22] Cable from Secretary of State Colin Powell to All Diplomatic and Consular Posts, August 27, 2003.

[23] Memo from Scott Carpenter to Paul Bremer, "State Department Staffing on Governance Teams," October 11, 2003.

[24] CPA Baghdad, "Personnel Assessment Team Report to the Secretary of Defense," February 11, 2004.

it was, 'Let's see what impediments we can put in their way. Let's see how long we can be in delivering this particular commodity or individual or amount of expertise. Let's see how long we can stiff 'em.'[25]

These allegations are certainly plausible, given human nature and the cavalier fashion in which Rumsfeld had rejected the initial slate of State nominees for ORHA in early 2003. Yet Bremer, McManaway, and Kennedy all recall State as one of the most responsive agencies in meeting the CPA's staffing needs.

Rumsfeld occasionally became directly involved in trying to encourage people to work for the CPA. In a memo to Douglas Feith, Rumsfeld said he had recommended a particular individual for consideration to go to Iraq. This person had banking experience, Rumsfeld noted, and possibly spoke Arabic. Rumsfeld was later informed that the individual's services were not needed by the CPA. "I would be curious to have someone check into that and explain to me what in the world is going on," Rumsfeld then wrote, obviously perturbed. "I thought we needed people out there."[26] Recalling the incident, Fred Smith, then the deputy head of the CPA's Washington office, said that the individual in question was about 80 years old, did not want to go to Baghdad, and had connections with some shady banking dealings in the Middle East. "This was the kind of issue that drove us nuts," Smith commented. "Rumsfeld would ask some stupid question about an individual or minor issue and we'd spend three or four hours chasing it down."[27]

Judge Daniel Rubini, CPA's senior advisor to the Iraqi Ministry of Justice, wrote in a scathing memo that all of the Ministry of Justice "objectives are directed towards instituting the rule of law in Iraq. The common denominator of these objectives is personnel support that has not heretofore been forthcoming." He continued that "any successes

---

[25] Packer, *The Assassins' Gate*, p. 396.

[26] Memo from Donald Rumsfeld to Doug Feith, CC: Reuben Jeffery and Larry Di Rita, "Subject: Kevin Woelflein," August 21, 2003.

[27] Author interview with Fred Smith, December 14, 2008.

to date in that regard we owe to the Herculean efforts of a very few" and that the office "has never had more than six non-military lawyers at a given time."[28] As Rubini explained in another email, he had "no announced replacements" for those in his office imminently departing. "For all existing vacancies, I have requested by name about 20 personnel . . . beginning November. I have info as to one arrival in mid January but I have no information through the system as to status of any other of my requests. Of course this continues to impact on mission accomplishment."[29] Other sectors were similarly impacted. "Getting personnel to Iraq was one of our most significant problems," argued Fred Smith.[30]

The CPA's inability to specify how many people it actually had at any moment detracted from its ability to ask for more. Senior CPA officials repeatedly acknowledged they did not have good estimates of how many CPA staff were in Iraq at any one time. As Fred Smith admitted, "We never got a good grip of how many people were in CPA. Sometimes people simply showed up in Baghdad. And sometimes people just left." In Smith's view, this inability to keep tabs on its own staffing represented one of the CPA's biggest shortcomings. "When trying to staff CPA from Washington," he recalled, "I never knew who was already in Baghdad performing the functions they were asking for, how long people would be there and what were the exact requirements. It was not helpful to be told 'We need 15 budget analysts.' What level, to do what tasks? But the people in Baghdad didn't have time to provide this kind of information."[31]

Too many staff also remained bottled up inside the Green Zone. "A lot of CPA officials never really got out of the palace," noted Ministry of Interior advisor Matt Sherman. "A lot of people felt more com-

---

[28] Daniel L. Rubini, "Coalition Provisional Authority Ministry of Justice: Prioritized Objectives (Excluding Prisons), Amman Justice Sector Conference," January 2004.

[29] Email from Daniel Rubini to Scott Norwood, Matthew Waxman, Carl Tierney, Michael Dittoe, Homer Cox, Ralph Sabatino, Lance Borman, and Bruce Fein, "Subject Revised Justice Sector Presence," December 30, 2003.

[30] Author interview with Fred Smith, July 23, 2008.

[31] Author interview with Fred Smith, July 23, 2008.

fortable interacting with other Americans or British than with Iraqis. The result was that CPA officials didn't coordinate enough with Iraqis, and often spent too much time writing plans rather than implementing them in the field."[32]

In addition to shortages in personnel, rapid turnover was a problem. It was difficult to get people to stay for any length of time, initially because the CPA's mission was thought to be of brief duration and later because of the danger, discomfort, and lack of career incentives. "In 90 days," remarked Bremer's press spokesman, Dan Senor, "people could barely figure out where the mess hall was. It was pointless to stay for such short tours."[33] Yet many did, and the CPA had, perforce, to depend heavily on such short-term help.

Remarkably, the U.S. military seems to have established a similarly poor record in staffing its highest headquarters in Iraq. In July, Sanchez reported to Central Command that "The overall fill rate for CJTF-7 is 37%" and "only one of thirty critical requirements has been filled." More than a year later, Rumsfeld acknowledged, in a meeting with Sanchez, that his headquarters had been staffed at below 50 percent of its intended size for most of its existence. "How could this happen, General? Why in hell didn't you tell someone about it?" To which Sanchez replied, "I did, Mr. Secretary. Every senior leader in the Pentagon knew the status of CJTF-7."[34]

The CPA may have contributed to its staffing woes by failing to accurately account for what it had and what it needed, but the fact that CJTF-7 was encountering the same difficulties suggest that the main problems were not in Baghdad but in Washington, where the Defense Department, the White House, and the administration as a whole were not making the extraordinary efforts needed to meet the unforeseen demands of their commitment in Iraq.

---

[32] Author interview with Matt Sherman, July 8, 2007.

[33] Author interview with Dan Senor, October 31, 2008.

[34] Sanchez, *Wiser in Battle*, pp. 209, 419.

Difficulties in Coordination

Coordination within the CPA imposed added difficulties. Individuals came from many agencies and walks of life. Some lacked relevant experience. Many stayed relatively briefly. Some senior staff also tended to travel frequently. They were, after all, volunteers who often still had responsibilities elsewhere. Marek Belka, for instance, had been deputy prime minister of Poland before coming to Baghdad, and became prime minister after leaving his post at the CPA. Other senior CPA staff were on leaves of absence from their day jobs and could not entirely ignore other obligations. It often took three to four days to get to Baghdad International Airport, catch a flight to Kuwait, and then get to one's final destination, so the briefest trip to Europe or the United States would soak up a week in travel time.

Ambassador Bremer was firmly in charge. However, a substantial amount of his time was spent interacting with Iraqis and giving speeches and press conferences to communicate CPA goals, decisions, and thinking to Iraqis, the citizens of coalition partners, and to the international community at large. He also spent a substantial amount of time communicating with his superiors, Secretary of Defense Rumsfeld and President Bush. Bremer held short (ten-minute) daily staff meetings and longer strategy meetings every Friday. In addition, his day was filled with meetings with senior administrators at CPA where functional and policy decisions were made. Individuals in charge of the "core foundations" and senior advisors who participated in these meetings did not always follow through with the directives they received, however. This failure was partly due to the organizational structure. Responsibility for the "nag function" was not clearly defined. While Bremer did not have time to write down decisions and ensure follow up, others in the organization lacked the authority to enforce discipline. McManaway partially filled this role, but he spent much of his time filling in for Bremer or troubleshooting special projects. The follow-up function was not built into the organization, at least not until after Bremer's November reorganization.[35]

---

[35] Author interview with Keith Crane, September 29, 2008.

Even within sectors, coordination could be difficult. "Security efforts were pretty stove-piped," acknowledged Walt Slocombe, CPA's senior advisor for defense and security affairs.[36] David Gompert, Slocombe's replacement, noted: "Based on the organizational structure, Walt had little influence over what happened with the police, justice, and intelligence efforts. There was no integration across the security sector."[37]

Each CPA office tended to have its informal lines back to Washington. Economics by and large reported to the Treasury Department, governance to the State Department, security to the NSC and Defense Department, and essential services interacted with the U.S. Army Corps of Engineers. Multiple reporting channels resulted in substantial duplication of reporting and confusion, further complicated by the fact that much of this back-and-forth with Washington was taking place informally, by cell phone or Yahoo accounts, not through record traffic that would have had a broader lateral distribution.

People worked very hard in Baghdad, and one cannot fault their diligence and dedication. Most, however, were not part of a career structure that could reward or punish them for good or poor performance. Their home offices might appreciate what they were doing or regret their absence, but for most of them it was a temporary assignment that would not influence their long-term career prospects one way or another. As a result, people sometimes went off on tangents of their own and gave less attention to direction and deadlines from the top than they might have in a more structured environment.[38]

Funding Constraints

If Washington was slow in providing the personnel, it was also remiss in providing the necessary funds for reconstruction in a timely manner. As the CPA representative in Najaf noted in January 2004, "Our big-

---

[36] Author interview with Walter Slocombe, May 5, 2008.

[37] Author interview with David Gompert, February 29, 2008.

[38] Author interview with Keith Crane, September 29, 2008

gest single source of influence is the ability to spend money quickly."[39] However, more often than not, this money was late in coming. On May 10, John Sawers cabled back to 10 Downing Street that "Money needs to be released by Washington. The clock is ticking."[40] Yet in July, former Deputy Secretary of Defense Hamre reported, on his return from an extended mission of inquiry to Iraq, that "CPA is badly handicapped by a 'business as usual' approach to the mechanics of government such as getting permission to spend money or enter into contracts," and warned that "Business as usual is not an option for operations in Iraq."[41]

In the early days of the CPA, when the primary source of funds was from vested Iraqi accounts, the main impediment to effective spending was the Office of Management and Budget. In CPA's first month, when immediate results were most important, it took OMB an average of ten days to approve CPA's funding requests.[42] Hamre told Secretary Rumsfeld and Bremer, "I was astounded to hear the constraints your lower level folks live with to get money and contracts."[43] And in his memoirs, Bremer recalled: "Since arriving in Iraq, I'd often run afoul of the bureaucrats in Washington who controlled our purse strings. It was a sad reality that, while the president had ordered me to act with decisive speed, I was often hamstrung by organizations like the OMB and the State Department's Agency for International Development."[44] Problems with OMB and its associate director, Robin

---

[39] Rick Olson to Paul Bremer, "Weekly GC Update—Najaf," January 19, 2004.

[40] Quoted in Gordon and Trainor, *Cobra II*, p. 472.

[41] Hamre et al., *Iraq's Post-Conflict Reconstruction*, pp. iii, 8.

[42] Memo from Dov Zakheim to Secretary of Defense, "Process and Average Time for the Office of Management and Budget (OMB) to Approve Funding Requests from the Office of the Coalition Provisional Authority," June 16, 2003; see also memorandum from John Hamre to the Secretary of Defense and the Administrator, Coalition Provisional Authority, "Subject: Preliminary Observations Based on My Recent Visit to Baghdad," July 2, 2003.

[43] Memo from John Hamre to the Secretary of Defense and the Administrator, July 2, 2003.

[44] Bremer, *My Year in Iraq*, p. 113.

Cleveland, would continue to slow the CPA's activities throughout its tenure.

As CPA's mission became more reliant on appropriated funds, Congress also began to raise obstacles to timely CPA action. In August, Bremer's legislative aide urged him to complain to Representative Jim Kolbe, Chair of the House Appropriations Committee (HAC) Subcommittee on Foreign Operations that committee staffers are "forever putting holds" on the "Congressional Notifications" CPA was required to submit on various funding requests, thereby delaying efforts to initiate various projects.[45] CPA agreed with OMB chief Joshua Bolten's assessment regarding the "need to eliminate the excessive reports the Congress is planning" to incorporate into the supplemental appropriation, noting that the increase in personnel with no functional role would drain CPA's already strained resources.[46] One senior administration official told a journalist that CPA's contracting officers "were scared to death to let that money go out because they already saw what was happening with some of the Iraqi money" as accusations of waste and corruption were being hurled at the CPA "and they were already being visited by congressional delegations."[47] Bolten also complained in a letter to the chairman of the Senate Appropriations Committee that "Both the House and Senate versions of the Supplemental contain numerous provisions that are not related to ongoing military operations in Iraq, Afghanistan, and elsewhere or relief and reconstruction activities."[48]

Appropriations decisions were often taken over advice based on CPA's experience on the ground in Iraq. For example, the House Appropriations Committee eliminated $90 million that was requested under

---

[45] Memo from Tom Korologos to Paul Bremer, "Kolbe, Hutchinson Meeting," August 16, 2003.

[46] Email from Colonel Scott Norwood to Joshua Bolten, "FW: Revised Priorities for Supplemental," October 21, 2003.

[47] Quoted in Packer, *The Assassins' Gate*, p. 243.

[48] Letter from Joshua Bolten to Senator Ted Stevens, October 21, 2003.

the heading of "Local Governance and Municipalities."[49] The money that was appropriated was often channeled through agencies with different interests or priorities from those of the CPA. The amount for construction that the CPA project management office controlled was reduced from $12 billion to $2 billion. Four different agencies would separately contract and manage the remaining programs, creating high overhead ratios for reconstruction projects, and hence less actual work completed. This also allowed for disparate agencies to determine spending priorities independent of CPA's overall strategy, which left CPA senior advisors feeling "burned."[50] The slow rate of spending that resulted from this bureaucracy and oversight meant that by the end of June 2004, when the CPA closed its doors, only 2 percent of the $18.6 billion supplemental appropriation had been spent.[51]

Inadequate Outreach

The CPA sought to establish a presence in each province, or governorate. The CPA's chronic staffing problems hit those governorate teams particularly hard. These offices, which were designed to have at least seven to nine people, sometimes consisted of one or two deployed officials in the early months of the occupation—if they were present at all. Bremer repeatedly stressed to Rumsfeld that the CPA had significant staff shortfalls and asked that his requests be treated with a sense of urgency and high priority. Despite these entreaties, it took the CPA six months to get officials into each of Iraq's 18 provinces; once there, the small staffs were often overwhelmed by the scale of their responsibilities. The teams engaged in weekly political reporting but interactions between the CPA governance team and the provincial offices were lim-

---

[49] Memo from L. Paul Bremer to Tom Korologos, "Iraq Supplemental Request," October 13, 2003.

[50] Memo from L. Paul Bremer to Donald Rumsfeld, "Effect of the Changes to the Implementation of the Supplemental," December 16, 2003. See also John Agresto to Paul Bremer Memo, "Problem with Funding of Ministry of Higher Education," December 3, 2003.

[51] Packer, *The Assassins' Gate*, p. 242.

ited. As a consequence, progress (or the lack thereof) at the provincial and local level depended largely on the initiative and improvisation of individual governorate coordinators and military commanders.

In early December, the CPA representative in Anbar Province complained that the "trend continues to be capturing staff/resources in central and regional offices before staffing governorates."[52] Although such frustrations regarding imbalances between the center and periphery were justified, they obscure the larger reality that virtually every element of the CPA was chronically understaffed. This problem was apparent as early as May 2003, when the regional coordinator for CPA-South, Danish Ambassador Ole Olsen, wrote to raise concerns about the staffing and equipping, warning that without significant resource increases over the next two to three weeks, "Our window for influencing the course of events in the South will be gone."[53] When Bremer returned to Washington in July 2003, he said that his number-one priority was to send more people. Yet a concern raised consistently at the August 15 regional coordination meeting was how poorly staffed the CPA regional headquarters remained.[54] As of September 21, only 558 of 1,258 slots (44 percent) for all of CPA had been filled, with governance headquarters also filling fewer than half of its 125 allotted positions. Consequently, CPA's Directorate of Operations warned Bremer in late September that "Lack of adequate staffing continues to hamper operations at the Central, Regional, and Governorate level."[55]

In its July report on Iraq, the CSIS team headed by John Hamre had recommended, "Decentralization is essential" because "the job facing occupation and Iraqi authorities is too big to be handled exclusively by the central occupying authority and national Iraqi Govern-

---

[52] Keith Mines to Paul Bremer, "Weekly GC Update—Anbar," December 6, 2003.

[53] Memo from Ambassador Ole Olsen to Ambassador Paul Bremer, "Performance Capability of CPA South," May 30, 2003.

[54] Chandrasekaran, *Imperial Life in the Emerald City*, p. 93; memo from Chappell to Bremer, August 16.

[55] Coalition Provisional Authority, "Personnel Status as of September 21, 2003," briefing slides, September 21, 2003; memo from Bearpark to Bremer, September 29, 2003.

ing Council."[56] However, throughout CPA's existence, regional officials complained about the overcentralization of decisionmaking on governance. In late June, a CPA official visiting from Baghdad was told by personnel in CPA–South Central, "The significance of the role of the local force commander/Government Support Team commander as the 'governor' does not seem . . . to be well-appreciated by the CPA staff in Baghdad. It appears . . . that the CPA staff routinely undertakes major policy decisions without benefit of wisdom from 'on-the-ground' local governors."[57] Similarly, the CJTF-7 Director of Operations wrote to Bremer stressing the need to tighten up coordination between civilian and military activity and between the center and the regions.[58] At the November regional coordination meeting, a near revolt broke out when it was learned that the CPA strategic plan had already been briefed to the "highest levels in Washington" without any input from the generals or regional coordinators.[59] In his read-ahead for the December Coordinators and Commanders Conference, Bremer was warned, "We continue to hear concerns that the national political process is effectively divorced from reality on the ground in the Governorates."[60] And in January, the new CPA representative in Kirkuk observed: "As a newcomer to the role of Governorate Coordinator, I am struck by a lack of information exchange between CPA in Baghdad and the Governorates." The official recounted that he had seen Bremer's briefing documents for his January 2 visit to Kirkuk only by the lucky coincidence of sitting next to him on the flight from Baghdad. He concluded, "It seems from here that the system for getting to Governorate Coordinators the accurate, timely information we need in order to represent CPA

---

[56] Hamre et al., *Iraq's Post-Conflict Reconstruction*, p. ii.

[57] Gerald B. Thompson, "Trip Notes: Hilla, Coalition Provisional Authority—South Central, 26–27 June 2003."

[58] Memo from Director of Operations, CJTF to Ambassador Bremer, "Directorate of Operations Tiger Teams," June 29, 2003.

[59] See Stewart, *The Prince of the Marshes*, pp. 110–111.

[60] Memo from Scott Carpenter to Paul Bremer, "Your Participation in the Coordinators and Commanders Conference," December 16, 2003.

policies on the ground is in urgent need of overhaul."[61] Although the CPA took some measures to provide greater flexibility to regional and governorate coordinators in implementing reconstruction projects, the complaints from provincial officials that CPA staff in Baghdad were making governance decisions in a vacuum persisted.

Reorganizing the CPA

Clayton McManaway departed Iraq in November for serious health reasons. This left Bremer without his closest and most trusted colleague, but it also provided an opportunity to designate two senior deputies, and thereby reduce the number of people reporting directly to him. In a November 18 memorandum to his staff Bremer announced a restructuring, the purpose of which was to "enhance the organization's ability to manage relief and reconstruction funding promptly, transparently, and wisely, while guiding Iraq's establishment of an internationally recognized, representative government and its development of a free market economy."[62] Under the new organization, Richard Jones (who was serving concurrently as Ambassador to Kuwait, but spending virtually all his time in Baghdad) became the senior of two deputy administrators and the director of policy. He had previously been the ambassador to Kazakhstan and Lebanon—and had extensive experience in the Middle East. Lieutenant General (Ret.) Joseph Kellogg became the second deputy and director of operations, with responsibility for managing most of the reconstruction effort.

Bremer also developed a more robust Office of Policy Planning, which led to an improvement in CPA's analytical capabilities. Its director, Andrew Rathmell, reported in a CPA memo that "by late 2003, CPA and CJTF-7/CENTCOM had developed a comprehensive, inte-

---

[61] Paul Harvey to Paul Bremer, "Weekly GC Update—Kirkuk," January 16, 2003.

[62] Memo from the Administrator to Deputy Administrator and Chief Policy Officer, CPA; Deputy Administrator and Chief Operating Officer, CPA; Chief Operating Officer, CPA; Directors, CPA; General Counsel, CPA, Senior Advisors, CPA; Commander CJTF-7; "Subject: Reorganization of Coalition Provisional Authority (CPA)," November 18, 2003.

grated Strategic Plan and were using this as a management information tool to drive and monitor progress. Effective coordination mechanisms had been developed within CPA and with CJTF-7 at both staff and senior leadership levels. While not perfect, the plan and the coordination mechanisms contributed to effective decisionmaking and forward planning."[63] CPA's policy planning staff collected information on—and evaluated—what was being done across CPA in the Iraqi ministries. It also conducted some policy analysis and worked on CPA's "Vision of Iraq" in the fall of 2003.[64]

Slocombe also left in the fall and was succeeded by David Gompert, who had worked with and for Bremer in several previous State Department positions. Gompert, who had also served in the National Security Council under George H.W. Bush, made a concerted effort to improve coordination among CPA's security sectors. "My responsibility was defense," he recalled. "But I spent a lot of time designing a system to better integrate the defense, police, intelligence, and justice components. I didn't have direct control over other sectors, but was responsible for coordinating them."[65]

In the fall of 2003 the focus of CPA efforts shifted toward quickly strengthening Iraqi institutions in preparation for a mid-2004 transfer of sovereignty. In a November 16 memo to his senior advisors, Bremer asked "that each of you review your strategic plan and consider which of the tasks you have identified must be completed or set well in motion before July." It continued by recommending that the advisors "discuss these with your Ministry and Interim Minister, and then inform the Planning Staff of the three most important tasks. The Plan-

---

[63] Draft Memo from Andrew Rathmell to the Incoming Chief of Mission, "Subject: Achieving Long-Term US goals in Iraq: Strategic Planning in the Mission," April 28, 2004.

[64] See, for example, memo from Executive Secretariat to Ambassador McManaway and Ambassador Kennedy, "Subject: Coordinated Responses for the Vision for Iraq Brochure from Amb. Kennedy, OPP and USAID," September 25, 2003; memo from Coalition Provisional Authority to Deputy Secretary of Defense, "Subject: Execution of Achieving the Vision Statement as of September 17, 2003," September 17, 2003.

[65] Author interview with David Gompert, February 29, 2008. Also see, for example, email from David Gompert to Frederick Smith, "Subject: Intell Coordination Help," May 25, 2004.

ning Staff will then prepare for us an overall strategic plan for these final months."[66]

Bremer also had to plan for the transfer of U.S. authority in Iraq from the CPA to a new U.S. embassy. An important element of this transition was the establishment of the Iraqi Reconstruction Management Office (IRMO), which would assume responsibility for all U.S. reconstruction matters. As Bremer outlined to Secretary of State Powell in a cable, "IRMO will be established as a temporary organization under the Chief of Mission's authority, direction and control, and will be comprised primarily of senior and technical advisors and supporting staff who will provide advice and operational assistance to Iraqi Ministries and operational coordination for the Iraq reconstruction effort."[67] IRMO's mission was to plan, execute, and manage the reconstruction initiatives of the country, employing the $18 billion from the congressional supplemental appropriation that had been passed in November 2003 but finally became available only beginning in February 2004.

State and Defense planned carefully for the transfer of responsibility from the CPA to a U.S. embassy in Iraq, and in most respects the handoff went smoothly. Some matters fell between the cracks, however. Gompert, for example, tried desperately to get DoD to send advisors to the Iraqi Ministry of Defense to succeed those the CPA had provided. He lobbied Secretary of Defense Rumsfeld, Deputy Secretary of Defense Paul Wolfowitz, Under Secretary of Defense for Policy Douglas Feith, and Under Secretary of Defense for Personnel and Readiness David Chu. "But we never got anyone," said Gompert. "I couldn't turn over the Iraqi Ministry of Defense to no one, so I asked the British to help out, even though Rumsfeld had specifically asked me not to."[68]

Another hiccup came in the late spring of 2004. "At the time," noted Fred Smith, "a number of people were trying to cut the umbilical cord and increasingly let the Iraqis run their own affairs. We often

---

[66] Info Memo from L. Paul Bremer, Administrator to All Senior Advisors, November 16, 2003.

[67] Cable from Ambassador Bremer to Secretary Powell, "Subject: IRMO Management Staff," April 1, 2004.

[68] Author interview with David Gompert, February 29, 2008.

called it 'graduating the ministries.'"[69] Lieutenant General Jeffrey Oster, CPA's deputy administrator and chief operating officer, ordered pink slips sent to a range of CPA's officials, including critical people advising in such areas as the Ministry of Interior. The notes thanked officials for their service, and many requested that individuals leave within a week. "It was unbelievable," noted David Brannan, CPA's director of security policy for the Ministry of Interior. "Virtually everyone in my office got a pink slip, even Jack Myers, who was running the entire fire department for Iraq. The impact on morale was devastating. Most people believed that no one in Washington cared how this thing in Iraq turned out."[70] With Bremer's involvement the order was quickly rescinded.[71]

If some agencies pulled back from supporting the CPA in its waning days, the State Department, which was due to inherit many of the CPA's responsibilities, went in the opposite direction and stepped up its support. By the spring of 2004 there were five Arabic-speaking former ambassadors on the CPA rolls, and a number of other experienced Foreign Service officers.

Conclusion

Whether or not the U.S. had adequate troop numbers in Iraq in 2003–2004 has been hotly debated for years. Less attention has been paid to the shortage of both civilian and military staff, about which there is no debate. According to one knowledgeable source, only seven people remained with the CPA throughout its 14-month lifespan.[72] Many agencies shared responsibility for the failure to fully man the CPA. The Defense Department supplied the largest number of personnel, about

---

[69] Author interview with Fred Smith, July 23, 2008.

[70] Author interview with David Brannan, July 20, 2008. Also see author interview with Matt Sherman, May 8, 2008.

[71] Author interview with L. Paul Bremer, August 12, 2008.

[72] SIGIR, "Lessons Learned Forum: Human Capital Management," January, 2006, pp. 73–74.

half the total, but never enough. Sanchez's charge that his headquarters was similarly short of personnel is even more shocking. The joint manning document for CJTF-7 showed that his staff was barely filled over the half-way mark through much of 2003. Senior leadership positions—general officers to lead staff sections—were not filled for many months.[73] In the end, it was the Secretary of Defense's responsibility to ensure that both organizations were adequately staffed, and the President's job to see that he did so. Between them they had all the authority and funding needed to ensure that all the relevant agencies and armed services did their part.

The CPA made do with what it had, improvising an organization and a set of policies and continuing to work extremely hard in the face of mounting challenges and against an approaching deadline for its own demise. The November reorganization relieved some of the pressure on Bremer and led to a more efficient delegation of responsibilities. Nevertheless, with his senior staff in near constant flux, Bremer naturally came to rely more heavily on those who had been with him the longest, leading some more experienced officers to feel underutilized. The CPA gradually extended its tentacles out into the provinces, but this process went too slowly and those so deployed never felt adequately connected to the center. The lack of a robust CPA presence outside Baghdad also left most American military commanders without regular direct contact with the civil authorities responsible for governing the country.

---

[73] Author interview with Catherine Dale, January 20, 2009.

CHAPTER NINE

Promoting Democracy

Iraq will have a representative form of government that protects the rights of all, promotes the rule of law, and is supported by a vibrant civil society. It will be underpinned by a democratically agreed constitution, transparent electoral processes, and strengthened political institutions. There will be an accountable and responsible system of local government. The effectiveness of elected officials will have been increased through training.[1]

In these words, the Coalition Provisional Authority laid out in the summer of 2003 its fundamental objective in Iraq. In the aftermath of decades of violent dictatorship, in a situation in which central authority had collapsed following the invasion, and in a society fractured along ethnic and sectarian lines, this was an awesome undertaking. As time went on, and with it the failure to find any weapons of mass destruction or former regime operational links to international terrorists, democratization became an ever more dominant rationale for the American presence.

Beyond the soaring rhetoric of freedom and liberty, however, the CPA's democratization program had to wrestle with the fundamental practical problem of how to create a transition to a stable and legitimate Iraqi government. Doing so required answering a series of intensely difficult policy questions: Who would rule Iraq? How would these rulers be chosen? Under what restraints would they exercise their power?

---

[1] "CPA Strategic Plan, 60 day report," August 1, 2003.

The story that emerges is one of a clash between the CPA's aspirations and the realities that it confronted in Iraq. CPA officials promulgated a deliberate, step-by-step democratization program that they hoped would achieve their goal of empowering secular, democratic forces in Iraq. This plan, which had a powerful internal consistency, did not fully conform to the political demands of Iraqis, most notably those of religiously based Shi'ite nationalists. Nor was it consistent with Washington's desire to hold troop levels in Iraq down and reduce them further as soon as conditions permitted. The push and pull of these contending factors defined the shifts in the CPA's democratization proposals. The dispute crystallized over the essential question of when and how to hold elections in Iraq. In the end, the CPA's plans had to give way to pressures from both Iraq and Washington for a more expeditious and less democratic process.

Seven Steps to Sovereignty

During the summer of 2003, the CPA's political team developed a plan for the transition to democracy and, with it, sovereignty. This plan closely paralleled Bremer's views as expressed in his first week and reflected the conclusions about democratization that the responsible CPA staffers had formed since their arrival in Baghdad. Iraq needed a new constitution, and Iraqis wanted one. Bremer determined that Iraqis should write and ratify the new constitution before holding national elections for a sovereign government. To that end, the CPA set in motion a number of initiatives. Its Office of Policy Planning produced a strategic plan that set out a timeline and a list of benchmarks in each area of reconstruction.[2] With respect to democratization, the plan included specific objectives: drafting the constitution, building an electoral apparatus, and developing political parties. Strikingly, while the CPA's security and economic goals were tied to specific dates, its political goals spoke only of the "short term," "medium term," and "longer term." Later drafts would incorporate more precise bench-

---

[2]   CPA, *Achieving the Vision: Taking Forward the CPA Strategic Plan for Iraq,* July 17, 2003.

marks. According to the CPA's director of policy planning, this initial vagueness at least partly reflected poor communication within the CPA: The governance team did not keep the policy planning group, responsible for recording such deadlines, particularly well informed of its plans.[3]

In early August, the CPA asked the Governing Council to form a constitutional preparatory committee (CPC) that would devise a system for writing a constitution—the CPA's hope was that the Governing Council would endorse a council of experts to write the document. In addition, the governance team working with USAID brought the International Foundation for Electoral Systems (IFES) to conduct a feasibility study for organizing elections.[4] Military commanders and regional CPA representatives in the provinces also set about creating local and provincial councils through a process of either neighborhood caucuses or direct selection by the coalition.[5]

In the first week of September, after months of consultation with U.S. and Iraqi leaders, Bremer went public with his plan in a radio address to the Iraqi people and in an op-ed in the *Washington Post*.[6] The seven steps to sovereignty he outlined were: (1) the creation of the GC, (2) the formation of the CPC to propose how to write the constitution, (3) increasing day-to-day responsibility for the GC, (4) writing the constitution, (5) ratifying the document, (6) national elections to choose a government, and ultimately (7) the dissolution of the CPA and the resumption of Iraqi sovereignty. Like the planning document that preceded it, this plan did not give a concrete timeline for this process but to most observers it appeared to entail at least a two-year

---

[3] Author interview with Andrew Rathmell, July 13, 2007.

[4] Lewis Lucke and Scott Carpenter to Paul Bremer, "Support for Pre-election Assessment," August 18, 2003.

[5] Paul Bremer to Commander, Coalition Forces, "Appointment of Interim Town Councils," May 31, 2003.

[6] CPA HQ to Secretary of Defense, Secretary of State, NSC, "Ambassador Bremer's 5 Sep 03 Radio Address," September 5, 2003; L. Paul Bremer, "Iraq's Path to Sovereignty," *Washington Post*, September 8, 2003.

occupation. As the CPA soon found out, it would not have the luxury of that much time.

From their first meetings with Iraqi political leaders, CPA officials started hearing indications that some Iraqis had very different ideas about how the country's political transition should be structured. At the May 16 meeting during which Bremer informed the leadership council of Iraqi exiles that they would not be taking power, several leaders emphasized the importance of Iraqi sovereignty and meaningful participation in governing the country. According to the CPA's account of the meeting, Hamid Bayati of SCIRI argued that "the longer Iraqis are not in control of their political life, the more problems would arise."[7] Chalabi specifically referenced the perceived promise that had been made the previous month about creating a transitional government within weeks. Another member of the council pushed Bremer to commit to a specific timeline for the transfer of power to Iraqis. This drumbeat of voices on the subject of sovereignty continued in CPA meetings with individual Iraqi politicians. SCIRI leaders conveyed to the CPA their concern that the U.S. sought to install itself as an occupying power indefinitely.[8] Even leaders who were open to the idea of delaying elections, such as Da'wa's Jaafari, emphasized the importance of some transition to local government and the need to communicate this clearly to the Iraqi people.[9] He quoted an Arabic proverb in response to the CPA proposal of an interim authority: "Sometimes, there is nothing longer than the interim."[10]

A key leader who would not meet with the coalition was Grand Ayatollah Ali Sistani, the most senior Shi'ite cleric in Iraq. Bremer was able to establish several channels to Sistani, and the CPA had to rely on accounts from them and other interlocutors to learn his views. On

---

[7] Meghan O'Sullivan, "Meeting with Members of the Iraqi Opposition Leadership Council," May 16, 2003.

[8] Meghan O'Sullivan to Paul Bremer, "Meeting with Abud Aziz Hakim [sic], Principal, SCIRI," May 23, 2003.

[9] Sam Rascoff, "Ambassador Bremer Meeting with Dr. Ibrahim Ja'afari (5/22/03)," May 23, 2003.

[10] Rascoff, "Ambassador Bremer Meeting with Dr. Ibrahim Ja'afari."

June 29, 2003, UN Special Representative to Iraq Sergio de Mello told Crocker that he had met with Sistani in Najaf on the previous day.[11] At that meeting, Sistani insisted that the CPA guarantee elections for the constitutional convention. If this did not happen within seven days, the cleric said, he would issue a *fatwa*, or religious ruling, declaring illegitimate any constitution produced by unelected authors. In the event, it appears that Sistani did not wait for CPA action and issued a *fatwa* on June 28 demanding elections for the constitutional assembly.[12] The CPA was slow to grasp the full weight of the ayatollah's pronouncement. In part, this was because it received contradictory accounts of his views: Mowaffak Rubaie emphasized Sistani's attachment to the idea of elections while Chalabi insisted that a process in which the CPA and the GC-created provincial assemblies would be acceptable to Sistani.[13] The CPA also discussed whether the cleric had issued merely an opinion, not a *fatwa*, but concluded that it was indeed the latter, more definitive judgment.[14] Beyond all this, Sistani's announcement came when the CPA was entirely focused on finalizing formation of the Governing Council and so simply did not dwell on it at the time.[15]

Even as Bremer was elaborating on his seven-step plan, some members of the GC had become increasingly strident in demanding an early restoration of Iraqi sovereignty. A CPA memo identified several people who were "agitating for immediate sovereignty," including Adnan Pachachi (who had served as Iraqi foreign minister and ambassador to

---

[11] Ryan Crocker to Paul Bremer, "June 29 Meeting with De Mello," June 29, 2003.

[12] Rajiv Chandrasekaran, "How Cleric Trumped U.S. Plan for Iraq," *Washington Post*, November 26, 2003, p. A01.

[13] Philip Hall to Ryan Crocker, "Mowaffak Al Rubaie: Constitutional Conference," June 27, 2003; CPA Political Team to Paul Bremer, "Ahmed Chalabi Meeting July 3," July 2, 2003. (This document lays out talking points for Bremer's July 3 meeting with Chalabi and references prior CPA conversations with Chalabi on the subject of elections.)

[14] Hume Horan to Paul Bremer, "Sistani stands by his fatwa," July 10, 2003.

[15] Author interview with Scott Carpenter, August 2, 2007; author interview with Roman Martinez, July 1, 2007.

the UN before the Ba'athists seized power), Chalabi, and Hakim.[16] The governance team worried about Chalabi in particular, fearing his ability to press the case for sovereignty in discussions at the United Nations and in Washington.[17] In addition, the CPA had received soundings that the CPC was going to recommend the use of elections to select the constitutional assembly, out of deference to Sistani's *fatwa*. As Ayad Allawi put it metaphorically: When the CPC meets, "Sistani is sitting in the room."[18] Adding to these political pressures, the insurgency was flaring up and the level of violence in the country was increasing. Despite all this, the CPA officials felt that changing course on sovereignty would be dangerous for a host of reasons. Carpenter laid these out in a memo to Bremer.[19] In the absence of a constitutional structure, the GC would have essentially unlimited power and could even halt the democratic process. Moreover, the Islamist parties continued to be very strong and the coalition's "pro-democratic friends" were not well-placed to fight them off.[20] The Kurds might take the opportunity to break away from Iraq. The possibility even existed that U.S. forces would be asked to leave by the GC, whose relations with the CPA had deteriorated over the course of the summer. Finally, Carpenter argued, a change in course would signal weakness. For these reasons, the CPA took a number of steps to fight for its initial plan.

The governance team tried to lobby the CPC and the GC directly, trying to persuade them of the difficulty of holding elections in Iraq.[21] The CPC released a report on October 1, 2003, that laid out two options for choosing the constitutional assembly: direct elections or a system of "partial elections" drawing from appointed councils. Despite

---

[16] Scott Carpenter to Paul Bremer, "Talking Points on the UNSCR and the GC's Constitutional Process," September 17, 2003.

[17] Scott Carpenter to Paul Bremer, "Engaging Ahmad Chalabi," September 20, 2003.

[18] Scott Carpenter to Paul Bremer, "Meeting with Ayad Allawi," September 22, 2003.

[19] Scott Carpenter to Paul Bremer, "Sovereignty," September 16, 2003.

[20] Scott Carpenter to Paul Bremer, "Sovereignty," September 16, 2003.

[21] Roman Martinez to Paul Bremer, "Meeting with CPC Chairman Fouad Massoum," September 18, 2003; Roman Martinez to Paul Bremer, "Approaching the GC on the Constitutional Process," September 13, 2003.

the efforts of the CPA, the report strongly recommended the direct election option.[22] The CPA then tried to persuade the GC to reject the CPC's conclusion by arguing that elections would only delay the sovereignty prized by some of the council's members. CPA officials even met privately with the IFES experts who had conducted an election feasibility study in Iraq and asked them to highlight to the GC their recommendation that elections be delayed by two years, although IFES had concluded that elections were technically possible in six to nine months.[23] At the same time, the CPA sought to achieve a "grand bargain" with the GC in which the council would gain expanded powers in return for quickly convening a constitutional assembly—something that could only be done by a selection process.[24] Events in Washington, however, soon upended these various efforts.

Stepping on the Gas

The September publication of Bremer's seven-point plan took some officials in Washington by surprise, due not so much to its content as to its timing—coming as it did just as the administration was beginning to have second thoughts about the sequence Bremer was espousing. Colin Powell felt Bremer's timetable was too slow. Robert Blackwill, who had recently been brought back from a post as ambassador to India to oversee Iraq policy within the National Security Council staff, later described his own reaction:

> [It was] a schoolbook solution, but a solution without . . . Iraqis. . . . So I immediately spoke with Condi [Rice] and Steve [Hadley] about this. We had a series of conversations about this in which I argued—and I think they were coming to the same conclusions . . . that this was just too long a timeline. . . . We dis-

---

[22] Scott Carpenter to Paul Bremer, "CPC Report to the GC," October 4, 2003.

[23] Roman Martinez to Paul Bremer, "Constitutional Discussion with the Governing Council," October 18, 2003.

[24] Meghan O'Sullivan to Paul Bremer, "A Grand Bargain with the GC," October 13, 2003.

cussed, well, what is a reasonable timeline for this? We decided that it is the following summer, that it is June of the following year.[25]

When Bremer traveled to Washington in late October, all the senior administration officials he met with conveyed to him their impatience with an extended timeline.[26] Various reasons were given, but the main concern was the rising tide of violent resistance to the occupation. The Department of Defense came forward with its own plan for a swift transfer of power to an expanded GC. The CPA's governance team found this proposal "deeply flawed" because it made no attempt to address the constitutional and political questions that they felt needed to be resolved before a handover of sovereignty.[27] Nevertheless, the emerging political reality was clear, and the team went on to note in a memo to Bremer that "if it is a political imperative to end the occupation by mid-2004," then an interim constitution and elected provisional government was the best solution.[28] In two follow-up memos, Bremer's aides argued that the CPA was in no position to overrule the GC on the issue of elections, largely because of the danger of an irreparable breach with the Shi'ites.[29] Instead, they recommended holding elections under an interim constitution for a new sovereign government while the process of drafting and ratifying a permanent constitution took place after the transfer of sovereignty.[30] The second memo went on to suggest that, by "operating stealthily," the CPA could

---

[25] Robert Blackwill, Interview with PBS Frontline, "The Lost Year in Iraq," July 25, 2006.

[26] Author interview with L. Paul Bremer, July 30, 2007.

[27] Meghan O'Sullivan and Roman Martinez to Paul Bremer, "Overview of Changing Timeline Proposals," October 28, 2003.

[28] O'Sullivan and Martinez to Bremer, "Overview of Changing Timeline Proposals."

[29] Meghan O'Sullivan and Roman Martinez to Paul Bremer, "Part I of Constitutional Strategy, Arguments against a Redline," November 4, 2003.

[30] Meghan O'Sullivan and Roman Martinez to Paul Bremer, "Part II of Constitutional Strategy: An Interim Constitution," November 4, 2003.

have a strong hand in drafting the interim constitution and therefore influence the future constitutional development of Iraq.[31]

Bremer accepted these proposals and communicated them to Rumsfeld, Rice, and Colin Powell on November 10. Rice asked Bremer to come to Washington immediately to brief the President.[32] The first plane available was a C-141 medical evacuation flight via Germany. Seated with badly wounded soldiers, Bremer flew all night and proceeded directly to the White House for a principals-only meeting in the Situation Room.[33] In discussions among the key national security figures of the Bush administration, a new plan emerged: By February 28, 2004, the Governing Council, with the advice of the CPA, would produce an interim constitution, the TAL; a Transitional National Assembly (TNA) would be chosen via local and provincial caucuses by May 31, 2004; and sovereignty would transfer to an interim government formed from the assembly by June 30, 2004. Elections would take place by March 31, 2005, for delegates to a constitutional convention and, following ratification of the constitution, a new government would be elected by December 31, 2005.

Crucially, this plan abandoned the initial governance team proposal to directly elect the interim government. On November 11, O'Sullivan reported that she had spoken with the election contractors, who said they could no longer confirm a six- to nine-month timeframe for holding elections.[34] Nevertheless, she followed this up with another memo the next day forcefully arguing for keeping elections to the transitional government. She wrote that a move away from the principle of transferring sovereignty to an elected body would constitute "a dramatic departure from our original vision."[35] "The key issue at stake," she went on, "is the legitimacy of the transitional parlia-

---

[31] O'Sullivan and Martinez to Bremer, "Part II of Constitutional Strategy."

[32] Bremer, *My Year in Iraq*, p. 218.

[33] Bremer, *My Year in Iraq*, pp. 219, 224.

[34] Meghan O'Sullivan and Irfan Siddiq to Paul Bremer, "Electing an Iraqi Transitional Assembly," November 11, 2003.

[35] Meghan O'Sullivan to Paul Bremer, "Keeping Direct Elections for the Transitional Parliament," November 12, 2003.

ment. There is no question that a directly elected parliament will be far more legitimate."[36] Moreover, she added, the move to an interim constitution could be defended as a necessary stopgap to a permanent settlement, but the move away from an elected government could "only be explained as (1) our anxiety about the security situation; (2) our fear over the outcome of democratic elections; and (3) our desire to leave Iraq at all costs."[37] Finally, she argued that a change in policy was unnecessary at that stage—if elections truly could not be organized in time for a June handover of sovereignty, the CPA could adjust its plans if it became apparent that the handover could not be accomplished by June. Bremer considered these arguments but still worried that holding elections would derail the timeline for sovereignty.[38] In particular, he worried that possiby finding out in May that elections had to be postponed would be highly disruptive to the political process.[39] Nevertheless, Bremer presented both options to the principals in the White House Situation Room. President Bush made the final decision to go with caucuses, citing the risk that elections posed to the July deadline.[40]

Before leaving for Washington, Bremer had prepared the way for the prospective timetable change with key Iraqi officials, including sending an emissary to Sistani.[41] Immediately on his return, he presented the new plan to the GC's nine-man presidency council. They were elated by the proposal and its commitment to an early handover of sovereignty.[42] However, the next day the SCIRI delegate, Adel Mahdi, objected to the caucus proposal—he worried that it would undermine the Shi'ite electoral majority.[43] Bremer forced a vote on the issue and

---

[36] O'Sullivan to Bremer, "Keeping Direct Elections."

[37] O'Sullivan to Bremer, "Keeping Direct Elections."

[38] Author interview with L. Paul Bremer, July 30, 2007.

[39] Bremer, *My Year in Iraq,* p. 226.

[40] Bremer, *My Year in Iraq,* p. 228.

[41] Author interview with Meghan O'Sullivan, January 29, 2008.

[42] Author interview with L. Paul Bremer, July 30, 2007.

[43] Chandrasekaran, *Imperial Life in the Emerald City,* p. 202.

the council, over the objections of SCIRI, adopted what became the November 15 Agreement.[44]

Building Iraqi Capacity

The following day, Bremer convened an all-hands meeting of CPA staffers to emphasize the need to scale back plans for further reforms and focus instead on "building capacity" among Iraqis to run their government.[45] He followed this up with a memo to all of CPA's senior advisors, stating,

> As you know we have reached agreement with the Governing Council for the transfer of sovereignty to an Iraqi Transitional Government by July 1, 2004. At that point, the CPA will dissolve, as will the Governing Council.
>
> I ask that each of you review your strategic plan and consider which of the tasks you have identified must be completed or set well in motion before July. Please discuss these with your Ministry and Interim Minister, and then inform the Planning Staff of the three most important tasks. The Planning Staff will then prepare for us an overall strategic plan for these final months.
>
> Please reply to Planning by December 1.
>
> Thank you all for your continued support.[46]

The third version of the CPA Strategic Plan, released in December, emphasized the need to work on four issues that cut across all issue areas and Iraqi ministries: (1) developing a professional civil service;

---

[44] Chandrasekaran, *Imperial Life in the Emerald City*, p. 204. See *The November 15 Agreement: Timeline to a Sovereign, Democratic and Secure Iraq.*

[45] Chandrasekaran, *Imperial Life in the Emerald City*, p. 217.

[46] Info Memo from L. Paul Bremer to All Senior Advisors, November 16, 2003.

(2) ministerial capacity-building; (3) anti-corruption; and (4) planning past the CPA.[47]

None of these priorities was entirely new. As early as June 2003, the CPA began to consider when ministries could be turned back to full Iraqi control; the political team reported that many ministries were likely to be ready quite soon.[48] Recognizing the need to train a cadre of senior civil servants to respond flexibly to, and to anticipate, the new requirements of a democracy and free market, on October 21 Bremer sent the Governing Council a note asking them to "think hard" about plans for the development of the Iraqi civil service.[49] Bremer had also initiated anti-corruption efforts well before the November shift in emphasis. On September 20, Jeremy Greenstock, the UK's special envoy to Iraq, wrote to the Governing Council recommending the formation of a body to oversee compliance with a defined set of standards of behavior for governmental officials.[50] The CPA took the first step in this direction itself by drafting an order for "Honest and Transparent Governance" that would impose basic requirements of ethical public service on members of the Governing Council, their deputies, and the interim ministers, as well as establish an Office of Public Integrity to assist compliance and enforcement of these requirements.[51] As one regional CPA official noted, "None of the organizational diagrams, regulations, and ultimately, the Constitution itself, will have any utility

---

[47] Coalition Provisional Authority, *Towards Transition in Iraq: Building Sustainability*, December 20, 2003.

[48] Memo from CPA Political Team to Paul Bremer, "Interim Administration Authorities," June 5, 2003.

[49] Memo from Mohamed Alhakim and David Kirk to Paul Bremer, "Planning for Civil Service and Management Development and Training," October 18, 2003.

[50] Memo from Ambassador Jeremy Greenstock to Governing Council, "Transparency in Government," September 20, 2003.

[51] Memo from Office of General Counsel to Paul Bremer, "Honest and Transparent Governance," October 8, 2003. Bremer did not sign the memo, however, which still appeared on the list of pending orders in January. As discussed below, elements of this order were eventually adopted through other measures.

if corrupt officials cannot be checked and rooted out before the final transition happens."[52]

However, building Iraqi capacity assumed a greater impetus after November 15. Training Iraq's civil service, preparing the Iraqi ministries for transition, and countering corruption became as central to CPA's governance mission as did the continued delivery of essential services.

Defining the CPA's priorities for 2004, the Office of Policy Planning and Analysis noted: "It has become evident that the primary risk to the achievement of CPA's vision for the reconstruction of Iraq is the lack of a coordinated, well-trained and well-managed civil service." A separate assessment concurred, stating, "Most ministries have very little indigenous senior management capability and depth, and certainly no concept of how a modern ministry should operate." Consequently, three critical tasks were established:

- Creation of a coordinated and restructured civil service, including a cabinet office secretariat and inter-ministerial committees
- Recruitment, education, and training of senior and mid-level management capacity in the civil service and security forces
- Establishment of sound management and financial systems.[53]

On March 31, Ambassador Bremer appointed 31 deputy ministers to 15 Iraqi ministries.

The CPA recognized that after 35 years of tyranny and mismanagement, there was only so much it could do during its remaining eight months to reform the Iraqi civil service.[54] CPA senior advisors recommended that 206 consultants should continue to be based in the min-

---

[52] Email from John Berry to Julie Chappell, "Re: Comments on Local Governance Strategy Paper," November 10, 2003.

[53] Memo from OPPA to Bremer, "CPA Priorities in 2004," December 18, 2003; CPA Baghdad, "Personnel Assessment Team Report to the Secretary of Defense," February 11, 2004.

[54] Memo from David Kirk to Paul Bremer, "Civil Service Capacity-building Program," December 18, 2003.

istries after the transfer to Iraqi sovereignty.[55] The CPA was not able to achieve all its self-identified critical tasks, acknowledging in June that "Little headway has been made on the more fundamental question of administrative support for the introduction of cabinet government to Iraq."[56] Looking back on this period, Fred Smith opined "that CPA may have tried to do too much in overseeing too many ministries. I recall an inordinate amount of staff time devoted to minor ministries. CPA would have benefited from concentrating exclusively on core ministries—Interior, Finance, Oil, Justice, Defense, maybe Foreign Affairs, the intelligence apparatus, maybe two or three more. In retrospect, CPA should have turned over 17 or 18 ministries in 2003 and moved on with the truly critical stuff."[57]

At a March 30 meeting with the Iraqi ministers, Bremer described the four criteria he would look at in deciding whether a ministry was ready for full authority: Does the ministry have short- and long-term strategies? Does the ministry structure and staffing support the goals? Have training needs been analyzed and training programs begun? And are fundamental management systems such as communications, personnel policies, financial and budgetary controls in place?[58] However, progress toward these goals was often difficult. Although the Ministry of Health was transitioned on March 28, as April dawned, the CPA's Director for Civil Affairs warned Ambassador Bremer that only seven of the remaining 26 ministries appeared to be ready for transition ahead of sovereignty. The three main problems cited were differences

---

[55] Memo from Giles Denham and Andrew Rathmell to Paul Bremer, "Ministry Transition Advisor Requirements—Final Report," March 29, 2004. This drastic reduction would prove problematic in some sectors. The senior advisor to the Ministry of Interior strongly protested the reduction to 27 personnel mandated by the transition planning teams, predicting it would have "disastrous consequences to our ability to complete the job." Memo from Steve Casteel to Chief Operating Officer, "Status of the CPA Ministry of Interior," April 23, 2004.

[56] Memo from Giles Denham to Paul Bremer, "Support for Cabinet Government in Iraq," June 10, 2003.

[57] Author interview with Fred Smith, December 14, 2008.

[58] Memo from Giles Denham to Paul Bremer, "Meeting with Ministers on April 20," April 16, 2003.

in approach between the CPA advisors and Iraqi ministers and officials; ineffectiveness of the ministers and their senior officials; and the lack of overall capacity in terms of staff, systems, and structures.[59]

Despite these deficiencies, Bremer approved a proposed timeline on April 15 by which two to three ministries would transfer to full Iraqi sovereignty each week until the end of June. CPA justified this decision to deemphasize its previous criteria by arguing that "Early transition is an incentive for the Ministers and ministries to ensure that CPA is satisfied with the ministerial process or risk delayed transition and the benefits or prestige it affords."[60] Consequently, over half the Iraqi ministries were transitioned by June 1 when the Iraqi Interim Government was announced. The remaining ministries that the CPA deemed unready for transfer in May were transitioned en masse by Prime Minister Ayad Allawi on June 24, a week before the formal transfer of sovereignty.

The CPA also invested major efforts in creating institutions to fight the corruption endemic to the Iraqi government. There were four key elements to the CPA's efforts to reduce and prevent corruption in the Iraqi government: a Commission on Public Integrity; the Board of Supreme Audit; the Judicial Review Committee; and the inspectors general. On November 17, Judge Daniel Rubini, the CPA senior advisor to the Ministry of Justice, presented a briefing on the CPA's "Iraq Government Integrity and Anti-Corruption Project" that advocated the rapid establishment of an anticorruption entity called the Office of Anti-Corruption and Integrity in Government.[61] Bremer presented this anticorruption program to the Governing Council on December 3, warning the council members that "Wherever it is found throughout the world, corruption inevitably wards off investment and foreign aid, impedes economic growth, and becomes a mechanism for extortion

---

[59] Memo from Giles Denham to Paul Bremer, "Transferring Full Authority to Ministries— Next Steps," March 31, 2004. See also memo from Giles Denham and Andrew Rathmell to Paul Bremer, "Ministry Transition Status Assessment," February 16, 2004.

[60] Memo from James Haveman to Paul Bremer, "Ministry Readiness to Transition Timeline," April 14, 2004.

[61] Rubini memos, November 17 and November 27, 2003.

from the people."[62] Bremer found a strong ally on the Governing Council in Mowaffak al-Rubaie, who was proactive in moving what became the "Commission on Public Integrity" (CPI) measure forward in the Governing Council, pushing for even more authority for the CPI than even the CPA had envisioned.[63] The Governing Council enacted the CPI measure on January 27, and on March 24, the CPA announced the formation of the CPI to enforce Iraq's anticorruption laws in cooperation with the Board of Supreme Audit, which Bremer sought to revitalize and strengthen.

Bremer signed the CPA order appointing inspectors general to the Ministries of Oil, Labor, and Social Affairs on January 22 and to the Ministry of Municipalities and Public Works on February 19. A number of the ministry senior advisors reported difficulty in identifying qualified candidates to serve as inspectors general; as of February 5, half the ministries had not yet identified an agreed nominee.[64] However, by the end of March all but two ministries had appointed inspectors general, and the final inspector general (Ministry of Defense) was appointed on May 8.

Bremer took a particular interest in developing the Iraqi bureaucracy's anticorruption capacity, which he held to be key to a successful transition to Iraqi sovereignty. On a November paper outlining the criteria by which senior advisors would be phased out, he hand-wrote, "What about corruption?"[65] Bremer agreed to make "Anti-Corruption" the first agenda item for his May 8 meeting with the Iraqi ministers, whom he told: "Unless you can rid the new Iraq of corruption, there

---

[62] Memo from L. Paul Bremer to the Governing Council, "Proposed Agenda for the December 3 Meeting," December 2, 2003.

[63] Memo from Candace Putnam to Paul Bremer, "Read Ahead for Your Meeting with Mouwaffak al-Rubaie," December 31, 2003.

[64] See memo from Giles Denham to Paul Bremer, "Establishing Inspector General Offices — Progress Report," January 31, 2004; and memo from Giles Denham to Paul Bremer, "Establishing Inspector General Offices—Progress Report," February 5, 2004.

[65] Unattributed author, "Senior Advisor Phase-Out Criteria," November 27, 2003.

will be no true democracy and future prosperity."[66] Consequently, even some of the CPA's harshest critics acknowledged the importance of Bremer's commitment. Ali Allawi, who served as Interim Minister of Finance during the CPA's tenure (and later Minister of Defense during the Iraqi Interim Government) concedes that "Bremer's order regarding the establishment of a Public Integrity Commission was a notable achievement, and created the framework for seriously tackling the exploding levels of high- and low-level corruption in the country."[67]

In the waning months of the CPA's existence, Gompert worked to establish a national security council modeled on the U.S. system. When the November 2003 decision was taken to accelerate the handover of sovereignty, no Defense Ministry existed, the old one having been disbanded by CPA Order Number 2. A decision had been made early on to hold up on forming a new ministry in order to maintain CPA control of the new military during its early months. This all changed in November 2003. "Before we could turn over the new army to an Iraqi government, we had to establish civilian control, one of the basic tenets of the government we were trying to establish. That meant the CPA had barely six months to organize, recruit, train, and put in place a civilian-led ministry. The other CPA senior advisors already had ministers with whom they were working, but Gompert had to work to create his from scratch."[68]

In addition to a new Defense Ministry, a National Security Cabinet Committee was organized and tasked with assisting the integration of national security ministries and developing a long-term counterinsurgency strategy. The CPA also established a "Situation Room" accessible to the prime minister and national security advisor, from where

---

[66] See memo from Giles Denham to Paul Bremer, "Meeting with Ministers on May 8," April 30, 2004; and memo from Giles Denham to Paul Bremer, "Agenda for May 8 Meeting with Ministers," May 7, 2004.

[67] Allawi, *The Occupation of Iraq*, p. 264. See also Diamond, *Squandered Victory*, pp. 150–151.

[68] Author interview with Fred Smith, December 14, 2008. In the fall of 2008, Smith moved from the CPA's Washington office to work for Gompert in Baghdad.

they could communicate directly with the other national security ministries and coalition military forces.[69]

The National Security Cabinet Committee was initially chaired by Bremer and then, once he had been named, by the Prime Minister designate, Ayad Allawi. Other members included the ministers of Interior, Defense, Foreign Affairs, Justice, and Finance. This framework improved the cooperation among Iraq's security ministries and allowed them to begin developing an interministerial counterinsurgency strategy. During the Fallujah crisis in April, the committee met regularly to coordinate civilian and military action, including military operations, patrols, intelligence briefs, media strategies, and humanitarian relief efforts.[70] Fred Smith, then an aid to Gompert, recalls that the arrangement did not function as well as it might, however, because of distrust among the Iraqi participants. "Given they all came from different groups and sects, this was something we were not going to be able to overcome in a few week's time." Ali Allawi, the Defense Minister, for instance, constantly told Smith that he did not trust the head of the Iraqi intelligence agency.[71]

Working at the Grassroots

While a great deal of the CPA's attention focused on negotiations with the GC over national electoral arrangements, the success or failure of the coalition's democratization program would ultimately depend on the people of Iraq. The CPA recognized this fact and took a variety of steps to promote democracy among the population. These included establishing programs to build Iraq's civil society, reaching out to the Sunni population in particular, and reforming the provincial councils.

---

[69] Info Memo from Fred Smith to the Administrator, "Subject: Additional Information from the 4 June 2004 MCNS Meeting," June 15, 2004.

[70] See, for example, memo from Roman Martinez through Scott Carpenter to the Administrator, "Subject: Iraqi Buy-In to Fallujah," April 20, 2004; Ministerial Committee for National Security, "Conclusions and Actions," April 26, 2004.

[71] Author interview with Fred Smith, July 23, 2008.

Scott Carpenter noted shortly after the November 15 agreement that "by whatever timeline or plan, history will judge the Coalition by how well (or poorly) it sets the Iraqi people on the road to a robust democracy."[72] To achieve that end, he laid out a strategy that emphasized investment in five areas: creating an election administration infrastructure, decentralizing the structure of government and aiding local bureaucracies in that transition, providing civic education, funding and training civil society organizations, and funding and training political parties. Significantly, he noted that it would take ten years of sustained effort to accomplish these goals. The political situation was complex and difficult to navigate, with early estimates of more than 100 new parties formed in the aftermath of the invasion.[73] A memo focusing on the issue of political party development noted that "the constituency the CPA would most like to support—a cross-ethnic, nonsectarian, secular, urban demographic—does not yet have an effective political voice within the current set of political parties."[74] Furthermore, religious and regional parties were by far the best placed in terms of means and organization to compete in elections. To remedy this situation, the memo recommended taking steps to "level the playing field" by supporting secular, pro-democratic parties and by crafting election laws that encouraged a consolidation in the number of parties. The CPA also opened "democracy centers" in the governorates to provide training and office resources to new political parties and sent experts to different cities to give lectures on democracy. In addition, the CPA oversaw the creation of women's groups and funded magazines and other civil society endeavors. While these efforts may have created positive effects among the groups that the CPA dealt with directly, Hume Horan struck a more melancholy note on his departure from Iraq: "The Iraqis will not be ready for the challenge of independence. Saddam left a deep psychological imprint on his subjects. It would take almost a

---

[72] Scott Carpenter to Paul Bremer, "Democracy Building Strategy," November 30, 2003.

[73] Judy Van Rest to Meghan O'Sullivan, "Political Parties in Iraq—interim response," October 27, 2003.

[74] Judy Van Rest to Paul Bremer, "Iraq Political Party Development," November 1, 2003.

generation of mandate-style colonialism to detoxify their politics and their psychology. But alas! There are no political dialysis machines."[75]

The biggest political challenge that the CPA sought to address was the lack of Sunni participation in the politics of the new Iraq. This issue had first cropped up when the CPA was putting together the Governing Council. Finding Sunni candidates was difficult because almost all politically engaged Sunnis had been members of the Ba'ath party.[76] The issue had only grown as the insurgency increased in scale and ferocity. As a governance team memo put it, "Sunni communities throughout Iraq, but especially within the Sunni triangle, are feeling politically and economically disenfranchised. . . . Sunnis do not feel like they are duly represented at the national level, often complaining that they have no representatives on the Governing Council."[77] Therefore, the CPA in the fall of 2003 sought to implement a strategy designed to intensify its efforts to bring Sunnis into the political process. This included senior-level CPA outreach to Sunni tribes, encouraging the GC to engage with the Sunni population, seeking to persuade the GC to limit its efforts at de-Ba'athification, and offering aid to the Sunni triangle.[78] In October 2003, the CPA created the Office of Provincial Outreach to focus entirely on Sunni and tribal outreach under senior British diplomat David Richmond and senior State Department Arabist Ronald Schlicher.[79] These efforts had only limited impact throughout the rest of the CPA's existence. The Sunnis, still chafing over their loss of position, privilege, and influence, resisted being drawn into the political process, and the insurgency raged on.[80]

---

[75] Hume Horan to Paul Bremer, "Some End of Tour Comments from Hume," November 22, 2003.

[76] Bremer, *My Year in Iraq*, p. 93.

[77] Scott Carpenter to Paul Bremer, "Sunni Strategy," September 30, 2003.

[78] David Richmond to Paul Bremer, "Strategy for Sunni Outreach," November 12, 2003.

[79] Office of Policy Planning and Analysis to Paul Bremer, "Office of Provincial Outreach establishment, milestones, and personnel plan," December 7, 2003.

[80] This critique was made by former CPA advisor Larry Diamond, among others; see Diamond, *Squandered Victory*, p. 295.

One Sunni complaint had to do with the poor quality and unrepresentative nature of the provincial and local councils that governed them.[81] This reflected a more widespread concern about these councils, which was shared by the Shi'ite population, as noted in Chapter Five. These councils had been formed in the summer of 2003, at the CPA's direction, either by direct appointment by local coalition commanders or by some sort of caucus process. Often, the leaders who initially rose to power had not been properly vetted and in some cases included senior Ba'athists, criminals, and other questionable characters.[82] The CPA initiated a "refreshment" process by which governorate coordinators and military commanders reviewed the membership of these councils and, where necessary, held provincial caucuses to select new members.[83] This process took place in late 2003 and early 2004. The Governing Council, which had been asked by the CPA to provide candidates, submitted lists in which "some of the names were completely unknown, some were of suspicious character, and others were completely unaware of their being named to the list or what their mission was."[84] In Najaf, daily demonstrations took place demanding the dissolution of the provincial council and the holding of general elections.[85] In Salah ad Din, the replacement of compromised provincial council members went smoothly, but the governorate coordinator reported that Sunni Arabs were deeply opposed to elections, which they felt would be easily manipulated at that stage by Shi'ites in the province.[86] In Al Anbar, the lead CPA official described an enthusiastic, hard-fought caucus process that successfully produced new council members.[87] Still,

---

[81] Scott Carpenter to Paul Bremer, "Sunni Strategy," September 30, 2003.

[82] Allawi, *The Occupation of Iraq*, p. 120.

[83] Meghan O'Sullivan and Roman Martinez to Paul Bremer, "Reforming Provincial Institutions," November 2, 2003.

[84] Dean Pittman to Paul Bremer, "Update on Governing Council Refreshment Teams," January 19, 2004.

[85] Rick Olson to Paul Bremer, "Weekly GC Update—Najaf," January 19, 2004.

[86] Bob Silverman to Paul Bremer, Weekly GC Update—Salah ad Din," January 19, 2004.

[87] Keith Mines to Paul Bremer, "From a Trickle to a Flood, Al Anbar Finds its Voice," January 20, 2004.

he added, "there is a clamoring for an open and transparent system that is not manipulated by the coalition and it is difficult to see anything short of an election that would deliver on this demand."[88] The caucuses were not executed in a uniform manner across Iraq's provinces, as governorate coordinators were not given a standard model and so improvised on the spot.[89] In some instances, such as in Dhi Qar province, the local CPA team simply disregarded directives from Baghdad and organized rudimentary ration-card elections, which were generally met with a positive response among the population.[90]

The Return of the United Nations

The November 15 Agreement had been pushed through the Governing Council over the objections of SCIRI, perhaps the most influential Shi'ite party. Eventually SCIRI was able to persuade other Shi'ite and non-Shi'ite groups to join its opposition to the plan. The heart of the dispute concerned the proposed caucus system for selecting the interim government that would rule after June 30, 2004. As a CPA memo put it, "The fundamental issue at stake is the GC's desire to exert more control over the process. Alongside (and connected to) this desire is the wish to ensure that the new institutions are representative (particularly of the Shi'ite majority)."[91] The governance team took steps immediately to try to allay these concerns and persuade Shi'ite members of the GC that the caucus process would not endanger the Shi'ite majority. SCIRI leaders Abdul Hakim and Adel Mahdi took matters into their own hands, however, and, a week and a half after the agreement, they trav-

---

[88] Mines to Bremer, "From a Trickle to a Flood," pp. 3–4.

[89] Author interview with Rory Stewart, June 21, 2007. See also Stewart, *The Prince of the Marshes*, p. 214.

[90] Anthony Shadid, "In Iraqi Towns, Electoral Experiment Finds Some Success," *Washington Post*, February 16, 2004, p. A01; Yaroslav Trofimov, "Iraqis Taste Democracy," *Wall Street Journal*, February 18, 2004.

[91] Meghan O'Sullivan to Paul Bremer, "Post-November 15 Agreement Concerns," November 20, 2003.

eled to Najaf to lobby Ayatollah Sistani against the caucus proposal. Shortly thereafter, in response to questions submitted by the *Washington Post*, Sistani decreed that the caucus system was unacceptable and demanded direct elections for the Transitional National Assembly.[92]

This demand was more than SCIRI or the CPA had bargained for. SCIRI's primary objective had been to secure greater control over the caucus process, and its leaders viewed elections with some uncertainty. The CPA, which had kept in touch with Sistani through various intermediaries, had not expected him to demand elections for the interim government, since his prior *fatwa* had dealt only with the constitutional assembly.[93] His decision came, therefore, as a major and most unwelcome shock. Within two weeks of its signing, the November 15 Agreement was in mortal danger and the CPA's pathway to sovereignty was about to be modified once again.

The CPA was initially inclined to resist this new intrusion by Sistani into the political process. The core issue from its perspective was "whether the CPA and the GC will allow Iraqi clerics to overrule and/ or nullify decisions made by Iraq's legitimate political authorities."[94] The problem for the CPA was that legitimacy lay in the eye of the beholder: To many Iraqis, Sistani and the Shi'ite religious hierarchy were far more legitimate than either the GC or the CPA. Nevertheless, the CPA embarked on an effort to persuade its Iraqi interlocutors to try to convince Sistani to accept the caucuses. As the governance team had experienced previously, GC members had various views on what Sistani was thinking. Chalabi assured the CPA that Sistani could be brought around; Rubaie insisted that Sistani's demand for elections was firm.[95] Bremer sent a letter to Sistani seeking to reassure him that the November 15 Agreement would not undermine the Shi'ite posi-

---

[92] Chandrasekaran, *Imperial Life in the Emerald City*, p. 205.

[93] Chandrasekaran, *Imperial Life in the Emerald City*, p. 205; author interview with Roman Martinez, July 1, 2007.

[94] Meghan O'Sullivan to Paul Bremer, "Update on Sistani and the November 15 Agreement," December 1, 2003.

[95] Meghan O'Sullivan and Irfan Siddiq, "Readout of Meetings of GC Members' Meeting Sistani," December 12, 2003.

tion.[96] The CPA also convened the governorate coordinators from the southern, Shi'ite majority provinces to get an on-the-ground appraisal of what ought to be done about the situation. The discussion reflected a variety of competing theories as to what Sistani wanted, but there was a strong sentiment from several CPA officials in the provinces that the coalition should not back down in the face of the cleric's demands.[97] At one stage, the CPA considered nine different options for selecting the transitional assembly, including sticking with the November 15 Agreement, conducting direct elections, employing a variety of caucus methods, convening a national conference, or having the CPA and the UN appoint the assembly.[98] The memo laying out these choices concluded, "Ultimately, the choice faced by the Coalition will be either to proceed with a nonelected TNA (either by persuading Sistani, or moving ahead without his support) or to allow the TNA to be directly elected. There are no intermediate options."[99]

A possible way to break the deadlock between the CPA and Sistani was to ask the UN to render a judgment on the feasibility of elections. Sir Jeremy Greenstock, the senior British representative to the CPA, was initially opposed to such a step:

> We should hold at bay the proposal that the UN should make a recommendation of an alternative to elections. That will take too long and will be divisive, not least with Washington. The better route is to get the IGC and the CPA to agree on a method for seeking popular approval of the [transitional assembly], and then to arrange for Annan to condone that, thus getting Sistani down from his ladder. We may have to offer a bit more by way of popular approval, but the cost will probably be worth it.[100]

---

[96] Paul Bremer to Ali Al-Sistani, December 2, 2003.

[97] Diamond, *Squandered Victory*, pp. 85–86.

[98] "Transitional National Assembly Options," January 20, 2004.

[99] "Transitional National Assembly Options," p. 5.

[100] Jeremy Greenstock to Paul Bremer, "Briefing for Your Return," December 28, 2003.

Other CPA officials felt that if the UN could certify that elections were not possible in the coming six months, Sistani might have a face-saving way to back down.[101] The issue was forced when Hakim wrote to UN Secretary-General Kofi Annan on December 28, 2003, and asked him to make a formal recommendation on whether or not elections could be held by the June 30 deadline. In Washington, Robert Blackwill, recently appointed deputy national security advisor with responsibility for Iraq, embraced the notion of bringing the UN back and promoted it within the administration.[102] After consultations with the United States and the United Kingdom, Annan dispatched Lakhdar Brahimi, who had helped midwife the new government of Afghanistan, to Iraq in early February. Brahimi concluded that elections could not be held for at least a year but also recommended scrapping the caucus system.[103] Sistani reluctantly accepted these findings.[104]

Political realities—in Iraq and in the United States—had again forced the CPA to revise its plan for democratizing Iraq. The CPA had accepted a firm deadline by which it would transfer sovereignty to Iraqis, accepted Sistani's demand for a directly elected constitutional assembly, invited the United Nations to take a significant role in Iraq's political process, and abandoned its new proposal for selecting an interim government. Despite all these changes, significant issues remained unsettled. Brahimi's report did not specify how the transitional government would in fact be chosen, and the interim constitution still had to be written.

Drafting an Interim Constitution

The CPA sought to make the most of its remaining months by helping to put in place an institutional structure that would last after the formal

---

[101] Author interview with Paul Bremer, July 30, 2007.

[102] Author interview with Scott Carpenter, August 2, 2007.

[103] Judy Aita, "Iraqi Elections Could Be Held by Year's End, Report Says," *Washington File*, U.S. Department of State, Bureau of International Information Programs.

[104] Allawi, *The Occupation of Iraq*, p. 230.

end of the occupation. The most important vehicle for doing so was the interim constitution, which came to be called the Transitional Administrative Law, or TAL, a name that had been chosen to avoid transgressing Sistani's dictate against unelected authorship of a constitution.[105] In the weeks after the November 15 Agreement, the Governing Council appointed a committee that set about writing this document. Two drafts emerged—Pachachi produced a constitution that was primarily a brief statement of liberal principles, and the Kurds produced an alternate version that granted them maximal autonomy.[106] The discussions soon bogged down, in part because of the Kurdish demands. In January, while the CPA was still mulling options for selecting the transitional government, a new, informal drafting committee convened consisting of Iraqi lawyers Feisal Istrabadi and Salem Chalabi and CPA officials Larry Diamond, Roman Martinez, and Irfan Siddiq.[107] They began a detailed process of writing and refining a new document. At one point, the State Department questioned whether such a detailed document would be necessary, suggesting instead a brief statement of principles. The CPA response emphasized that the intention had always been to produce a detailed document for two reasons. First, clearly defining the separation of powers for the sovereign government was necessary to prevent chaos in its functioning. Second, the TAL would provide a blueprint for the permanent constitution, thus making it one of the most important means for the CPA to influence Iraq's political development.[108]

The CPA had three principal objectives for the TAL: (1) creating a democratic system of checks and balances; (2) preserving Iraq's viability as unified, federal state (in contrast to Kurdish demands for near-total autonomy); and (3) preventing Islam from becoming the controlling basis for law (in contrast to Shi'ite Islamist demands for a

---

[105] Noah Feldman and Roman Martinez, "Constitutional Politics and Text in the New Iraq: An Experiment in Islamic Democracy," *Fordham Law Review*, Vol. 75, No. 2, November 2006, p. 895.

[106] Author interview with Scott Carpenter, August 2, 2007.

[107] Diamond, *Squandered Victory*, pp. 142–145.

[108] Roman Martinez to Richard Jones, "DC Meeting Tonight," January 30, 2004.

more prominent role for *sharia*). The CPA played a critical role as an intermediary to the Kurds. To make the Kurdish region more palatable to Arab Iraqis, the CPA proposed language by which any group of governorates could create "regional blocs of common interest."[109] This provision sought to demonstrate that "federalism is for all of Iraq, and is not merely a favor for the Kurds."[110] In addition, the CPA sought to avoid potentially explosive territorial disputes, most notably in Kirkuk, a city claimed by both Arabs and Kurds that controlled access to nearly 40 percent of Iraq's oil reserves. Another controversial issue was the threshold for ratification of treaties by the transitional government. According to CPA constitutional advisor Larry Diamond, the U.S. government pushed for the lowest bar possible to enable the interim government of Iraq to reach a status-of-forces agreement covering future U.S. military activity and presence in Iraq.[111]

In mid-February, Pachachi circulated a Chairman's Draft of the document to the broader GC, which then inaugurated two weeks of day-and-night negotiation among the council members on the final text. A number of contentious issues were addressed in the last two weeks of discussion, including the manner in which the TAL could be amended, the role of Islam, the structure of the presidency, and the method of ratification.[112] After an all-night session, the GC unanimously approved a final draft early in the morning of March 1, 2004.[113] The result was a progressive document that enshrined a federal, parliamentary system of government with a three-person presidency council. It referred to Islam as "a source" but not *the* source of law, guaranteed an extensive set of basic rights, and sought to reserve a quarter of the seats in the parliament for women. It provided for the election of the TNA by January 31, 2005, the ratification of a permanent constitution

---

[109] Meghan O'Sullivan and Roman Martinez, "Follow Up to January 29 PC," January 30, 2004.

[110] O'Sullivan and Martinez, "Follow Up to January 29 PC."

[111] Diamond, *Squandered Victory*, pp. 158–160.

[112] CPA Governance Team to Paul Bremer, "TAL Drafting Strategy," February 21, 2004.

[113] Feldman and Martinez, "Constitutional Politics and Text," p. 896.

by October 15, 2005, and the election of a new government by the end of that year.

Just as the CPA thought that matters had been settled, word came a few days later that Sistani was objecting to Article 61(c) of the TAL, a clause that had been added late in the negotiations at the Kurds' insistence. It allowed the rejection of a permanent constitution if two-thirds of the voters of any three provinces chose to vote against the document. To Sistani and much of the country's Shi'ite population, this last provision was an unacceptable "Kurdish veto" on the TAL process. Sistani explained his opposition in a letter to Bremer with a unique analogy to American constitutional law: "Can the blacks of America veto the vote of the American people? Can the Spanish people of America veto the entire will of the American people?"[114] The TAL signing ceremony, which had been scheduled for March 5, was abruptly postponed. However, Shi'ite members of the GC, unlike as in previous instances, were eventually able to prevail on Sistani to change his mind. The TAL was signed into law by the Governing Council on March 8, 2004.[115]

The TAL signing did not entirely dampen popular discontent with the document. Handbills started appearing in mosques and bazaars decrying the TAL as a document that, in the words of one flyer, was "made behind doors under pressure of the occupiers . . . so as to finish it before the election campaign of Bush."[116] The handbill detailed a variety of specific objections to the TAL's provisions, including the so-called Kurdish veto. The CPA noted evidence of "orchestrated opposition to the TAL" and developed a strategy to fight back, including releasing pro-TAL pamphlets, political outreach by Bremer and the GC, and communication with Sistani and senior Shi'ite politicians.[117] The memo detailing this strategy emphasized that broad acceptance of the TAL was essential to the coalition's strategic objective of a demo-

---

[114] Bremer, *My Year in Iraq*, p. 304.

[115] "Law of Administration for the State of Iraq for the Transitional Period," March 8, 2004.

[116] Diamond, *Squandered Victory*, p. 183.

[117] Meghan O'Sullivan to Paul Bremer, "Combating the Anti-TAL Campaign," March 21, 2004.

cratic Iraq.[118] As this process went forward, the CPA also sought to address the pressing and unresolved question of how the interim government would be chosen.

With time running short, the governance team briefly considered convening a national conference to expand the GC into an interim government.[119] The impetus for doing so was resistance within the GC to Washington's proposal that the UN be invited to send Brahimi back to Baghdad to help select the interim government.[120] The Shi'ite members of the council in particular distrusted Brahimi, both because he was a secular Sunni and because the UN as an institution was deeply unpopular in Iraq by reason of its association with the sanctions regime.[121] Nevertheless, Rice and Bremer felt that there was no effective option other than having the UN return, and Bremer pressured the GC to agree.[122] On March 17, 2004, the GC sent a letter to New York inviting the return of Brahimi. The UN team had its own concerns about returning. O'Sullivan reported a conversation with Brahimi's top political advisor, Jamal Benomar, who emphasized the difficulty the UN would have in accessing the intricacies of Iraqi politics, given its months-long absence from the country, and emphasized the need for cooperation from the GC.[123] In late March, the UN dispatched Carina Perelli, the head of the organization's Electoral Assistance Division, and Brahimi arrived a week later.

---

[118] O'Sullivan to Bremer, "Combating the Anti-TAL Campaign," p. 2.

[119] Irfan Siddiq to Paul Bremer, "Next Steps on the Political Process," March 10, 2004.

[120] Siddiq to Bremer, "Next Steps."

[121] Author interview with L. Paul Bremer, July 30, 2007.

[122] Bremer, *My Year in Iraq*, p. 311.

[123] Meghan O'Sullivan to Paul Bremer, "March 19 Conversation with Brahimi Advisor Jamal Benomar," March 19, 2004.

Conclusion

It took nearly six months from the onset of the war for the United States to fully articulate its plan for the return of sovereignty to Iraq. By then, the original American project for Iraq was clearly failing. Violent resistance to the occupation was rising, and most Iraqis blamed the United States for the damage. As a result, the seven-step plan outlined by Bremer on September 3 was abandoned in favor of a more expeditious approach that moved both elections and the adoption of a permanent constitution to after this transfer.

In retrospect, it would have been better to have clarified American intentions at an earlier date. It might also have been better to have begun with the more expedited process that was eventually adopted. Doing so would have recognized that the United States was not going to deploy enough of the assets needed—in terms of troops, civilian officials, and money—to effectively secure and govern Iraq for the extended period necessary to draft a permanent constitution and then hold national elections before empowering an Iraqi government. But that conclusion was not evident at the time, and Bremer's more deliberate plan was consistent with the advice he was getting from experts in the fields of democratization and postwar reconstruction. Indeed, most such experts were urging him to adopt a slower rather than a faster move toward the restoration of sovereignty than he intended.

The CPA successfully adjusted to the new and much accelerated timetable agreed on in November, and set in train the various steps needed to effectuate the transition by the mid-2004 deadline. These included the elaboration of a liberal interim constitution, a strengthened bureaucracy, and the beginnings of a more coherent interagency structure for managing Iraq's national security apparatus. The Iraqi government that took over was weak, divided, corrupt, and incompetent. In this, it was not much different from many other postwar regimes—in Bosnia, Kosovo, or Afghanistan, for instance. In Iraq, however, the conflict was not over, and the new interim government proved even less capable than the CPA of stemming the rising tide of sectarian violence. But the Transitional Administrative Law that the CPA had been instrumental in fostering did form the basis for a democratic constitution subsequently adopted by the Iraqi people, and the election of a new government that has, with considerable American assistance, gradually improved security throughout the country.

CHAPTER TEN

Disarming Militias and Countering Insurgents

The CPA's closing months were dominated by mounting opposition from two groups: Sunni insurgents and Shi'ite militia. These two threats came together in the spring of 2004 in a manner that almost derailed the upcoming transfer of power.

Muqtada al-Sadr

Among the militia, the most militant were adherents of the radical young cleric and demagogue, Muqtada al-Sadr. The CPA had first become concerned about al-Sadr in the early summer of 2003. Judge Don Campbell, the CPA senior advisor for the Ministry of Justice, told Bremer that an Iraqi judge, Raad Juhi, had found convincing evidence of al-Sadr's direct involvement in the April murder of Ayatollah Abd al-Majid al-Khoei. The magistrate had issued a warrant for the arrest of al-Sadr and several of his senior officials. In July, Clay McManaway sent a memo to Secretary Rumsfeld noting that there was "concern over the likelihood that Muqtada al-Sadr will attempt to repeat last weekend's performance in Najaf this Friday" and that "armed elements will again infiltrate the city and the shrine."[1] The newspaper *Voice of al-Sadr* published a list of 124 names under a headline, "Long ages in hell await the tyrants." One of those on the list was shot to death two weeks later. The newspaper prefaced the list by stating: "Sooner or later the

---

[1] Memo from Ambassador Clayton McManaway to Secretary Rumsfeld, "Re: CPA Highlight," July 23, 2003.

hands of the people will reach out to this list of the agents of Saddam's tyranny. Woe betide them! They shall be cursed by the worst torture in this world and in the hereafter!"[2] Bremer sent a memo to Secretary Rumsfeld in August, noting that Najaf's religious leaders, including the two Grand Ayatollahs, were concerned by al-Sadr because he was trucking thousands of activists into Najaf every week from Baghdad's poorest neighborhoods.[3]

Juhi told the CPA that he had two eyewitnesses prepared to testify that they heard al-Sadr give the order to kill al-Khoei. But before asking the Iraqi police to make the arrest, the magistrate wanted to conduct an autopsy of the body. He needed the permission of Grand Ayatollah Sistani because al-Khoei was buried at the shrine of Ali, Shi'ite Islam's holiest site. Juhi said he expected this process to take several days. The CPA came up with a detailed list of reconstruction projects they could put into effect quickly in Najaf and in the Shi'ite neighborhoods of Baghdad following the arrest. Indeed, senior CPA officials strongly supported arresting al-Sadr. In a meeting on August 9, for example, CPA officials concluded that "if we still do nothing" when the investigation of al-Sadr was completed, "we'll have an even bigger al-Sadr problem on our hands."[4] CPA officials argued that the most effective option would be to work with Iraqi police and justice officials to issue an arrest warrant, and then target al-Sadr.

But CPA's push to arrest al-Sadr unraveled in mid-August. McManaway received a call and memo from Doug Feith with a list of concerns about arresting al-Sadr.[5] Some in the U.S. military were beginning to get nervous about arresting as-Sadr because it might lead to unrest in the south and in the Shi'ite sectors of Baghdad. This

---

[2]   Info Memo from Hume Horan to the Administrator, "Subject: Muqtada al-Sadr's Published Threats," July 31, 2003; and memo from Secretary Rumsfeld to L. Paul Bremer, III, "Re: CPA Issues," August 4, 2003.

[3]   Memo from Rumsfeld to Bremer, "Re: CPA Issues."

[4]   Memo from Scott Carpenter, Dan Senor, and Hume Horan to the Administrator, "Subject: Our Discussions on al-Sadr," August 9, 2003.

[5]   Author interview with Clayton McManaway, December 5, 2007.

anxiety was compounded following riots in the southern Shi'ite city of Basra on August 11 and 12. The 1st Marine Expeditionary Force began to actively campaign against al-Sadr's arrest, possibly because it was due to rotate out of Iraq in a few weeks and didn't want trouble.[6] As Lieutenant General Ricardo Sanchez acknowledged, and the U.S. military concurred:

> When my staff and I evaluated the situation, we agreed with the Marines, who didn't like the idea of trying to arrest al-Sadr. They were three weeks away from going home and they didn't want to create any instability. In addition, all the multinational forces, led by the Poles and Spaniards, were actively flowing into the region. . . . The fact that most of the coalition nations would not engage in offensive operations also created a problem. I concluded that we were simply too vulnerable while in transition, and that this was the wrong time to undertake the mission.[7]

General John Abizaid also agreed, noting that Sanchez was "absolutely right" and "we can't fracture the coalition even before we put it together."[8] Neither Sanchez nor Abizaid shared these reservations with Bremer, however.[9] CIA headquarters in Langley sent President Bush an assessment about the "risks of action" against al-Sadr, noting that the United States should ignore him and that his support base was weakening. The CIA was also concerned about a Shi'ite uprising if al-Sadr was seized or killed.[10] On August 18, McManaway spoke to Douglas Feith on the phone. "Secretary Rumsfeld requested that I send a series of questions to Bremer and McManaway," noted Feith. "They were questions that the Secretary felt needed to be answered before action was taken."[11] The questions included the following:

---

[6] Author interview with Clayton McManaway, December 5, 2007.

[7] Sanchez, *Wiser in Battle*, p. 246.

[8] Sanchez, *Wiser in Battle*, p. 247.

[9] Author interview with L. Paul Bremer, July 30, 2007.

[10] Author interview with Frank Miller, June 6, 2008.

[11] Author interview with Douglas Feith, November 4, 2008.

- Who will arrest Sadr? If an Iraqi, will it be a Sunni or a Shi'ite?
- Who would detain Sadr and where? What is the process there after—how long before trial, then what, etc.?
- What is your plan to inform and guide Iraqi public opinion about the arrest? International opinion?
- Have you consulted with the Shi'ite clerical leadership in Howza, in an-Najaf?
- What would be the role—if any—of the Governing Council?
- Is this something that must be done now? Or can it wait for the results of your campaign to inform Shi'ite opinion?
- How do you plan to consult with the UK?[12]

McManaway interpreted these questions as foot-dragging, since Bremer had been in contact with the Pentagon and with Rumsfeld personally, about arresting al-Sadr, repeatedly answering such questions. On August 15, for example, Bremer had written a memo answering many of the same questions. Bremer felt he had already gone over these matters with Rumsfeld, in writing and on the phone, during the run-up to the proposed action. "Feith told me, 'Don't do it now, we still have questions about this,'" said McManaway. "I was furious. We didn't know whether it had come from Rumsfeld or the White House, but it was communicated to us through Feith. It was a huge mistake regardless."[13] Bremer responded with a memo titled "Muqtada al-Sadr" to Feith on August 19, providing answers to the questions.[14] To CPA officials, it appeared that one of the most significant concerns from Washington was that targeting al-Sadr might trigger a Shi'ite revolt against the United States and that the benefits of capturing him were heavily outweighed by the potential costs. And there appeared to be little stomach throughout the U.S. government for conducting action against al-Sadr, although, Feith contends, "as far as I am aware, Rums-

---

[12] Memo from Ambassador L. Paul Bremer to Douglas J. Feith, Under Secretary of Defense for Policy, "Subject: Muqtada al-Sadr," August 19, 2003.

[13] Author interview with Clayton McManaway, July 22, 2008.

[14] Memo from Bremer to Feith, "Subject: Muqtada al-Sadr."

feld never nixed the plan to go after Sadr."[15] In any case, after the August 19 call, Bremer recalled, "the focus on al-Sadr faded away for the time being."[16]

But al-Sadr was just getting started, and he and his Mahdi Army continued to present a serious threat to security. Bremer wrote in his memoirs that on the night of October 12, 2003, he was at his desk when a firefight erupted across the river. "Big shoot-out in Sadr City, sir," Scott Norwood reported.[17] Al-Sadr's Mahdi Army took advantage of the pilgrimage to Karbala to assert itself. It stole five municipal vehicles at gunpoint from the municipal garage, occupied a mosque in the center of town, and set up a court. Sheikh Khalid al-Kazemi, a Shi'ite cleric in Karbala, announced that he did not recognize the CPA or the local government. CPA assessments indicated that the governor "clearly feels intimidated, wishes that the CPA would take decisive action across the board, but is hoping that any confrontation with Jaish al-Mahdi (Mahdi Army) will take place elsewhere than in his city."[18]

On October 15, al-Sadr took over the state-owned Samir Hotel in Najaf and renamed it his ministry of defense.[19] Fears in Karbala were acute. The governor of Karbala told CPA officials that "the people of the city feel they are experiencing a 'civil war' and that action has to be taken on a national scale to strip all the parties and all individuals of weapons."[20] The previous week, al-Sadr had named the coalition "terrorist occupiers." He had started wearing a white burial cloth instead

---

[15] Author interview with Douglas Feith, November 4, 2008.

[16] Author interview with L. Paul Bremer, November 15, 2007. Author interview with Clayton McManaway, December 12, 2007.

[17] Bremer, *My Year in Iraq*, p. 190.

[18] Email from John Berry to Mike Gfoeller, Amb Patrick Kennedy, Charles Heatly, Catherine Dale, Simon Cholerton, and Douglas Brand, "Subject: Security Situation in Karbala Following Moqtada Sadr's Announcement of a New Government," October 13, 2003.

[19] Email from Julie Chappell to Baghdad Governance, "Subject: Regional Sitrep Highlights," October 20, 2003.

[20] Email from John Berry to Mike Gfoeller, Scott Carpenter, Maura Connelly, Paul Bremer, Patrick Kennedy, Catherine Dale, David Richmond, Scott Norwood, and Jessica LeCroy, "Subject: Karbala Security: Sitrep 1700 Tue, Oct. 14," October 14, 2003.

of a dark imam robe, a symbol that he welcomed martyrdom. Al-Sadr was also collaborating with a radical Sunni cleric, Ahmed al-Kubaisi, and was busing Sunni extremists from central Iraq to the south to augment his militia.

As in August, there were major differences about how to respond. People in the first camp, which included some in the U.S. military and other coalition forces, were hesitant to arrest al-Sadr. The Marine battalion commander in Najaf argued that "Sadr is playing out a string, and in the end no one will care about him if we just leave him alone."[21] Jeremy Greenstock similarly argued in a video teleconference with Deputy Secretary of Defense Paul Wolfowitz that "Sadr has been inert for some time and . . . his political influence is declining."[22] Still others, including Spanish military officials in Najaf, argued that disarming militias was not their responsibility. As CPA official Robert Ford reported: "I had a long, slightly contentious conversation with the Spanish Lt. Colonel here and the Spanish Major who works with him this morning . . . [They] don't particularly care about disarmament."[23]

The second camp, which included most CPA officials, pushed for action against al-Sadr. As Bremer noted in a meeting with Clay McManaway, Douglas Brand, and Sanchez, "We can't afford to let him get away with it again. We've got to get the Iraqi police to arrest him and the others named on the August arrest warrants."[24] Their plan was to have coalition forces back up the Iraqi police to prevent al-Sadr from slipping back to Sadr city, where rooting him out would be bloody and destabilizing. In his nightly email to his wife, Bremer noted: "We have to stop Muqtada now or risk a much wider conflict in the Shi'a

---

[21] Email from Robert Ford to CPA Officials, "Subject: Najaf Sitrep," September 21, 2003.

[22] Email from Scott Norwood to Paul Bremer, "Subject: Notes from DSD SVTCs," December 30, 2003.

[23] Email from Robert Ford to Mike Gfoeller, "Subject: Political Tidbits," October 6, 2003. The email was forwarded to senior CPA officials. Pat Kennedy subsequently replied, noting that "I have talked to Jerry and then have got to CJTF-7; Gen. Sanchez has been told to send a message to the Spanish to include Ford in any and all meetings." Email from Amb. Patrick Kennedy to Mike Gfoeller and Scott Carpenter, "Subject: Further Deterioration in Najaf; Spanish Position Harmful," October 7, 2003.

[24] Bremer, *My Year in Iraq*, p. 191.

heartland as he continues to advance. But of course there will be the usual voices for delay, compromise, and attenuation. One is reminded of Churchill's comment on the period from Munich to the Polish invasion when the voices of moderation led directly to the 'bull's-eye of disaster.'"[25]

Other senior CPA officials agreed. According to Robert Ford, al-Sadr was "muscling people to demonstrate that he matters, and he is making us look irrelevant."[26] In a poignant and somewhat prophetic assessment, Mike Gfoeller, the CPA Regional Coordinator for Center South, argued that "we can no longer postpone action in this matter. In particular, we need to act before he undertakes a dangerously destabilizing action, such as seizing control of the Shrine of Ali in Najaf." Gfoeller continued by noting that the failure to arrest al-Sadr would have devastating repercussions for Iraq:

> The credibility of our dedication to establishing the rule of law in Iraq is being undermined by the widespread perception here that we have failed to move against Sadr out of fear. The vast majority of public opinion in the Shi'ite Heartland, both educated and uneducated, believes that Sadr ordered the murder of Ayatollah al-Khoei. When we respond to our interlocutors that this is a matter for the Iraqi police to investigate, they shoot back that we, as the occupying power, are responsible for law and order under international law. Hence, they say, solving the al-Khoei case and dealing with Sadr's other criminal activities are our responsibility. . . . The longer we wait to take action against Sadr, the more this corrosive atmosphere of fear and repression will spread.[27]

CPA officials argued that al-Sadr's support was growing, not diminishing. As Gfoeller noted in an email exchange among CPA officials, "Sadr's baleful and destructive influence is now felt in all five

---

[25] Bremer, *My Year in Iraq*, p. 193.

[26] Email from Robert Ford to CPA Officials, "Subject: Najaf Sitrep," September 21, 2003.

[27] Email from Mike Gfoeller to Paul Bremer, "Subject: Dealing with Muqtada al-Sadr," September 30, 2003.

provinces of the Shi'ite Heartland, from al-Wasit to Najaf. . . . It is no exaggeration to say that his growing influence threatens to undo all the progress we have made here since May." He continued by noting that "unless justice is seen to be done with regard to Sadr, we will never be able to establish true stability in this region of Iraq."[28] The Office of General Counsel at the CPA argued that there were clear legal grounds to arrest al-Sadr, in addition to the murder of Ayatollah Abd al-Majid al-Khoei. Al-Sadr had seized control of the Kufa mosque, broken the locks, and taken money belonging to the mosque. He had attempted to break into the Imam Ali holy shrine and steal money on two occasions: September 2 and October 3, 2003. And al-Sadr erected barricades on public streets, which violated paragraph 358 of the Iraqi Penal Code. The Office of General Council concluded that "the Administrator [should] authorize the apprehension of Muqtada al-Sadr by Iraqi and Coalition Forces police, at a time and place that they determine to be appropriate from a law enforcement perspective."[29]

In mid-November, CPA planned another effort to arrest al-Sadr despite opposition among some in the U.S. military. But Bremer eventually postponed the effort because of the November 15 Agreement, which spelled out Iraq's path to sovereignty. "I decided to hold off this time," Bremer noted, "because the big story was the November 15 Agreement. I didn't want the Muqtada ordeal to take center stage."[30]

Al-Sadr surfaced again in January. While on a trip to Washington, Bremer got word from Richard Jones in Baghdad that al-Sadr's

---

[28] Email from Mike Gfoeller to Paul Bremer, "Subject: Achieving Victory in South-Central Region: Course Corrections," October 18, 2003. As Scott Carpenter argued, any arrest of al-Sadr would have to be followed by a public affairs strategy. See Info Memo from Scott Carpenter to Administrator, "Subject: Moqtadeh al-Sadr: Political and Public Affairs Strategy," October 12, 2003.

[29] Action Memo from Office of General Counsel to the Administrator, "Subject: Request for Approval to Apprehend Muqtada Sadr," October 7, 2003. The office wrote an additional memo several days later. See Info Memo from the Office of General Counsel to the Administrator, "Subject: Response to Questions on Muqtada Sadr Action Memo," October 11, 2003.

[30] Author interview with L. Paul Bremer, November 15, 2007, and author interview with Clayton McManaway, December 5, 2007.

men had marched on the mosque of Imam Ali in Najaf. The local police had chased them off, but the mob returned in larger numbers and announced their intention to establish a sharia court in the sacred precinct to "try" four Iraqi policemen they had kidnapped several days before. The Spanish troops responsible for Najaf were getting nervous and talking about a "dialogue." Al-Sadr's timing was perfect: The governor of Najaf and the Custodian of the Shrine were in Saudi Arabia for the hajj, as was much of Najaf's elite. In an email to Bremer, Rick Olson painted an alarming picture and argued that al-Sadr would be more difficult to deal with for at least three reasons.

The first was immunity. Since neither the coalition nor Iraqis were comfortable using force within the walls of the sanctuary, al-Sadr's Mahdi Army would be accountable to no one. The second was money. The shrine, like the great cathedrals of Europe, was a repository of holy treasures, and, in Olson's words, was "a cash cow because of pilgrims' alms." The third reason was prestige. Al-Sadr's control of the Imam Ali mosque would put him in charge of two of Shi'ite Islam's holiest sites, since he already controlled the mosque at Kufa, the site where Ali was murdered. The local leadership, Olson told Bremer, was simply too weak and divided to confront al-Sadr. Olson's recommendation was blunt: "The Coalition must act to detain Muqtada Sadr for the murder of Abdul Majid Al-Khoei in April of 2003. It is now clear that unless we take action against him personally, the outrages will continue. We have fired too many shots over his bow to no effect."[31] Bremer agreed with Olson's suggestion, noting that he asked Jones to convene a meeting with CJTF-7 to get a plan of action into place.[32]

At Bremer's request, Jones sent a cable recommending that the CPA support the Iraqi police in arresting al-Sadr. The cable provoked CIA headquarters to send the President a paper warning that al-Sadr's

---

[31] Email from Rick Olson to Paul Bremer, "Subject: Situation in Najaf," January 25, 2004; also see email from Rick Olson to Richard Jones, "Subject: RE: Situation in Najaf," January 25, 2004.

[32] Email from Paul Bremer to Rick Olson, "Subject: RE Situation in Najaf," January 24, 2004. Mike Gfoeller strongly supported Olson's recommendation. Email from Mike Gfoeller to Paul Bremer, "Subject: Re: Situation in Najaf," January 25, 2004.

arrest would spark major unrest among the Shi'ites.[33] Bremer then asked Mike Gfoeller to provide his assessment of the pros and cons of acting against al-Sadr. Scott Carpenter passed Bremer's request to Gfoeller but could not resist adding, somewhat sardonically: "as if we're actually going to do it."[34] Gfoeller's response was explicit: "My recommendation is unchanged from before with regard to Muqtada al-Sadr and his henchmen. I recommend strongly that they should be arrested under the extant warrants issued last summer, issued for the murder of Abdul Majid al-Khoei."

Gfoeller argued that al-Sadr was also strongly suspected of involvement in numerous killings, including the assassination of senior judge Muhan al-Shuwayri in Najaf and the local representative of the Ministry of Science and Higher Education in Diwaniyya, Abdul Adheem Aziz. Al-Sadr also continued to maintain sharia courts in Najaf, complete with jails for holding those convicted. As Gfoeller warned: "Our failure to enforce the outstanding warrants against Sadr and his chief henchmen has greatly weakened the influence and prestige of the CPA in the Shi'ite Heartland. Our failure to respond to his continuous challenges to the legitimate authorities is having the same effect."[35] He argued that the arrest of al-Sadr would likely trigger several days of unrest in Kufa, Najaf, Karbala, Amara, Baghdad, and other cities, but it would likely involve a small percentage of the population and would eventually calm down.

Again, however, there was deep reluctance within the White House, Pentagon, and some coalition allies to confront al-Sadr. On February 2, Bremer told Sanchez bluntly: "We need to arrest Sadr." But Sanchez responded, "I'm waiting for orders from Washington." Rice also informed Bremer that Secretary of State Colin Powell was uneasy about an operation against al-Sadr because the UN was planning to dispatch an election team to Iraq in March, and he didn't want it to

---

[33] Bremer, *My Year in Iraq*, p. 284.

[34] Email from Scott Carpenter to Mike Gfoeller, "Subject: MAS," January 6, 2004.

[35] Email from Mike Gfoeller to Scott Carpenter, "Subject: Recommendation to Arrest Muqtada al-Sadr Now," January 6, 2004. Carpenter forwarded the email to Bremer the following morning.

undermine the fragile political situation.[36] The acting Iraqi governor in Najaf was dismissive, indicating that the situation was under control and Iraqis would negotiate to resolve issues with al-Sadr. The chief of the Shrine Police also challenged the CPA, saying there was no crisis.[37] Brigadier General Fulgencio Coll, the Spanish commander in Najaf, argued that al-Sadr had "no real support" and felt that any attempt to arrest him would increase his support base by making him a martyr.[38] As one CPA assessment cogently noted, the "Spanish are loathe to confront Sadr."[39] Finally, the U.S. military resisted the effort to capture al-Sadr, and he slipped away again.

Fallujah

On March 31, a small convoy of sport utility vehicles carrying Blackwater security guards was ambushed in the center of Fallujah. The gunmen raked the cars with AK-47 fire and set one of them ablaze. A frenzied mob of locals dragged the corpses from the burning vehicles and beat the charred bodies with shovels; two blackened bodies were hung on the city's main bridge across the river. Stuart Jones, CPA's Al Anbar governorate coordinator, reported that the hatred of U.S. forces was palpable: "The message and posture throughout the city remain defiantly anti-coalition. We have no real partners."[40]

This provocation in Fallujah caused the United States to enter into a simultaneous confrontation with its two main adversaries in Iraq,

---

[36] Author interview with L. Paul Bremer, August 12, 2008.

[37] Email from Darian Arky to Alastair Totty, John Public, Curt Whiteford, David Ballard, Irfan Siddiq, Joseph Adamczyk, Judy Van Rest, Meghan O'Sullivan, Michael Adler, Michael Gfoeller, Michael Whitehead, Milton Kinslow, Peter Wilkinson, Ronald Schlicher, Scott Carpenter, Sallay Kakay, "Subject: Sadr Update," January 26, 2004.

[38] Coalition Provisional Authority, "Report on January 20 Sadr Meeting," January 20, 2004.

[39] Email from Rick Olson to Scott Carpenter, Howard Pittman, and Mike Gfoeller, "Subject: MND Meetings Today," January 20, 2004.

[40] Memo from Stuart Jones to the Administrator, "Subject: Weekly GC Update—Al Anbar," April 2, 2004.

insurgents and militias, and to do so at precisely the moment when the process of selecting a new Iraqi government was coming to a head. General John Abizaid, commander of U.S. Central Command, argued that the U.S. Marines, who had only recently arrived in the region around Fallujah, were not yet ready to execute a major operation there. "The timing is not right and they haven't had time to implement their engagement program," he argued. "We should wait." But Rumsfeld overruled him, responding that "we have to attack." He continued that "we must do more than just get the perpetrators of the Blackwater incident. We need to make sure that Iraqis in other cities receive our message."[41] The mission in Fallujah, codenamed Operation Vigilant Resolve, involved eliminating the city as a safe haven for Sunni insurgents, finding and destroying weapons caches, capturing or killing the perpetrators of the Blackwater incident, and preparing the ground for long-term law and order.

In preparation, coalition and Iraqi officials dropped fliers over sections of Fallujah, English-language versions of which are illustrated in Figure 10.1. The CPA was involved in the information operations campaign, which included video footage and compiling interviews on humanitarian aid, convoys, and civil affairs operations for distribution to local citizens and the media. The CPA was also involved in briefing Arab media and television stations and conducting interviews with Western outlets such as Fox and CNN International.[42] But several CPA officials were subsequently skeptical that the information campaign could have overcome massive local resistance to the operation. As Ronald Schlicher later told Richard Jones, for example, the CPA should "not labor under the illusion that very high casualties in Fallujah can be offset by the endless laundry list of scattershot measures included in DoD/OSD's 'strategy' paper circulated so rapidly this morning . . . we should not kid ourselves that they will prevent or mitigate a severely negative reaction."[43]

---

[41] Sanchez, *Wiser in Battle*, p. 332.

[42] Coalition Provisional Authority, Fallujah Information Operations Update, May 6, 2004.

[43] Email from Ronald Schlicher to Richard Jones, "Subject: Fallujah—Possible Political Impact," April 21, 2004.

Figure 10.1
Mock-Up Fliers for Fallujah Campaign

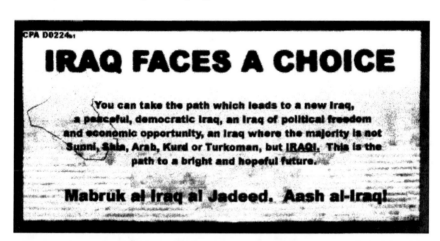

SOURCE: Memo from Rob Tappan to the Administrator, Subject: Proposed Fallujah Fliers, April 25, 2004.
RAND MG847-10.1

The tense environment in Iraq was compounded by the U.S. decision to respond to al-Sadr's provocations, which had become more brazen, as he sought to extend his power and influence. With a heavily armed cohort of bodyguards, he took over the main mosque in Kufa, where the Prophet Muhammad's son-in-law had led worshippers in the 7th century. According to one CPA assessment: "Since last summer, his Mahdi Army militia has grown from 500 to perhaps 6,000 members. Sadr's forces have arrested policemen in Kufa and Najaf. He has invaded the Mosque of Ali in Najaf. He operates an illegal sharia court there." It continued that the failure to arrest al-Sadr had a serious impact on the CPA:

> Our failure to arrest him so far has convinced the population in my Shi'ite provinces that we are not serious when we speak of equal justice for all and the rule of law. . . . Moreover, our failure so far to shut down the illegal, Sadr-controlled shari'a court in Najaf has radically reduced our prestige there, while increasing his. Without question, we need to implement the warrants and arrest Muqtada al-Sadr and his henchmen as soon as possible, if only to restore our credibility among the Shi'a and build support for the rule of law.[44]

A CPA report noted that United States military police driving past al-Sadr's Kufa Mosque observed 50–75 Mahdi Army militia around the mosque, displaying weapons such as AK-47s and rocket-propelled grenades. "We believe the show of force," it noted, "reflects militia muscle-flexing in the wake of the Ashura bombings in Karbala/Baghdad."[45]

On March 18, reacting to a sermon by al-Sadr that was printed in his *Al Hawzah* newspaper in which he praised the 9/11 attacks on New York and Washington as "a blessing from God," the CPA closed down the paper. On April 3, coalition forces arrested Mustafa Al Yacoobi for his alleged involvement in the murder of Abd al-Majid al-Khoei. As National Security Advisor Mowaffak Al-Rubaie argued, the "ineffectiveness of the Governing Council," the "continuing presence of coalition troops in Iraq (principally U.S. forces)," and the "continuous lack of security 1 year after the fall of Saddam" were leading factors of the

---

[44] Email from Mike Gfoeller to Paul Bremer, "Subject: South Central Region: Progress, Opportunities, and Risks," March 2, 2004.

[45] Coalition Provisional Authority, "Spot Report: Moqtada Sadr Show of Force in Kufa; Shari'a Court Activity," March 8, 2004. Also see Coalition Provisional Authority, "Najaf Security Update," March 25, 2004.

crisis.[46] On April 4, large-scale violence broke out in Najaf when supporters of al-Sadr attacked coalition forces. The next day, CPA's Office of General Counsel sent Bremer, at his request, an action memo seeking the detention of al-Sadr "for inciting violence in disregard of CPA Public Notice Regarding Public Incitement to Violence and Disorder, 5 June 2003."[47] Bremer put the memo aside pending a decision to take such a move. The CPA composed a political plan for dealing with al-Sadr and his militia. Its goals were to bring al-Sadr to justice and defeat the Mahdi Army by applying increasing pressure to separate him from the Iraqi population. It concluded that an aggressive political campaign would help bring justice, and coalition military and intelligence activities "will destroy Muqtada's military capability."[48] The only way to convince the Mahdi Army to disband through a demobilization and demilitarization process, David Gompert argued, was "to be attacked so vigorously and relentlessly that they are willing to negotiate a cease-fire. This will involve capturing or killing Sadr and his key lieutenants, and delivering decisive tactical defeats to at least some of the forces."[49]

On the night of April 6, the U.S. Marines commenced Operation Vigilant Resolve, designed to secure control of Fallujah. Iraqi security forces were supposed to work alongside the U.S. Marines during the operation, but many deserted when it began. The Marines encountered stiff resistance from insurgents, who were fighting from fixed defensive positions and armed with machine guns and mortar pits protected by snipers. The fighting caused a massive uproar among Iraqis. "To say that the Fallujah offensive angered the Sunni Muslims of Iraq,"

---

[46] Mowaffak Al-Rubaie, "Comprehensive Approach for the Muqtada al Sadr Crisis," April 28, 2004.

[47] Action Memo from Office of General Counsel to the Administrator, "Subject: Detention of Muqtada al Sadr for Incitement," April 5, 2004; Coalition Provisional Authority, "Directive for the Detention of Muqtada Al-Sadr for Public Incitement to Violence," April 5, 2004.

[48] Coalition Provisional Authority, "A Political Plan for Muqtada and Muqtada's Militia, Version #4," "May 3, 2004.

[49] Info Memo from David C. Gompert to the Administrator, "Subject: Demobilization and Demilitarization (D&D) of Mahdi's Army," April 5, 2004.

remarked Sanchez, "would be a gross understatement."[50] Iraqi Governing Council members were irate, and several threatened to quit unless
the operation was immediately called off. Brahimi, who was in Baghdad trying to put an Iraqi government together, threatened to abort
his efforts and leave. As CPA officials Meghan O'Sullivan and Roman
Martinez wrote in an assessment to Bremer, the "collapse of the Governing Council and fall of the Cabinet are close at hand, especially if
the Coalition opts not to show flexibility or a willingness to negotiate
within the next few days."[51]

By April 8, 48 hours after Operation Vigilant Resolve began, the
White House had called it off. Sanchez recalls in his memoirs a con-
frontational meeting with Bremer and Abizaid. "Ric, it's been decided
that you've got to stop your offensive operations and withdraw from
Fallujah immediately," Bremer told him. Sanchez objected to with-
drawing under fire. "We can't stop now. If we don't finish the mission,
we're going to have to come back and do it later." After a further tense
exchange between the two, Bremer repeated, "You've got to withdraw!
The transfer of sovereignty is in danger!" At this point Sanchez and
Bremer were shouting at each other. Abizaid interjected to calm the
discussion, and eventually Sanchez agreed to order the withdrawal.[52]

Bremer's account of the meeting is less dramatic but equally
revealing. His notes of the occasion state, "We (Jones and Bremer) had
an hour and a half with Abizaid and Sanchez. We achieved a high
degree of agreement on the way forward. Which is to prosecute vigorous offensive operations against MAS (Muqtada al-Sadr's) people and
facilities everywhere but Najaf. Abizaid had an idea of starting to go
after targets in and around Najaf right away based on his reading of
the President's guidance yesterday and his instructions from Rumsfeld. We pushed back only on the timing noting that we need to leave

---

[50] Sanchez, *Wiser in Battle*, p. 350.

[51] Memo from Meghan O'Sullivan and Roman Martinez to the Administrator, "Subject: P9
Meeting on the Insurgency," April 9, 2004. Also see Info Memo from Meghan O'Sullivan
to the Administrator, "Subject: Options Paper for Keeping the Political Process Alive," April
10, 2004.

[52] Sanchez, *Wiser in Battle*, pp. 354–355.

Najaf alone until after the Arba'een pilgrimage. And we all agreed that outside Najaf the military should prosecute the war against MAS with great vigor and speed. We must convince people that we are serious. All of us are concerned, however, that this delay in dealing with the heart of the matter will cost us support among the moderate Shi'a. There are increasing signs of their preparing to side with the MAS, which, we also agreed, is our worst nightmare."[53]

At this point the focus of CPA and U.S. military attention shifted from Fallujah and the Sunnis to al-Sadr and the Shi'ites. Lieutenant General Sanchez had ensured that the military mapped where al-Sadr lived, documented his patterns of travel, staked out the eight- to ten-mile route he took from his home to the Kufa mosque, and concluded that the best course of action would be to launch the operation while al-Sadr was traveling so that they could spare collateral damage. Sanchez then briefed Bremer, who agreed with the decision to target al-Sadr but needed to get permission from the President.

Permission was refused. Sanchez recalls Bremer telling him, based on instructions from Washington, "Your guidance is the following. Do not create any condition where it will even remotely bring the possibility of having an encounter with Muqtada al-Sadr. Such an operation will endanger the transfer of sovereignty."[54]

Blackwill confirms that Washington decided against an effort to capture al-Sadr because this would upset the political environment and trigger a Shi'ite revolt leading up to the July handover.[55] By this point, Bremer himself opposed targeting al-Sadr, who had fled to the main mosque in Kufa. As he explained to Condoleezza Rice on April 6, "We can't conduct an assault on a mosque. But we can go after others close to him."[56]

The combined offensives against Fallujah and al-Sadr triggered a massive backlash among both Sunni and Shi'ite Iraqis. A CPA paper

---

[53] Author interview with L. Paul Bremer, August 12, 2008.

[54] Sanchez, *Wiser in Battle*, pp. 364–365.

[55] Author interview with Robert Blackwill, November 16, 2007.

[56] Author interview with L. Paul Bremer, August 12, 2008.

presented to the U.S. National Security Council noted that "the civilian casualties" from U.S. military operations "have outraged Sunnis (and other Iraqis) and heightened resentment against the Coalition."[57] U.S. and Iraqi reports noted that coalition forces were increasingly being viewed by the Iraqi population as "forces of occupation," and that the U.S. military should increasingly put an Iraqi face on patrol operations.[58] Reports from other provinces also noted that "locals increasingly insist that resistance to the occupation comes directly from the excesses of Coalition Forces."[59] CPA advisors dealt with several crises at Iraqi universities after U.S. military forces "kicked open doors and broke windows because they heard that the Sadr army militia was on campus." As John Agresto, CPA's higher education advisor, told Bremer, these incidents have "the potential to ignite students against us in serious ways."[60]

Support for al-Sadr continued to grow in the aftermath of these pull-backs. According to a May public opinion poll that circulated through the CPA, 81 percent of Iraqis in Baghdad, Basra, Mosul, Babel, Diyala, Ramadi, and Sulaymaniyah acknowledged that their opinion of al-Sadr was "better" or "much better" than three months earlier. In a memo to Bremer, Don Hamilton noted that al-Sadr's popularity in the poll was "almost certainly an artifact of his anti-Coalition stance."[61]

---

[57] Coalition Provisional Authority, "Elements of a Sunni Strategy: CPA Paper for NSC Meeting," April 17, 2004.

[58] Ministerial Committee for National Security, "Security Situation—Updates and Analysis," May 6, 2004; Info Memo from Andrew Rathmell to the Administrator, "Subject: Rapid Reconstitution of New Iraqi Security Forces Capabilities," April 9, 2004.

[59] Memo from Mark Kennon to the Administrator, "Subject: Weekly GC Update—Salah ad-Din," April 21, 2004.

[60] Info Memo from John Agresto to the Administrator, "Subject: Military on Campuses," April 15, 2004.

[61] Info Memo from Don Hamilton to the Administrator, "Subject: Further Results from IIACSS Poll 14–23 May," May 28, 2004; and Independent Institute for Administration and Civil Society Studies, "Public Opinion in Iraq: First Poll Following Abu Ghraib Revelations," May 2004, briefing slide 14.

Disarming Militias

Muqtada al-Sadr's Mahdi Army was not the only armed group of concern to the CPA. The largest such group was the Kurdish Peshmerga, which had fought alongside coalition forces to defeat Saddam. The other militias were Shi'ites attached to various political parties. One was SCIRI's Badr Corps, which had been supported by Iran during its fight against Saddam's regime. The Da'wa, Iraqi National Congress, and Iraqi National Accord parties also had small militia.

The increasing strength of militia forces presented a challenge to the CPA's ability to establish for the Iraqi government what the German philosopher Max Weber has referred to as "the legitimate use of physical force within a given territory."[62] By the end of 2003, the CPA estimated there were over 30 known militias in Iraq with between 30,000 and 60,000 armed supporters.[63] Efforts to disarm these groups were complicated by the history of tolerance for the Peshmerga. One CPA assessment stated that because of its support in the war to overthrow Saddam's regime, "the Peshmerga are authorized, per the weapons policy, to retain heavy weapons under the supervision of the coalition." It continued that "excepting the Peshmerga, who must remain North of the Green line, we cannot accept any other militant formations."[64] In a memo to Rumsfeld, Bremer similarly explained that the national weapons policy "will apply to all of Iraq, although in practice, Peshmerga will be permitted to bear arms north of the Green Line."[65] Rumsfeld supported this position, encouraging Bremer to employ Peshmerga in

---

[62] Max Weber, "Politics as a Vocation," in H. H. Gerth and C. Wright Mills, eds., *From Max Weber: Essays in Sociology* (New York: Oxford University Press, 1946), p. 78.

[63] Coalition Provisional Authority, *Iraq: Integrated Security Sector Development* (Baghdad: Coalition Provisional Authority, Office of Policy Planning and Analysis, December 2003), p. 18.

[64] Coalition Provisional Authority, "Talking Points for Security Meeting with SLC," May 22, 2003. The "green line" refers to the de facto boundary established in 1991 that separates Kurdish areas from the rest of Iraq. After conducting military operations in Kurdish areas, Iraqi forces withdrew to this boundary in response to the threat of U.S. airstrikes.

[65] Memo from L. Paul Bremer to Secretary Rumsfeld, "Subject: National Weapons Policy Principles," May 21, 2003.

a variety of ways, including "to provide security for the pipeline going into Turkey."[66]

Massoud Barzani argued to Bremer that the Peshmerga "were not a militia force like the other parties had, but a symbol of Kurdish dignity and a force that had helped liberate Iraq." There was a need, he continued, "to prevent a power vacuum that the Peshmerga now filled."[67] Barzani said the same thing to other CPA officials, noting that it was "fundamentally wrong to compare the Peshmerga to the party militias that exist in other parts of the country," and that the Kurds "would insist on having some armed guarantee of their safety."[68]

But the CPA's exception for the Peshmerga had a serious cost, legitimizing as it did the existence of other such groups. Abdul Aziz Hakim of SCIRI told the CPA that while he believed in principle "the time for militias in Iraq is over," U.S. policy was inconsistent.[69] Hakim incredulously asked: "Why were the Badr Corps being treated differently to the Peshmerga?"[70] And he noted that the Badr Corps "have received nothing in return while the Peshmerga are looked at with pride and given plenty of opportunities for retraining."[71] In several meetings with Meghan O'Sullivan, Hamid Bayati repeatedly complained about the "preferential treatment the Peshmerga were receiving

---

[66] Memo from Donald Rumsfeld to Jerry Bremer, CC: Gen. Dick Myers, Paul Wolfowitz, Doug Feith, Gen. John Abizaid, "Subject: Security for Pipeline," August 18, 2003.

[67] Cable from HQ Coalition Provisional Authority Baghdad to SECDEF WASHDC, SEC-STATE WASHDC, "Subject: KDP's Barzani Pledges Support for the Coalition Effort on Iraq's National Political Process," June 30, 2003.

[68] Memo from Political Team, Civil Affairs to Special Presidential Envoy L. Paul Bremer, "Subject: Amb. Crocker's July 4 Meeting with Barzani," July 5, 2003.

[69] From the Political Team to Bremer, "Subject: Your meeting with Abdul Aziz Hakim, SCIRI," July 10, 2003. Also see email from Hume Horan to Scott Carpenter, Meghan O'Sullivan, Roman Martinez, Ethan Goldrich, Lydia Khalil, Philip Hall, Martin Hetherington, Jonathan Cohen, Joanne Dickow, and Irfan Siddiq, "Subject: Ambassador Bremer's Meeting with Sayyid Abdul Aziz al Hakim, SCIRI," July 1, 2003.

[70] Memo from Julie Chappell to Ambassador Sawers, "Subject: Political Process; Call on Abdul Aziz al Hakim," June 20, 2003.

[71] Email from Joanne Dickow to Scott Norwood and Patrick Kennedy, "Subject: Background for Meeting with Abdul Aziz Al-Hakim," July 1, 2003.

vice [sic] other militias in Iraq." And he noted that U.S. forces were humiliating members of SCIRI's Badr Corps with repeated raids that confiscated SCIRI's money and weapons. Bayati summed up his discussion by threatening that if U.S. attacks against SCIRI supporters continued, it would be impossible for SCIRI leaders like himself to work with the United States.[72]

In addition, senior SCIRI representatives told the CPA that the deteriorating security environment and low quality of Iraqi forces made disbanding the Badr Corps virtually impossible in the near term. Adel Abdel Mahdi argued, for example, that there was "a security gap between the incompetent police and the over-armed Coalition military, one which caused the assassination of Hakim and the general worsening security situation." He continued that while there was some progress on building Iraq's security institutions, "we cannot be expected to rely on them now. They have proven ineffective at the current time. Instead, all political parties and civil organizations, especially those with security experience like the Badr Corps, should carry the burden of responsibility for providing security."[73]

In an email to McManaway, Mike Gfoeller argued that "we would need to deal with both SCIRI and Da'wa with care, since they are essentially political parties with militias. While we are unwilling to see their armed units usurp the role of the police on the streets of Najaf, we welcome their participation in the GC in Baghdad. Hence the need for political sensitivity in this matter."[74] By January, however, several senior CPA leaders, including David Gompert, were pushing for a much stronger line to disarm all militia. Gompert wrote a series of talking points for Bremer's discussion with Kurdish leaders, noting that "there can be no armed forces outside the Iraqi state's control."

---

[72] Memo from Meghan O'Sullivan, "Meeting with Hamid Bayati, SCIRI," May 22, 2003. Also see memo from Meghan L. O'Sullivan through Scott Carpenter to Ambassador L. Paul Bremer, "Re: Meeting with Abdul Aziz Hakim," May 23, 2003.

[73] Info Memo from Meghan O'Sullivan through Scott Carpenter to the Administrator, "Subject: Readout of 19 October Meeting with Slocombe, Brand, SCIRI," October 19, 2003.

[74] Email from Mike Gfoeller to Clayton McManaway, "Subject: Conversation with Gen. Kelly on the Security Situation in Najaf," September 8, 2003.

Otherwise, he warned, "the U.S. cannot support your political and economic goals." Gompert argued that the CPA would prefer seeing Kurdish units transferred into the Iraqi Civil Defense Corps based on several guidelines: Transfers should be made in small units; the units should be available for active service anywhere they were needed in the country; and they would come under national command (which meant, for the time being, U.S. command).[75]

Overall, the strengthening of militias had a destabilizing impact on Iraqi security. According to one CPA assessment: "The Badr Corps, Mahdi Army, and Da'wa Party militia collectively represent a ticking time bomb under the still fragile foundation we are laying for Iraq's democratic future. Before June 30, we need to take energetic, vigorous, and decisive steps to disarm and disband the three Shi'ite militias in South Central. Otherwise, it is quite likely that the situation in our AOR will deteriorate rapidly thereafter. Already the people of Najaf are anticipating a power grab by SCIRI after June 30."[76]

This assessment was supported by other senior CPA figures, including Gompert, who wrote a blunt memo to Bremer noting that the U.S. military was unwilling to act against the Badr Corps, Mahdi Army, and Da'wa militia. Meghan O'Sullivan similarly told Bremer that "the continued ability of militias—mainly Shi'a ones—to operate freely is not only demoralizing to average Iraqis, but sending a message to Sunnis that the Coalition will turn a blind eye to Shi'a vigilantism."[77] The U.S. military's strategy of leaving disarmament to the Iraqi police and Iraqi Civil Defense Corps left a vacuum that "provides SCIRI/ Badr, Mahdi Army and Da'wa with self-justification to keep and even strengthen armed capabilities."[78] Secretary Rumsfeld actually asked Bremer in a March 2004 memo, "What do you think about having

---

[75] Info Memo from David Gompert to the Administrator, "Subject: Militia Dissolution," January 4, 2004.

[76] Email from Mike Gfoeller to Paul Bremer, "Subject: South Central Region: Progress, Opportunities, and Risks," March 2, 2004.

[77] Memo from Meghan O'Sullivan to the Administrator, "Subject: Readahead for March 30 PC Meeting," March 30, 2004.

[78] Memo from David Gompert to L. Paul Bremer, February 9, 2004.

one of the militias become the guards for the Shia holy sites? Think of it like the Swiss Guard at the Vatican, that has guarded the Vatican for decades."[79]

The CPA did not become serious about demobilizing militias until early 2004, when Gompert and CPA official Terry Kelly began intensive negotiations with the major political parties. Their goal was to disarm, demobilize, and reintegrate all Iraqi militias. Roughly one-third would transition into Iraqi security forces, such as the police and army; another third would be eligible to retire and receive pensions; and a final third would be given job training for the civilian market. "The chances of this working were slim from the beginning, especially given the security environment," noted Kelly. "But it was worth a shot."[80] Gompert and Bremer also recognized that, at this late date, the CPA was going to be able to do no more than chart a path for the successor Iraqi regime to follow if it wanted.[81]

As Gompert and Kelly quickly discovered, each party was cognizant of the status of other militias. SCIRI was suspicious that the CPA might preserve the Peshmerga while dissolving the Badr Corps, and refused to disband as long as the Mahdi Army remained strong in the Shi'ite heartland.[82] Disarmament became more difficult as the handover loomed. Tempers flared in June between Ayad Allawi, a prominent Shi'ite and the interim prime minister, and Jalal Talabani from the Patriotic Union of Kurdistan, after Allawi tried to enforce the demobilization of militias. Talabani's response was emotional, warning Allawi that he "better not think he can become a dictator."[83] Gompert and Kelly helped establish a database for monitoring the process and a system for screening and counseling militia members and families to

---

[79] Memo from Donald Rumsfeld to Jerry Bremer, CC: Gen. Dick Myers, Paul Wolfowitz, Doug Feith, and Reuben Jeffery, "Subject: Militia Guard Holy Sites," March 8, 2004.

[80] Author interview with Terry Kelly, February 27, 2008.

[81] Author interview with L. Paul Bremer, August 12, 2008.

[82] Info Memo from David C. Gompert to the Administrator, "Subject: Badr Corps," January 31, 2004.

[83] Email from Meghan O'Sullivan to L. Paul Bremer, Brian McCormack, Nicolaidis Constantinos, and David Noble, "Subject: Call from Jalal Talabani," June 9, 2004.

identify who was eligible for benefits. They also helped establish training programs so that skills could be translated immediately into jobs.[84] They were successful in obtaining agreements from the major parties for what the CPA called "transition and reintegration." By late May, CPA secured formal agreements with all but one of the major militia groups, then totaling roughly 90,000 fighters.[85] The Mahdi Army was not included in these negotiations both because it was thought unlikely to cooperate and because, unlike the other militias, it was not represented by a then-recognized political party.

In preparation for the late June handover, Gompert and Kelly set in motion a transition and reintegration strategy, but this effort quickly ran into roadblocks. USAID refused to participate because it was pessimistic about the possibility of success. As a senior USAID official in Baghdad noted to Kelly: "We believe the risk of institutional failure is so large that we don't want to get involved."[86] In any case, the militias ultimately refused to disband because of the deteriorating security environment, a desire for power, and a lack of faith in Iraq's security forces to establish security. After the CPA dissolved, neither the U.S. military nor the State Department pursued the transition and reintegration program that the CPA had initiated.

Conclusion

The failure to provide timely and consistent guidance to both the CPA and CJTF-7 on how to respond to months of repeated provocations from Muqtada al-Sadr was primarily Washington's problem. Most agencies were opposed to offensive operations against al-Sadr. Both Bremer and Sanchez sought the go-ahead from Washington on

---

[84] Coalition Provisional Authority, *Transition and Reintegration Strategy* (Baghdad: Coalition Provisional Authority, May 2004); and Action Memo from David C. Gompert to the Administrator, "Subject: Militia Transition and Reintegration," January 28, 2004.

[85] Info Memo from Fred Smith to the Administrator, "Subject: Militia Transition and Reintegration Next Steps," May 22, 2004.

[86] Author interview with Terry Kelly, February 27, 2008.

a number of occasions, although not always in tandem, but they never received a clear decision from Rumsfeld or the President until the final order to stand down in mid-April.

The decision to respond to Sunni and Shi'ite provocations by launching simultaneous offensives against both the Fallujah insurgents and Muqtada al-Sadr's militia was ill conceived. The effort to retake Fallujah was pushed by Washington despite reservations on the part of the American civilian and military leadership in Iraq. Fallujah had to be retaken, and some months later it was, but reoccupation was not essential just as tensions with the Shi'ite militia were cresting, and negotiations to form an interim Iraqi government were approaching the decisive phase. Abizaid argued against the Fallujah operation but was overruled by the President. He subsequently expressed reservations to Sanchez against going after Fallujah and al-Sadr simultaneously. Sanchez responded that he had consulted with his commanders and was sure that they could handle both jobs.[87]

It would have been better to have confronted al-Sadr in August 2003, when he first emerged as a major threat. Better yet, it would have been preferable to have disbanded all the militias coincident with the decision to dissolve the Iraqi army in May 2003. Disarming and disbanding militias is necessarily a military task, even if some degree of acquiescence could be negotiated with some or all of them. In May of 2003, however, Washington was still hoping to reduce the American troop strength in Iraq to 30,000 soldiers by year's end and had no stomach for the confrontations that such a course would have required.

The CPA's 2004 plan for disbanding the militias had some degree of Iraqi buy-in, but these arrangements came too late. They could only have been implemented if they had been backed by more military muscle than Washington was prepared, at that time, to deploy and use for the purpose.

---

[87] Sanchez, *Wiser in Battle*, pp. 333–334.

CHAPTER ELEVEN

Exit and Appraisal

Washington's decision to back off the operations against Fallujah and al-Sadr calmed the Governing Council and allowed Brahimi to proceed with the selection of a new government. Assisted by Bremer and Blackwill, Brahimi began a marathon set of consultations with members of the Governing Council and other notable Iraqis, including tribal leaders, with a view to securing agreement on the composition of an Iraqi interim government. The UN envoy's initial preference for the post of prime minister was Shi'ite nuclear scientist Hussein Shahrastani. Bremer and Blackwill, however, felt that he was too pro-Shi'ite to be a unifying figure and Brahimi soon soured on him as well.[1] After canvassing various other names, Brahimi suggested Ayad Allawi, who was enthusiastically embraced by the Americans because of his secular orientation, perceived toughness, and pro-Western attitude (as an exile, he was known to have had a long association with the CIA). On May 28, 2004, the GC, in an apparent attempt to take credit for the choice and preempt Brahimi, itself announced Allawi's selection as prime minister.[2] After another series of negotiations, an interim cabinet was drawn up. Brahimi then offered the presidency to Pachachi, who turned it down, leading to the ascension of Sunni Sheikh Ghazi al-Yawar.[3] The Iraqi Interim Government announced itself on June 1, 2004.

---

[1] Author interview with L. Paul Bremer, July 30, 2007.

[2] Allawi, *The Occupation of Iraq*, pp. 284–285.

[3] Bremer, *My Year in Iraq*, p. 376.

In his last month in power, Bremer used his authority to promulgate "orders" with the force of law to build up the institutional framework for Iraq's transition to sovereignty. Order Number 97 defined the conditions under which political parties could contest the upcoming elections. The primary authority for regulating the process lay with the Independent Election Commission of Iraq, whose makeup Carina Perelli had announced on June 4, 2004. The most fateful electoral decision Bremer made was Order Number 96, which defined Iraq as a single electoral district. This decision came after weeks of debate within the CPA and the UN about how best to structure elections in Iraq. The core question was whether Iraq should be subdivided into electoral districts that would elect members of parliament or whether parties would put forth national lists on which all Iraqis would vote. A memo laying out the two options noted that a national list would promote party discipline and allow votes to count proportionally in electoral outcomes rather than being "wasted," as happens to losing votes in a first-past-the-post system.[4] On the other hand, the memo went on to note, such a system would create a weak link between voters and their representatives and make it "difficult to ensure that all regions and interests are adequately represented."[5] A multiple-district approach would allow the election of locally recognized leaders and offer an opportunity to smaller parties and even individual candidates.[6] These political actors would stand no chance in a national-list system, which would be dominated by the established political parties, notably SCIRI, Da'wa, and the Kurdish parties. The chief drawback of a multiple-district system would be that it would take some time to actually draw up the electoral districts, and some argued that it would not be possible to do so by January 2005. Nevertheless, the memo did suggest that existing census data could be adapted to the task or even that satellite imagery could be used in drawing the boundaries.[7] The memo closed by noting

---

4   Thomas Warrick to Christopher Ross, "Electoral Systems," undated, p. 1.

5   Warrick to Ross, "Electoral Systems," p. 2.

6   Warrick to Ross, "Electoral Systems," p. 2.

7   Warrick to Ross, "Electoral Systems," p. 3.

deep divisions on the correct choice: The UN elections team, the CPA's electoral contractor, and the governance team favored a national list; the Office of Provincial Outreach (which focused on Sunni areas), the governorate coordinators for southern Iraq and Baghdad, the State Department's Bureau of Near Eastern Affairs, and constitutional advisor Larry Diamond favored multiple districts.[8]

Faced with these divisions, Bremer referred the issue to Washington, outlining the alternatives in a message to Rice and recommending a single electoral district with a national-list system. His recommendation was approved and duly promulgated.[9] In the event, most Sunnis later chose to boycott the election and consequently had few representatives in the new parliament. This was a result that the multiple-district approach could have ameliorated but probably at the cost of delaying the election in a manner unacceptable to Sistani and the Shi'ite majority.

On June 28, 2004, two days before sovereignty was supposed to be handed over, Bremer stood before a hastily gathered contingent of media, opened a blue Morocco leather folder, and read out a letter transferring sovereignty to the Iraqi people and their government. It said: "We welcome Iraq's steps to take its rightful place of equality and honor among the nations of the world." He then handed the letter to chief justice Medhat al-Mahmoud, who was flanked by Prime Minister Allawi, president Sheikh Ghazi al-Yawar, and Deputy Prime Minister Barham Saleh.

In Ankara, Turkey, President Bush was attending a NATO summit meeting. Secretary of Defense Rumsfeld handed him a note from Condoleezza Rice that read, "Iraq is sovereign. Letter was passed from Bremer at 10:26 a.m., Iraqi time." Bush smiled, scrawled "Let Freedom Reign!" on the note with a black marker, and passed it back to Rumsfeld, who grinned broadly.[10]

---

[8]   Warrick to Ross, "Electoral Systems," p. 3.

[9]   Author interview with L. Paul Bremer, July 30, 2007.

[10]   Agence France Press, June 28, Istanbul.

Mission Accomplished or Mission Impossible?

Following Bremer's departure, the CPA was quietly dismantled. Neither the Defense Department nor the State Department was eager to claim its legacy. DoD had no desire to repeat such a foray into the political and economic aspects of reconstruction, nor was State inclined to look to the CPA experience for positive lessons. For both agencies, as for the rest of the world, the now-defunct CPA simply became a convenient repository for blame about everything that had gone wrong during its 14-month existence.

In the course of that relatively brief period, the CPA had restored Iraq's essential public services to near—or in some cases beyond—their prewar level, instituted reforms in the Iraqi judiciary and penal systems, dramatically reduced inflation, promoted rapid economic growth, supported and helped broker what became the largest debt relief package in history, put in place institutional barriers to corruption, began reforms of the civil service, promoted the development of the most liberal constitution in the Middle East, and set the stage for a series of free elections. All this had been accomplished without the benefit of prior planning or major infusions of U.S. aid and despite Washington's inability to fill more than half of the CPA's positions at any time. The CPA failed to hand over power to a competent, united, and honest government, but doing so may never have been a viable option. Measured against progress registered over a similar period in more than 20 other American-, NATO-, and UN-led postconflict reconstruction missions of the past 60 years, the CPA's accomplishments in most of these fields bear respectable, in some cases quite favorable, comparison.[11]

What the CPA did not do is halt Iraq's descent into civil war. With the return of sovereignty, violent resistance to the occupation devolved into an even more violent conflict between Sunni and Shi'ite extremist groups. With respect to security, arguably the most important aspect of any postconflict mission, Iraq comes near the bottom in any ranking of modern postwar reconstruction efforts.

---

[11] For comparisons among post-conflict reconstruction missions, see Dobbins et al., *America's Role in Nation-Building, The UN's Role in Nation-Building,* and *Europe's Role in Nation-Building.*

The CPA thus did reasonably well, by historical standards, in most areas for which it had the lead responsibility, but it failed in the most important task, for which it did not. The degree to which one judges the CPA's overall performance, therefore, must depend heavily on how one assesses its contribution to the deteriorating security situation.

The United States went into Iraq with a maximalist reform agenda—standing up a model democracy that would serve as a beacon for the entire region—and a minimalist application of money and manpower. In particular, it deployed only enough troops to topple the old regime, but not enough to deter the emergence of violent resistance or to counter and defeat the resultant insurgency. The difficulties it encountered owe much to this disjunction between the scope of America's ambitions and the scale of its commitment.

The failure to deploy a sufficiently large force, to quickly assume responsibility for public safety on the collapse of the former regime, and to institute appropriate counterinsurgency tactics once an insurgency emerged, cannot be laid at the CPA's door. Bremer raised the question of troop levels with Rumsfeld and President Bush within days of being named to head the CPA. He embraced the public security mission, assigned it high priority, and sought military support in executing it.

The CPA has been accused of compounding, if not actually creating, the security challenge by disbanding the Iraqi army and barring thousands of senior Ba'athist officials from public office. Both of these actions did further antagonize a Sunni minority that was, in any case, facing a wrenching loss of power and prerogative occasioned by the move toward representative government. It is not clear whether the old Iraqi army would have proved any more effective at domestic policing than did the Iraqi police, which was not disbanded. Neither was retaining senior Ba'athists in top government positions entirely feasible, even if it had been desirable, given the intense public antipathy toward that party and the not entirely unreasonable desire of Iraq's new leaders to fill these jobs with their own adherents, rather than Saddam's. Any representative government in Iraq was going to shift power, wealth, and influence from the Sunni to the Shi'ite and Kurdish communities. Within limits, this was justified, given that the Sunni community had been disproportionately favored under Saddam. Contrasted with the massive ethnic cleansing that eventually took place under the Iraqi governments that succeeded the CPA in 2004–2005, Bremer's measures seem rather mild. Would sectarian passions have been cooled if the CPA had retained the army and a larger number of the Ba'athist elites? Perhaps, but the opposite seems equally possible. Bremer came to regret having turned the administration of de-Ba'athification over to Iraqi political leaders, but this does not mean that he could have safely resisted their pressures altogether. Perhaps a more limited measure would have been wiser, but the CPA could not afford to lose the support of the Shi'ite and Kurdish as well as the Sunni communities, and most of their leaders wanted an even deeper purge.

What does seem clear is that the CPA moved too slowly to rebuild an army and reform the police. Earlier and larger-scale efforts to train, equip, and then oversee the subsequent development of both forces might not have prevented the emergence of an insurgency, but they would certainly have made it easier to counter. Bremer was ready to turn military training over to the U.S. military in early 2003, but lower levels of the CPA seem to have resisted the shift, and Bremer and Sanchez could not bring the issue to a conclusion until Rumsfeld made the decision for them in March 2004. Bremer fought efforts to shift police training to the U.S. military in 2004, but was overruled. It is not clear whether any of this made much difference. The U.S. military had made a complete hash of military training in Afghanistan in 2002 (nearly all the new recruits deserted once their initial training was complete). In Iraq, the shift to U.S. military leadership eventually produced larger numbers of police and soldiers but little early improvement in quality. In fact, the Iraqi police actually became much more abusive in 2005–2006. It seems that in 2003, neither the civilian nor military components of the U.S. government were equal to recruiting, training, mentoring, and monitoring forces on the scale required. This does not excuse the CPA's failure to have started earlier and more expansively, but it does put them in some perspective. Formally dissolving the army was probably an unnecessary and counterproductive gesture; failing to recall a larger number of former soldiers more quickly was the more costly mistake.

Given what the CPA had to work with in the way of extant plans, human resources, and funding, it now seems apparent that its larger mission—to establish a peaceful, functioning, secular democracy within a united Iraq that could serve as a positive model for the region—could never have been achieved with the manpower, money, and time available to it.

Having been set what in retrospect seems an impossible task, Bremer and his team were then left bereft of adequate support, backstopping, and oversight. For the first six months of the occupation of Iraq, Washington seemed to be barely paying attention to developments there. Bremer's reports were initially not being circulated beyond the Department of Defense; the White House was not pushing for an interagency process; and Bremer, while subjected to copious advice, was receiving little direction. Bremer's two most controversial decisions, disbanding the army and firing thousands of Ba'athist officials, had been carefully reviewed and fully approved by his superiors, but these measures had not been adequately debated or fully considered by the rest of the national security establishment. President Bush himself was sometimes surprised by Bremer's decisions, although inclined to back him and defer to his judgment as the man on the spot.[12] If senior levels of State, Defense, and the White House were occasionally surprised and displeased by decisions made in Baghdad, the failure lay principally with Washington for not establishing a clear and transparent channel for disseminating Bremer's reports and for giving him instructions.

The decision to assign oversight of the CPA to the Defense Department was an important contributing factor to this lack of support and supervision. Whatever sense it may have made in the abstract to shift responsibility for the nonmilitary aspects of nation-building in Iraq to DoD, doing so only a few weeks before the invasion imposed immense start-up costs on the operation. DoD, despite its wealth of resources, had no modern experience with setting up, supporting, and running a branch office of the U.S. government half a world away. The result was a series of heroic, but in many cases unnecessary, improvisations, as

---

[12] Author interview with Andrew Card, November 15, 2001.

new arrangements had to be established to handle tasks long familiar to the Department of State but new to DoD.

One of the administration's more serious conceptual errors was to model its efforts in Iraq on the post–World War II occupation of Germany and Japan. Those two countries were both highly homogenous societies, with no likely proclivity toward sectarian conflict. They were first-world economies whose populations did not need to be taught how to successfully run a free-market system. And both countries had surrendered unconditionally. Iraq in 2003, by contrast, looked a lot more like Yugoslavia in 1995—ethnically and religiously divided, with an economy wrecked by war and sanctions and a pattern of historic sectarian grievances. The deceptive ease with which a democratic transition had been arranged in Afghanistan 15 months earlier encouraged an underestimation of the costs and risks of nation-building on this scale. Had the administration recognized that it was taking on tasks comparable to those NATO had assumed only a few years earlier in Bosnia and Kosovo, but in a society ten times bigger, it might have scaled up its initial military and monetary commitments and scaled back its soaring rhetoric. (Alternatively, of course, such a realization might have caused the administration to reconsider the entire enterprise.)

It was also a mistake for the United States to have premised so much of its appeal to the Iraqi people on an improvement in their economic circumstances. This emphasis on the economic aspects of reconstruction derived, in some measure, from a very inaccurate reading of history, in particular of the post–World War II German and Japanese occupations. Germany did not receive significant reconstruction aid until 1948, and Japan never did. In both societies, democratic political reforms had been put in place well before their economic revitalization. Growing prosperity helped consolidate democracy, but it did not precede or even accompany its introduction. In Iraq it would have been better to confine American promises to (1) liberating the Iraqi people, (2) protecting them, and (3) allowing them to choose their own government. The above promises, if fulfilled, would also have promoted more sustained economic growth than did pumping in large amounts of aid into the midst of a civil war. This is not to argue that the United States should not have helped rebuild Iraq, as it did, but rather that it should have put security first and employed economic assistance as a contributing element in a larger strategy focused on public safety and political reform.

Given the circumstances in which they found themselves, Bremer and his team performed credibly. Senior levels of the CPA staff were generally competent and experienced, although the turnover was too rapid. Not every decision was optimal, but choices were made in an orderly fashion on the basis of professional advice, despite the hectic pace of events. Bremer was restrained and judicious in the use of his extraordinary powers, sometimes resisting or ignoring ill-considered advice from Washington. Most CPA policies were consistent with best practices that had emerged during postconflict reconstruction missions over previous decades. The results, in most spheres other than security, bear favorable comparison with those of earlier such operations.

On the negative side, the CPA structure was overly centralized, particularly during the first six months. Staff turbulence exacerbated this problem, leading Bremer, not unnaturally, to rely increasingly on the few key staffers who stayed for the duration. The frustration of CPA officers assigned to the provinces was particularly acute because their capacity to communicate with and influence the center was limited both practically, as a result of inadequate communications, and organizationally, as a result of this centralization of decisionmaking.

Preparation for the occupation of Iraq has been rightly criticized. Of planning, there was a good deal, in the State and Defense Departments and in several military commands. However, these disparate streams were never fully integrated into a national plan that could have been given to the CPA leadership when it deployed. This meant that the CPA, chronically understaffed, had to create a strategic plan "on the fly."

Even if it had had a plan, the CPA would have lacked the resources to implement it because Washington agencies had not made the necessary preparations. Contrast the planning and preparation for the conventional battle that toppled Saddam with those for its aftermath. The former plans represented more than a year of intellectual work on the part of the administration's top military and civilian leadership. Equally important, the planning process had been accompanied by the movement of hundreds of thousands of men and tens of thousands of machines into position for battle and by the allocation of tens of billions of dollars for its execution. By contrast, the United States did not mobilize more than a couple of hundred officials in preparation to govern Iraq. The result was to leave ORHA and then the CPA bereft not just of a plan, but of the money and manpower needed to carry it out.

In any postconflict situation, some degree of improvisation is inevitable, no matter how good the prewar planning. But many of the demands placed on the CPA in its early months were foreseeable, because American-led coalitions had faced similar situations in Somalia, Haiti, Bosnia, Kosovo, and Afghanistan over the previous decade. For instance, all societies emerging from conflict have too many soldiers and too few police. Given that the United States had had more than a year to prepare for the stabilization and reconstruction of post-Saddam Iraq, U.S officials should have set aside money and recruited personnel to manage the disarmament, demobilization, and reintegration of excess soldiers into the civilian society and to reform, train, and reequip the police force. Yet both these essential programs had to be designed, funded, and manned largely from scratch only after the CPA was established.

It has been rightly said that no war plan survives first contact with the enemy. It is also true that no postwar plan is likely to survive first contact with the former enemy. The true test of any planning process is not whether it accurately predicts each successive turn in an operation, but whether it provides the operators the resources and flexibility to carry out their assigned tasks. This the planning process for postwar Iraq signally failed to do.

It is unlikely that American officials will again face decisions exactly like those required of the CPA in the spring of 2003. Second-guessing those decisions can take one only so far in preparing for future challenges. But it is certain that the United States will again find itself assisting a society emerging from conflict to build an enduring peace and establish a representative government. Learning how best to pre-

pare for such a challenge is the key to more successful future operations. In this regard, Iraq provides an object lesson of the costs and
consequences of unprepared nation-building.

Bibliography

Aita, Judy, "Iraqi Elections Could Be Held by Year's End, Report Says," *Washington File*, U.S. Department of State, Bureau of International Information Programs. As of January 7, 2009:
http://www.cpa-iraq.org/pressreleases/20040224_brahimi_report.html

Allawi, Ali A., *The Occupation of Iraq: Winning the War, Losing the Peace*, New Haven, Conn.: Yale University Press, 2007.

Amnesty International, *Iraq: Memorandum on Concerns Relating to Law and Order*, London: Amnesty International, July 2003.

Bayley, David H., *Changing the Guard: Developing Democratic Police Abroad*, New York: Oxford University Press, 2006.

Bensahel, Nora, Olga Oliker, Keith Crane, Richard R. Brennan, Jr., Heather S. Gregg, Thomas Sullivan, and Andrew Rathmell, *After Saddam: Prewar Planning and the Occupation of Iraq*, Santa Monica, Calif.: RAND Corporation, MG-642-A, 2008. As of December 22, 2008:
http://www.rand.org/pubs/monographs/MG642/

Bent, Rodney, interview with United States Institute of Peace, September 14, 2004. As of February 13, 2009:
http://www.usip.org/library/oh/sops/iraq/rec/bent.pdf

Bremer, L. Paul, "Iraq's Path to Sovereignty," *Washington Post,* September 8, 2003. As of January 7, 2009:
http://www.pbs.org/wgbh/pages/frontline/yeariniraq/documents/bremerplan.html

———, interviews with PBS Frontline: "The Lost Year in Iraq," June 26 and August 18, 2006. As of February 14, 2009:
http://www.pbs.org/wgbh/pages/frontline/yeariniraq/interviews/bremer.html

———, *My Year in Iraq: The Struggle to Build a Future of Hope,* New York: Simon and Schuster, 2006.

———, "What We Got Right in Iraq," *Washington Post,* May 13, 2007.

————, "How I Didn't Dismantle Iraq's Army," *New York Times*, September 6, 2007.

Blackwill, Robert, interview with PBS Frontline, "The Lost Year in Iraq," July 25, 2006. As of March 1, 2009:
http://www.pbs.org/wgbh/pages/frontline/yeariniraq/interviews/blackwill.html

Brookings Institution, Iraq Index/State Department, *Iraq Weekly Status Report*. As of February 27, 2009:
http://www.brookings.edu/saban/iraq-index.aspx

Cha, Ariana Eunjung, "In Iraq, the Job Opportunity of a Lifetime," *Washington Post*, May 23, 2004.

Chandrasekaran, Rajiv, "How Cleric Trumped U.S. Plan for Iraq," *Washington Post*, November 26, 2003, p. A01.

————, "Iraq's Barbed Realities," *Washington Post*, October 17, 2004.

————, *Imperial Life in the Emerald City: Inside Iraq's Green Zone*, New York: Alfred A. Knopf, 2006.

Cockburn, Patrick, *The Occupation: War and Resistance in Iraq*, London: Verso, 2006.

Diamond, Larry Jay, *Squandered Victory: The American Occupation and the Bungled Effort to Bring Democracy to Iraq*, New York: Times Books, 2005.

Dobbins, James F., *After the Taliban: Nation-Building in Afghanistan*, Dulles, Va: Potomac Books, 2008.

Dobbins, James, Seth G. Jones, Keith Crane, Andrew Rathmell, Brett Steele, Richard Teltschik, and Anga R. Timilsina, *The UN's Role in Nation-Building: From the Congo to Iraq*, Santa Monica, Calif.: RAND Corporation, MG-304-RC, 2005. As of December 22, 2008:
http://www.rand.org/pubs/monographs/MG304/

Dobbins, James, Seth G. Jones, Keith Crane, and Beth Cole DeGrasse, *The Beginner's Guide to Nation-Building*, Santa Monica, Calif.: RAND Corporation, MG-557-SRF, 2007. As of December 22, 2008:
http://www.rand.org/pubs/monographs/MG557/

Dobbins, James, Seth G. Jones, Keith Crane, Christopher S. Chivvis, Andrew Radin, F. Stephen Larrabee, Nora Bensahel, Brooke Stearns Lawson, and Benjamin W. Goldsmith, *Europe's Role in Nation-Building: From the Balkans to the Congo*, Santa Monica, Calif.: RAND Corporation, MG-722-RC, 2008. As of December 22, 2008:
http://www.rand.org/pubs/monographs/MG722/

Dobbins, James, John G. McGinn, Keith Crane, Seth G. Jones, Rollie Lal, Andrew Rathmell, Rachel M. Swanger, and Anga R. Timilsina, *America's Role in Nation-Building: From Germany to Iraq*, Santa Monica, Calif.: RAND

Corporation, MR-1753-RC. As of December 22, 2008:
http://www.rand.org/pubs/monograph_reports/MR1753/

Fallon, Richard H., Jr., "'The Rule of Law' as a Concept in Constitutional Discourse," *Columbia Law Review*, Vol. 97, No. 1, 1997, pp. 1–56.

Feith, Douglas J., *War and Decision: Inside the Pentagon at the Dawn of the War on Terrorism*, New York: HarperCollins, 2008.

Feldman, Noah, and Roman Martinez, "Constitutional Politics and Text in the New Iraq: An Experiment in Islamic Democracy," *Fordham Law Review*, Vol. 75, No. 2, November 2006, pp. 883–920.

Ferguson, Charles H., *No End in Sight: Iraq's Descent into Chaos*, New York: PublicAffairs Books, 2008.

Foote, Christopher, William Block, Keith Crane, and Simon Gray, "Economic Policy and Prospects in Iraq," *Journal of Economic Perspectives*, Vol. 18, No. 3, Summer 2004, pp. 47–70.

Garner, LTG (Ret.) Jay, interview with PBS Frontline: "Truth, War & Consequences," July 17, 2003. As of February 13, 2009:
http://www.pbs.org/wgbh/pages/frontline/shows/truth/interviews/garner.html

———, interview with PBS Frontline: "The Lost Year in Iraq," August 11, 2006. As of February 13, 2009:
http://www.pbs.org/wgbh/pages/frontline/yeariniraq/interviews/garner.html

Gordon, Michael R., "The Strategy to Secure Iraq Did Not Foresee a 2nd War," *New York Times*, October 19, 2004.

———, "Fateful Choice on Iraq Army Bypassed Debate," *New York Times*, March 17, 2008.

Gordon, Michael R., and General Bernard E. Trainor, *Cobra II: The Inside Story of the Invasion and Occupation of Iraq*, New York: Pantheon Books, 2006.

Hamre, John, Frederick Barton, Bathsheba Crocker, Johanna Mendelson-Forman, and Robert Orr, *Iraq's Post-Conflict Reconstruction: A Field Review and Recommendations.* July 17, 2003. As of December 30, 2008:
http://www.csis.org/media/csis/pubs/iraqtrip.pdf

Hayek, Friedrich A., *Law, Legislation, and Liberty, Volume 1: Rules and Order*, Chicago, Ill.: University of Chicago Press, 1973.

Henderson, Anne Ellen, "The Coalition Provisional Authority's Experience with Economic Reconstruction in Iraq," United States Institute for Peace Special Report 138, April 2005.

International Monetary Fund, *Iraq: Third and Fourth Reviews Under the Stand-By Arrangement, Financing Assurances Review, and Requests for Extension of the Arrangement and for Waiver of Nonobservance of a Performance Criterion,*

Washington, D.C.: International Monetary Fund, Country Report No. 07/115, March 2007.

"Iraq: Three Years, No Exit: Rebuilding Iraq Has Been Tougher Than Expected," CBS News Online, March 13, 2006. As of February 27, 2009: http://www.cbsnews.com/stories/2006/03/13/eveningnews/main1397666.shtml

Jones, Seth G., Jeremy M. Wilson, Andrew Rathmell, and K. Jack Riley, *Establishing Law and Order After Conflict*, Santa Monica, Calif: RAND Corporation, MG-374-RC, 2005. As of December 22, 2008: http://www.rand.org/pubs/monographs/MG374/

Kaufmann, Daniel, Aart Kraay, and Massimo Mastruzzi, *2007 Governance Matters VI: Governance Indicators for 1996–2006*, Washington, D.C.: World Bank, 2007.

"Law of Administration for the State of Iraq for the Transitional Period," March 8, 2004. As of January 7, 2009: http://www.cpa-iraq.org/government/TAL.html

Miller, T. Christian, *Blood Money: Wasted Billions, Lost Lives, and Corporate Greed in Iraq*, New York: Little, Brown, and Company, 2006.

Natsios, Andrew, interview with ABC News Nightline: "Project Iraq," April 23, 2003. As of February 14, 2009: http://www.mtholyoke.edu/acad/intrel/iraq/koppel.htm

The November 15 Agreement: Timeline to a Sovereign, Democratic and Secure Iraq. As of February 13, 2009: http://www.cpa-iraq.org/government/AgreementNov15.pdf

Oakley, Robert B., Michael J. Dziedzic, and Eliot M. Goldberg, eds., *Policing the New World Disorder: Peace Operations and Public Security*, Washington, D.C.: National Defense University Press, 1998.

Packer, George, *The Assassins' Gate: America in Iraq*, New York: Farrar, Straus and Giroux, 2005.

"Paper Says Bush Talked of Bombing Arab TV Network," *Washington Post,* November 23, 2003.

Perito, Robert M., *The American Experience with Police in Peace Operations*, Clementsport, Canada: Canadian Peacekeeping Press, 2002.

———, *Where is the Lone Ranger When We Need Him? America's Search for a Post Conflict Stability Force*, Washington, D.C.: U.S. Institute of Peace, 2004.

Phillips, David, *Losing Iraq*, New York: Westview Press, 2005.

"President Names Envoy to Iraq: Remarks by the President in Photo Opportunity After Meeting with the Secretary of Defense," Washington, D.C.: Office of the Press Secretary, May 6, 2003. As of March 1, 2009:

http://georgewbush-whitehouse.archives.gov/news/releases/2003/05/20030506-3. html

"Putting Cruelty First: An Interview with Kanan Makiya," Parts 1 and 2, Democratiya.com, December 16, 2005. As of December 22, 2008: http://www.democratiya.com/interview.asp?issueid=3 http://www.democratiya.com/interview.asp?issueid=4

Rathmell, Andrew, Olga Oliker, Terrence K. Kelly, David Brannan, and Keith Crane, *Developing Iraq's Security Sector: The Coalition Provisional Authority's Experience,* Santa Monica, Calif.: RAND Corporation, MG-365-OSD, 2005. As of January 5, 2009: http://www.rand.org/pubs/monographs/MG365/

Ricks, Thomas E., *Fiasco: The American Military Adventure in Iraq,* New York: Penguin Press, 2006.

Rubin, Michael, "Iraq in Books: Review Essay," *Middle East Quarterly,* Vol. 14, No. 2, Spring 2007.

Sanchez, Ricardo S., with Donald T. Philips, *Wiser in Battle: A Soldier's Story,* New York: HarperCollins, 2008.

Senor, Dan, and Walter Slocombe, "Too Few Good Men," *New York Times,* November 17, 2005.

Shadid, Anthony, "In Iraqi Towns, Electoral Experiment Finds Some Success," *Washington Post,* February 16, 2004, p. A01.

Slocombe, Walter B., "To Build an Army," *Washington Post,* November 5, 2003.

Snyder, Jack, *From Voting to Violence: Democratization and Nationalist Conflict,* New York: W. W. Norton, 2000.

Special Inspector General for Iraq Reconstruction, *Hard Lessons: The Iraq Reconstruction Experience,* Washington, D.C.: U.S. Government Printing Office, 2009.

Stewart, Rory, *The Prince of the Marshes: And Other Occupational Hazards of a Year in Iraq,* Orlando, Fla: Harcourt, 2006.

Tenet, George, *At the Center of the Storm: My Years at the CIA,* New York: HarperCollins, 2007.

Transparency International, *Corruption Perceptions Index 2003,* Berlin: Transparency International, 2003.

Trofimov, Yaroslav, "Iraqis Taste Democracy," *Wall Street Journal,* February 18, 2004.

Tyler, Patrick, "Aftereffects: Postwar Rule: Opposition Groups to Help Create Assembly in Iraq," *New York Times,* May 6, 2003.

United Nations Security Council Resolution 1483, S/RES/1483, May 22, 2003. As of January 5, 2009:
http://www.iamb.info/pdf/unscl483.pdf

U.S. Department of Defense, *Iraq: The Path to Democracy*, Washington, D.C.: Office of the Secretary of Defense for Legislative Affairs, July 23, 2003.

Ward, Celeste J., *The Coalition Provisional Authority's Experience with Governance in Iraq*, United States Institute of Peace, Special Report 139, May 2005.

Weber, Max, "Politics as a Vocation," in H. H. Gerth and C. Wright Mills, eds., *From Max Weber: Essays in Sociology*, New York: Oxford University Press, 1946.

Woodward, Bob, *State of Denial: Bush at War, Part III*, New York: Simon & Schuster, 2006.

Wright, Donald P., and Timothy R. Reese, *On Point II: Transition to the New Campaign: The United States Army in Operation Iraqi Freedom, May 2003–January 2005*, Fort Leavenworth, Kan.: Combat Studies Institute Press, 2008.

Index